T0341206

Advanced Biological, Physical, and Chemical Treatment of Waste Activated Sludge

Advanced Biological, Physical, and Chemical Treatment of Waste Activated Sludge

Antoine Prandota Trzcinski

CRC Press
Taylor & Francis Group
Boca Raton London New York

CRC Press is an imprint of the
Taylor & Francis Group, an **informa** business

CRC Press
Taylor & Francis Group
6000 Broken Sound Parkway NW, Suite 300
Boca Raton, FL 33487-2742

International Standard Book Number-13: 978-1-138-54118-4 (Hardback)

Library of Congress Cataloging-in-Publication Data

Names: Trzcinski, Antoine P. (Antoine Prandota), author.
Title: Advanced Biological, physical, and chemical treatment of waste activated sludge /Antoine Prandota Trzcinski.
Description: Boca Raton: Taylor & Francis, a CRC title, part of the Taylor & Francis imprint, a member of the Taylor & Francis Group, the academic division of T&F Informa, plc, 2019.
Identifiers: LCCN 2018024175 | ISBN 9781138541184 (hardback: alk. paper)
Subjects: LCSH: Sewage--Purification--Activated sludge process. | Sewage sludge.
Classification: LCC TD756 .T79 2018 | DDC 628.3--dc23
LC record available at https://lccn.loc.gov/2018024175

**Visit the Taylor & Francis Web site at
http://www.taylorandfrancis.com**

**and the CRC Press Web site at
http://www.crcpress.com**

Contents

Preface

A wastewater treatment plant is a facility in which a combination of various processes (e.g., physical, chemical and biological) are used to treat industrial wastewater and remove pollutants. The waste residue generated during these treatment processes is known as sludge. Sludge is potentially hazardous because it contains adsorbed residual organic pollutants from treated wastewater. The treatment of sludge is thus considered one of the most significant issues in wastewater treatment, due to higher energy demands and treatment costs. Sludge production currently results in serious environmental issues in many developed, and developing, nations. Rapid industrialization, in conjunction with the extensive growth of urban zones, has also raised concerns in relation to sludge disposal. Sludge disposal processes can account for 60% of total operating costs and 40% of total greenhouse gas emissions from wastewater treatment plants. Furthermore, sewage sludge is rich in pathogenic microorganisms and toxic pollutants, with the potential to cause serious risks to health. In order to acquire "Class A" solids, and fulfill the demands of the Environmental Protection Agency, sludge must be stabilized and detoxified prior to its final disposal or use for land application. Previously, the disposal of excessive sludge was undertaken through traditional methods, including incineration, land filling or ocean dumping. However, an increase in related environmental concerns and stringent environmental laws have led to these disposal options being replaced by biological methods, i.e., composting and aerobic and anaerobic digestion. These biological processes are now widely accepted and are employed for the following: the removal of toxic compounds and pathogenic organisms; to reduce the total volume of sludge; and to transform sludge into stable biosolids. A number of advanced techniques have been developed for sludge treatment and minimization. These consist of physical, chemical and biological technologies, or a combination of all three. This book documents these advanced processes from theoretical and practical perspectives.

In the activated sludge process, biodegradable organic material in the influent wastewater is used by microbes, with some converted to cellular material for the growth and reproduction of the microbes and some used as energy sources to maintain the microbes' metabolism, with by-products, such as carbon dioxide and water. In activated sludge systems, the waste activated sludge yield is between 0.7 and 0.8 kg/kg biochemical oxygen demand. The treatment process therefore produces an increase in biomass, although the yield is less than the value of the influent organic load. The organic material contained in the biomass is slowly biodegradable, largely due to the time required to break down the cell walls. In mesophilic digestion, this may take 8 days or more. Some sludge minimization technologies operate by increasing the consumption of the biomass as a food source, thus reducing the overall production of biomass. This consumption may take place within the activated sludge process, in a downstream digester or in an additional secondary treatment stage.

Extended aeration is a proven process for reducing sludge yields and has been implemented in many wastewater treatment plants worldwide. This process operates by allowing a sufficient solids retention time in the activated sludge process to allow for natural cell death and the breakdown of cell walls. Extended aeration processes can operate at solids retention times of 20 to 25 days, and some plants have operated at significantly longer solids retention times up to 60 days. High mixed liquor suspended solids concentrations are required to provide these long solids retention times without excessive tankage and this can affect the stability of the secondary treatment process and impact the performance of the secondary clarifiers.

Anaerobic and aerobic digestion systems for sludge are now commonly integrated into wastewater treatment plants, specifically to stabilize the waste activated sludge. However, this leads to the efficiency of biological treatments being highly compromised due to a number of factors, including: complex structural components in sludge, rate-limiting cell lysis, the presence of extracellular polymeric substances and various inhibitory compounds (i.e., ammoniacal nitrogen). Bacterial cells and cell walls/membrane form strong barriers to the penetration of hydrolytic enzymes to degrade the intracellular organic components in waste activated sludge. The component cell walls are either very hard or barely dissolve because they are composed of recalcitrant complex compounds, e.g., lignin, cellulose and hemicellulose. In the case of anaerobic digestion, the efficiency of sludge degradation is generally affected by hydrolysis. This is the first stage in anaerobic digestion, which is known as the rate-limiting step, due to the complex structural components present in the sludge. For biological stabilization, anaerobic digestion is a more accepted technology as compared to aerobic digestion and composting. This is due to the requirement for a lower energy footprint and low cost. Anaerobic digestion is specialized for the production of methane, which could be used as biofuel and also assist in greenhouse gases emission reduction. On the other hand, aerobic digestion and compositing processes are energy demanding and require a workforce due to the continuous supply of air and mixing requirements. Although the by-product (compost) of composting can be applied on fields as organic fertilizer, there are concerns associated with the lower quality and toxicity of the biosolids, especially if it is produced from industrial sludge.

The newer generation of sludge minimization technologies aims to increase the biodegradability and the rate of degradation of the biomass produced in the wastewater secondary treatment process. Cell lysis technologies aim to increase the rate of hydrolysis by lyzing the cellular material and releasing the organic material into the bulk medium. Some technologies rely primarily on physical lysis, such as ultrasound, while others such as MicroSludge™ combine chemical and physical processes. Other technologies rely on changing reactor environments and the availability of oxygen to affect the biological processes and the biodegradability of sludge. This is accomplished through cyclic environments that alternate oxic, anoxic and facultative conditions, such as in the Cannibal™ process. The Biolysis® "O" process creates an extreme oxidation condition through the contact of solids with ozone, improving the biodegradability of the biomass when it is returned to the activated sludge system. The mechanisms by which changing reactor environments and causing microbial stress affect the biodegradability of sludge is an emerging area of scientific development.

The efficient stabilization of sludge can also be improved through the application of pre-treatment to the sludge prior to biological digestion. Various methods of pre-treatment have been reported in the literature in the past 30 years, e.g., ozonation, thermal hydrolysis, ultrasound, enzymatic lysis, acidification, alkaline hydrolysis and mechanical disintegration. Moreover, more advanced research has been undertaken over recent years to apply these pre-treatments in a number of combinations, aimed at enhancing the process efficiency to a high level, e.g., alkaline-thermal, thermal-H_2O_2, microwave-alkaline and Fenton-ultrasonic. The primary aim of sludge disintegration using the pre-treatment option is to rupture the microbial cell wall, leading to the release of both extracellular and intracellular organic compounds. This can accelerate the subsequent biological treatment (aerobic/anaerobic digestion) of the sludge and reduce the solids retention time required during digestion.

This book describes the details of the sludge treatment process through the digestion processes, along with key factors impacting the efficiency of the digestion process. Furthermore, in order to enhance the efficiency of the biological treatment of sludge, there is a discussion of the use of various advanced methods of pre-treatment, i.e., biological, physical and chemical methods either as a single process or in combination. Some of these methods can be applied to raw wastewater (also known as mainstream treatment) in the return activated sludge recycle loop, as a pre-treatment prior to the anaerobic digester or even as a post-treatment. In each case, a schematic flow sheet is provided for clarity.

The chapters provide an overall evaluation of various advanced treatment methods with respect to their advantages, disadvantages, energy (temperature, electric or power) and chemical requirements. Most of the pre-treatment methods, such as microwave and ultrasound, demand high energy in the form of electricity. A net return of this energy could be achieved if the subsequent stabilization is performed by anaerobic digestion. The energy from biogas would lead to a lower footprint; however, the net profit would be decreased because of the energy requirement. Some pre-treatments have the capability to enhance biogas yields by 40%. However, in typical biogas engines, 15% of the biogas energy is dissipated and from the remaining, only 35% is converted to electricity while 65% is converted to thermal energy. Thus, the potential for pre-treatment requiring electricity is relatively limited. In contrast, the viability of thermal hydrolysis and heat pre-treatments is comparatively better because the potential of biogas to produce heat energy (65%) is normally sufficient for this purpose. In combined pre-treatments, the energy demand can be lowered, which could ultimately have a positive effect on the energy balance. Combined pre-treatments make it possible to achieve a good balance either way, by decreasing the initial energy requirement or by increasing the conversion of sludge from soluble chemical oxygen demand to methane with a similar energy requirement as in the single pre-treatment. Combined pre-treatments are more favorable with respect to energy requirements and the efficiency of the process; however, operating two treatments simultaneously is difficult at a full-scale plant and requires more research particularly at the large scale. The selection of pre-treatment method is also influenced by other factors. Biological pre-treatment can be efficient but takes a longer time. Microwave and thermal pre-treatment can achieve a high temperature in a short time but is less favorable due to its high-energy consumption and potential to inhibit aerobic and anaerobic bacteria present in the sludge. Comparatively, ultrasonic is a widely applied technology even at the full scale. The shear forces of ultrasonication have the ability to remove pathogens, increase sludge solubilization and decrease clogging and odor problems.

Author

Dr. Antoine Prandota Trzcinski graduated with a bachelor's degree in engineering and a master's degree in chemical engineering from the University of Louvain-la-Neuve in Belgium in 2005. In 2009, he was awarded a PhD from the Department of Chemical Engineering at Imperial College London. During his PhD, he developed a novel process for producing biogas from municipal solid waste and for the treatment of landfill leachate. He gained practical experience in aerobic and anaerobic processes including membrane bioreactors, carbon and nitrogen removal, wastewater and organic solid waste treatment.

In 2009, he carried out research and development at the pilot scale on the production of value-added products from algae at Manchester University. He gained experience in designing a plant for various projects (mass and energy balance, unit sizing and economics) in a biorefinery context in collaboration with Shell. He worked extensively on liquid and solid-state fermentation processes using biomass such as sugarcane bagasse, rapeseed meal, coffee waste, waste glycerol, wheat bran and soybean residues for the production of sugars, ethanol, enzymes and biodiesel using integrated biorefinery concepts.

Dr. Trzcinski joined Nanyang Technological University in 2011 as a senior research fellow in the Nanyang Environment and Water Research Institute and continued working on solid waste treatment such as waste activated sludge and wastewater treatment in anaerobic membrane bioreactors. In particular, he developed novel combinations of pre-treatments of waste activated sludge that result in greater biogas production. He generated three patents from this work in collaboration with the Public Utilities Board of Singapore. He was also actively involved in the preparation of pilot and full-scale anaerobic digestion projects involving municipal solid waste, food waste and waste activated sludge with governmental agencies and private companies. His long-term vision is to transform the conventional waste activated sludge process into its anaerobic counterpart, which would result in an energy factory with significantly lower sludge production. To prove the concept, in 2014 Dr. Trzcinski secured a grant from the Ministry of Education, looking at concurrent carbon and nitrogen removal with biogas production in membrane bioreactors. His interests also include fouling mitigation in membrane bioreactors, the characterization of soluble microbial products, the identification of bacterial and archaeal strains, pharmaceutical and antibiotics removal from wastewater, the fate of nanoparticles in the environment and bioelectro stimulation of microbes to improve bioprocesses through interspecies electron transfer.

In 2016, he joined the University of Southern Queensland as a lecturer and teaches environmental engineering, environmental engineering practice, hydraulics, solid and liquid waste treatment and applied chemistry and microbiology as well as continuing his research in these fields.

1 Conventional Waste Activated Sludge Process

1.1 DESCRIPTION OF MAIN UNITS IN THE PROCESS

The purpose of wastewater treatment plants (WWTPs) is to convert raw sewage into acceptable final effluent and to safely dispose of solids removed in the process. The treatment plant should ideally be constructed at minimum capital cost, have economic maintenance and operating costs and be able to provide a reliable service capable of handling design flows. The system should have low offensive odors.

The goals of wastewater treatment can be outlined as follows:

- Reliable and economical operation.
- Protect public health from contamination of water supplies and water bodies from the outbreak of waterborne diseases, e.g., cholera, typhoid and dysentery; viral diseases, e.g., infectious hepatitis, polio, etc.; worm infections, e.g., tapeworm and roundworm; toxic and heavy metals contamination, e.g., mercury, copper, cadmium and lead; chlorinated organic compounds, e.g., pesticides, herbicides and insecticides.
- Maintain the aesthetics of natural water bodies from visual degradation, odors and foul taste.
- Protect the ecology of water bodies. The contamination of nitrate and phosphorus can lead to eutrophication of lakes and waterways. Effluent with high biochemical oxygen demand (BOD) values will deplete dissolved oxygen content resulting in fish kills, and toxic chemical spills can kill aquatic life and birds.

For more than 100 years (Arden and Lockett, 1914), the conventional waste activated sludge (WAS) process has been used worldwide for the treatment of municipal wastewater. Complete oxidation and nitrification were observed during experiments on settled sewage (Arden and Lockett, 1914). The deposited solids resulting from the oxidation of sewage were designated "activated sludge."

Wastewater treatment processes can be broadly grouped into three main classes (i.e., primary, secondary and tertiary) of treatment. These classes define the extent of wastewater treatment. The selection of appropriate treatment processes is dependent on the nature and strength of the pollutants in the wastewater, the quantity of flow, the discharge license for the treatment plant and other practical considerations.

The municipal WWTP (Figure 1.1) has traditionally been designed to remove floating, settleable and suspended solids, e.g., oil, grease, paper, rags, scum and other non-filterable residues (NFR); soluble organics, i.e., BOD, chemical oxygen demand (COD) and total organic carbon (TOC) of organic materials; and pathogenic bacteria.

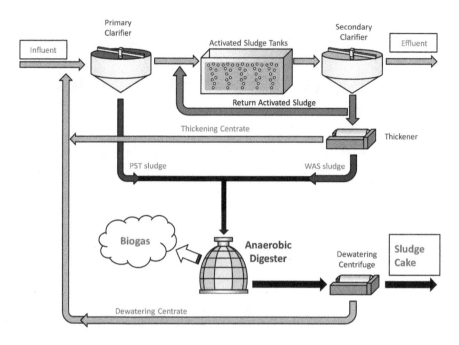

FIGURE 1.1   Overview of a conventional waste activated process.

In recent times, treatment plants have also been designed to remove nutrients, e.g., nitrates and phosphorus, and toxic chemicals (organic and inorganic).

The primary treatment involves physical processes designed to remove floating material and settleable matter of organic and inorganic origin. The influent wastewater passes through screens, grit chambers and primary sedimentation tanks. The objective of primary sedimentation is to remove readily settleable solids and those floating material that escaped the upstream screens. As the suspended solids are mostly organic matter, they tend to be flocculent by nature. Depending on the type of tank, about 50%–65% of suspended solids can be removed by primary sedimentation. Some reduction in the BOD5 value is also achieved by the removal of suspended organic material (25%–40%). The process also involves the collection and removal of settled solids or primary sludge (PS), which is then fed by gravity or pumped to a digester. This is referred to as primary settling tank (PST) sludge in Figure 1.1.

The PS is collected into hoppers by bottom scrapers. Surface scrapers remove floating particles, e.g., fats, mineral oils, grease and other floating matter. The partially clarified liquid is then passed to the next stage of the treatment by overflowing into peripheral weirs or surface-collecting troughs. The primary sedimentation tanks also act as flow/strength equalization basins. They may be subjected to additional loads if biological sludge from the secondary clarifiers is recirculated into the primary sedimentation tanks, which is usually the case for sludge thickening.

The effluent from the primary treatment is not acceptable for discharge into inland waters, although sometimes it may be discharged with approval from the appropriate state pollution control authority into ocean outfalls where large volumes of water provide the assimilation capacity. Primary effluent still has about 30%–50% of the original suspended solids and nearly all of the dissolved organic and inorganic matter.

The biological treatment of wastewater significantly reduces the soluble organic matter through microbial-mediated processes. The biodegradation by-products (in the form of sludge) are then separated from the clarified effluent. Almost all wastewater can be biologically treated if proper analyses have been carried out on the influent water and it is processed under the appropriate environmental conditions.

The biological process comprises an active and mixed microbial population that is sustained by organic waste matter or substrates (food to microorganisms) of proteins, carbohydrates and lipids. The microbial population may be present as a fixed film media or dispersed in a mixed medium. The former refers to an attached growth process where the microorganisms are immobilized or fixed to an inert support structure such as rocks in a trickling filter. A suspended-growth process uses microorganisms dispersed in a mixed medium such as those found in the activated sludge system and in anaerobic digestion (AD).

Activated sludge tanks are the "heart" of the process where activated sludge is formed and microbiological degradation of the organic matter (BOD, COD or TOC) takes place through processes of biochemical transformation. The conventional activated sludge process is wholly aerobic and requires the maintenance of 0.5–2 mg/L of dissolved oxygen in the settled sewage. Some aeration tanks are also designed and operated to achieve nitrification.

The rate of return sludge from the final settling tank to the aeration tank is expressed as a ratio (R) of the wastewater influent. The rate of return sludge or recycling should be such that the mixed liquor suspended solids (MLSS) concentration at the designed value is maintained. To monitor this important operating component of the activated sludge system, the sludge volume index (SVI) is often measured by operators on-site.

The main purpose of the final sedimentation tank is to provide clarification and thickening of the MLSS from the aeration tank. WAS exhibits hindered settling, which is different from the settling behavior of PS.

Sludge refers to the solids that settle and are removed from primary and secondary settling tanks. Raw or primary sludge from the PST has a high volatile organic content of 65%–75% and appears as a grey slimy suspension with an offensive odor. The sludge contains wastes of enteric origin and may contain significant numbers of pathogenic microorganisms. Sludges from humus tanks of trickling (biological) filters and final clarifiers of activated sludge have lower volatile solids (VS) of 45%–70% in a flocculent suspension with a characteristic inoffensive earthy odor when fresh. Secondary sludge is essentially a microbial biomass consisting mainly of bacteria, fungi and protozoa, entrained with non-biodegradable suspended and colloidal solids, and owing to its flocculent properties tends to settle less readily than PS.

The presence of enteric microorganisms requires sludge to be treated through some transformation process before release to the environment for reuse (e.g., soil conditioner) or disposal in landfill or incineration. The cost of the facilities for stabilizing, dewatering and disposing of sludge can be about one-third of the total investment in a treatment plant. The operating costs in sludge processing and handling can even be a larger fraction of the total plant operating expenditure depending on the system employed. Sludge treatment can sometimes account for 90% of the operational problems. Therefore, for these reasons, a properly designed and efficiently operated sludge treatment process and disposal is essential.

It is difficult to separate solids from water in an untreated sludge and being putrescible the solids will readily undergo decomposition. However, in comparison with secondary biological solids, PS thickens and dewaters better because of its fibrous and coarse nature. A mixture of sludge from the primary and secondary settling tanks typically will have a dry solids content of 4%–6% with a specific gravity of about 1.01. After AD, the VS are reduced to 32%–48% with a dry solids content of 8%–13% and a specific gravity of 1.03–1.05. The digested sludge tends to be black in color and the odor is tar-like. Water will readily separate from the solids, which at this stage are stable and will not significantly degrade further.

Sludge from conventional systems is digested anaerobically in sealed tanks or aerobically in the case of extended aeration processes. In the process of sludge digestion, the organic matter is broken down by microorganisms to form digested sludge (sometimes known as biosolids), which are relatively stable and inert solids. Prior to digestion, the sludge is usually gravity thickened in a thickening tank, which is similar to a circular clarifier except that the bottom slope is steeper. Thickening is a physical process used to increase the solids content of the sludge. Thickening reduces the required size of the digester and the amount of supernatant liquor. Typically, fresh sludge with about 0.8% solids content is thickened to about 4%. This represents a five-fold decrease in sludge volume. Sludge enters the center and solids settle to form a sludge blanket. The thickened sludge blanket is gently stirred by a rake mechanism to release gas bubbles, and to push the sludge to a center sump for removal. Supernatant passes over an effluent weir around the outer peripheral of the tank and is returned to the primary treatment unit or to the secondary treatment process. Other means of thickening include flotation, centrifugation and gravity belts.

AD uses microorganisms that thrive in an environment devoid of molecular oxygen and there is an abundance of organic matter. In nature, the process occurs in anaerobic environments such as peat bogs, marshes and lake sediments and in the rumen of cud-chewing animals. The biological process converts organic matter to methane (CH_4), which is the major end-product of AD and carbon dioxide (CO_2). For thousands of years, anaerobic fermentation has been exploited by humans for the fermentation of beer and wine, and more recently in the treatment of waste solids in sewage plants.

The major disposal routes for digested sludge may be to land application for agriculture, landfill and incineration. The choice is influenced by the quantity and composition of the sludge, and the location of the source relative to the potential disposal sites. Environmental protection requirements must ensure that the disposal method does not cause environmental damage.

Prior to disposal, the sludge is dewatered. Digested sludge may be dewatered by air-drying on sand beds, vacuum filters, centrifuges, filter presses and sludge lagoons. Air-drying is dependent on climatic conditions and may achieve a high solids content of 30%–45%. The mechanical means of dewatering is dependent on the initial solids content and the nature of the sludge, and often requires the addition of a chemical additive to enhance the dewatering process. Filter presses can produce a 45%–50% solids content sludge.

Digested sludge from a WWTP can be a valuable soil amendment. Its application to soil completes a natural cycle where the produce from the land is returned to the soil. However, although the environment in a digester will significantly reduce the microorganism population, any remaining pathogens in sludge will pose a health risk. The fecal coliform count provides a good indicator of pathogens and some regulators specify $<2 \times 10^6$ fecal coliform per gram of total solids (TS) in sludge for land application.

The presence of heavy metals and toxic organics such as polychlorinated biphenyls (PCBs) and pesticides, which are not affected by biological treatment processes, are also a concern in the land disposal of sludge. Such contaminated sludges are generally not accepted for land application.

Composting is another viable and cost-effective means of treating wastewater sludges. The process also minimizes the potential for odor. Aerobic composting generates temperatures in the pasteurization range of 50°C–70°C that will destroy enteric pathogenic microorganisms. Subject to the limitations for heavy metals and other toxic compounds, a properly composted sludge may be used as a soil conditioner in agriculture or horticulture applications.

1.2 RATIONALE FOR ADVANCED SLUDGE TREATMENT

Biological WWTPs have been employed throughout the world in the treatment of municipal wastewater. Despite the fact that they are efficient in removing organics, large amounts of excess sludge are generated. For example, the average annual production of excess sludge is 3 million wet tons in Australia, and 240 million wet tons in Europe, the United States and China combined (Pritchard et al., 2010). The existing WWTPs in the United States, for instance, generate over 6.5 million tons dry solids (Mt DS) annually; it is estimated that around 3 and 2 million tons per year is produced in China and Japan, respectively. The main methods for sludge disposal have been and still are landfill, agricultural use and incineration, all incurring very large costs (e.g., $30–$70 per wet ton in Australia and €30–€100 per wet ton in Europe) (Batstone et al., 2011). In developed countries such as the United States, the reuse and disposal rate reaches 94% and it is roughly 97% in Japan, where more than half (52%) of sewage sludge is being recycled to produce building materials and 12% is anaerobically digested for bioenergy recovery (Zhen et al., 2017). Comparatively, in China over 80% of sewage sludge is dumped improperly. Even for landfill, which is the most commonly used method in China, most of the sludge is disposed of directly after mechanical dewatering with higher than 80% moisture content and very low compressive strength. This simple disposal not only causes wasted resources but also brings about a series of secondary disasters (e.g., landslides and environmental pollution). Sewage sludge management is highly complex and costly, representing a stern global challenge. Therefore, reducing sludge production in WWTPs has become a hot topic for both practitioners and researchers.

A large amount of surplus biological sludge is generated during the activated sludge process. The cost of sludge treatment and disposal may be as high as 50% of the total cost for a WWTP (Zhang et al., 2007a). AD is commonly accepted as an ideal method to stabilize sludge for safe disposal and utilization (Nickel and Neis, 2007). It has the advantages of a low biomass yield, a high stabilization degree as well as production of methane gas (McCarty, 1964). It is known that the digestible organic fraction in WAS is only about 30%–45% (w/w) of the biomass in a conventional anaerobic treatment, while methane production can improve markedly by disintegrating the WAS cells to release the intracellular organics using chemical or mechanical disruption methods (Carrère et al., 2010).

Sludge is produced during physical, biological and chemical treatment in a conventional WWTP. There are a number of possible sources of sludge at a WWTP. Sludge from primary sedimentation is PS. PS typically has 2%–8% readily settleable solids with 60%–80% organic content (Davis, 2011). Sludge from the secondary clarifier after the activated sludge basins is WAS. WAS is mainly microbial flocs formed in the activated sludge process. WAS TS content is typically 0.5%–2% with 60%–85% organic content (Davis, 2011). Sludge collected after the addition of chemicals is chemical sludge and this is often associated with phosphorous removal.

The secondary sludge is primarily composed of biomass and microbial cells, produced during the activated sludge process. Overall, the biomass contains approximately 30% protein, 40% carbohydrate and the remaining 30% is lipids in particulate forms (Lin et al., 1999; Sahinkaya et al., 2015). Table 1.1 provides the chemical characteristics of various types of sludge (e.g., COD, VS, TS, pH and total and ammonia nitrogen). Municipal wastewater sludge contains high concentrations of organic matter, e.g., COD and BOD, phosphorous, nitrogen and heavy metals such as Zn, Cu, Ni, Pb, Hg and Cr (Tao et al., 2012). Industrial sludge contains a variety of specific toxic pollutants, which can vary according to the type of industry and product process, e.g., pharmaceutical industrial sludge contains a high concentration of endocrine disrupters and toxic antibiotics, such as sulfonamides (García-Galán et al., 2012). Oil refineries generate a considerable amount of oil sludge containing hundreds of organic compounds that are highly toxic, carcinogenic and mutagenic. These pollutants are classified as priority pollutants by the Environmental Protection Agency, leading to strict controls being required for their release into the environment (Anjum et al., 2016).

PS and WAS, due to their high organic content, can be odorous. Sludge stabilization is needed to reduce the organic fraction, odors and the quantity of solids requiring follow-on management (Bitton, 2005). AD is often used to stabilize sludge before safe disposal and utilization (Nickel and Neis, 2007), while methane, a valuable energy source, is generated (Wang et al., 1999).

In the activated sludge process, organics (indicated by parameters such as BOD and COD) are consumed by the microbial consortium and converted into more biomass while a smaller portion is released as carbon dioxide. The microbes (largely bacteria) may secrete high-molecular weight compounds referred to as extracellular polymeric substances (EPS). EPS are composed of polysaccharides and proteins and are not readily biodegradable. EPS bind the bacteria together and provide a matrix for the microbial flocs (Figure 1.2). EPS, originating from the microbial activity (secretion and cell lysis) and/or from the wastewater itself, i.e., from the adsorption of organic matter (e.g., cellulose and humic acids), are the major constituent

TABLE 1.1

Chemical Characterization of Various Types of Sludge

Type of Sludge	pH	TS (g/L)	VS (g/L)	COD (mg/L)	TN (mg/L)	AN (mg/L)	TP (mg/L)	References
Secondary sewage sludge	6.46	53.2	39.8	4,195 (s)	109 (t)	88	334	Jin et al. (2015)
Dewatered activated sludge	7.82	220.8	108	158,312 (t) 1,518 (s)	9239 (t)	431	NR	Zhang et al. (2015a)
Mix of primary and waste activated sludge	6.38	59.7	40.2	62,000 (t) 5,500 (s)	2840 (t) 3400 (s)	150	NR	Souza et al. (2013)
Sewage sludge (sludge thickening tank)	6.5	57.6	42.2	1,615 (s)	173 (t)	NR	NR	Liu et al. (2011)
Secondary sewage sludge	6.30	55.7	41.2	2,310 (s)	141 (t)	104	301	Xu et al. (2013)
Sewage sludge (sludge thickening tank)	6.9	55.2	34.6	425 (s)	105 (t)	98	NR	Liu et al. (2012b)
Aerobic granular sludge	NR	18.8	14.5	23,600	NR	NR	NR	Palmeiro-Sanchez et al. (2013)
Waste activated sludge	6.38	22.87	17.26	NR	1523 (t)	197.7	NR	Shao et al. (2013)
Secondary sewage sludge	6.5	13.1	9.4	11,000 (t) 60.7 (s)	NR	NR	NR	Tyagi and Lo (2012a)
Mixed primary and secondary sludge	6.61	NR	35.5 (wb)	71.35 (t) 7.96 (s)	4510 (t)	900	1810	Jang et al. (2014)
Thickened sewage sludge	6.83	15.4	10.2 (wb)	NR	NR	NR	NR	Liu et al. (2015)
Excess sludge	6.82	12	NR	17,868 (t)	98.8 (t)	NR	268.8	Gong et al. (2015)
Sewage sludge	7.72	138	96 (wb)	128,000 (t)	8335 (t)	NR	4400	Serrano et al. (2015)
Slaughtering processing sludge	7.4	10.6	74 (% of TS)	4,239 (t)	NR	NR	NR	Erden (2013)
Primary sludge	7.2	20	89 (% of TS)	30,500 (t)	NR	NR	NR	Yenench et al. (2013)
Waste activated sludge	6.7	7.48	5 (wb)	18,249 (t) 1,752 (s)	NR	5006	NR	Liu et al. (2013)
Waste activated sludge	6.83	7.48	5 (wb)	16,249 (s)	NR	132.6	NR	Liu et al. (2012a)

Source: Reproduced from Anjum et al., 2016, with permission from Elsevier.

Note: TN, total nitrogen; AN, ammonia nitrogen; TP, total phosphorus; wb, wet basis; db, dry basis; s, soluble; t, total; NR, not reported.

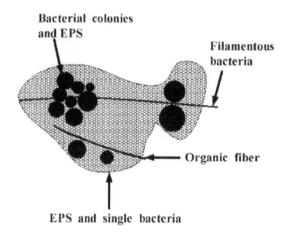

FIGURE 1.2 Representation of a biological floc in waste activated sludge. (Reproduced from Keiding and Nielsen, 1997, with permission from Elsevier.)

of sludge organic fraction, mainly composed of proteins, polysaccharides, nucleic acids, humic substances, lipids, etc.

Depending on the spatial distribution within sludge floc matrixes, EPS are usually divided into three categories: slime EPS (S-EPS), loosely bound EPS (LB-EPS) and tightly bound EPS (TB-EPS). S-EPS are evenly distributed in the aqueous phase and LB-EPS extend from TB-EPS and are characterized by a highly porous and dispersible structure; comparatively, TB-EPS adhere to the surface of the bacterial cells inside the sludge flocs. The presence of these three-dimensional, gel-like and negatively charged biopolymers govern the surface physicochemical properties of sludge matrixes. EPS provide a protective shield and prevent cell rupture and lysis, thereby influencing sludge functional integrity, strength, flocculation, dewaterability and even biodegradability. Non-biodegradable particles may also be included in these flocs (Figure 1.3).

In the secondary clarifier, these microbial flocs settle and are then withdrawn as WAS. WAS mainly consists of intact microorganisms and their secretions forming particles larger than 0.1 μm that cannot be directly assimilated by the microorganisms. Besides protection from EPS, microbial cells themselves possess a hard cell envelope composed of glycan strands cross-linked by peptide that presents physical and chemical barriers to direct AD. In consequence, sewage sludge with a high EPS and cell content has a stiff structure and will be more difficult to hydrolyze. Hydrolysis of the cells must first take place before the soluble materials released can be converted to methane gas in the anaerobic digester. Cell lysis of the microorganisms limits the rate of hydrolysis, which further limits the rate of the whole anaerobic process (Lafitte-Trouqué and Forster, 2002). This is because the cell wall and the membrane of prokaryotic organisms are composed of complex organic materials such as peptidoglycan, teichoic acids and complex polysaccharides, which are recalcitrant to biodegradation. The low digestibility, therefore, requires a long retention time in the range of 30–60 days during biological treatment.

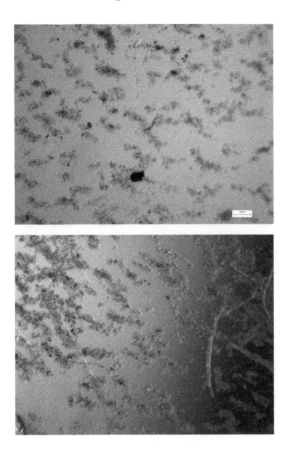

FIGURE 1.3 Microscopy pictures of waste activated sludge showing bioflocs. The bar shows 100 μm. Typical bacteria have a size of 0.1–10 μm. The bottom picture includes filamentous bacteria.

Sludge is a by-product of wastewater treatment, which requires further treatment and disposal. AD is often used given its advantages of a high degree of organics stabilization and methane production (Grönroos et al., 2005). AD is, however, a relatively slow process. Much of the sludge organics are particulate macromolecules and can only pass through cell membranes and be utilized by microorganisms when hydrolyzed into soluble simple organics (Pavlostathis and Giraldo-Gomez, 1991). The hydrolysis of particulate macromolecules is the rate-limiting step in sludge AD (Eastman and Ferguson, 1981; Pavlostathis and Giraldo-Gomez, 1991). The hydrolysis of WAS is especially slow because of its composition. WAS comprises intact microbial cells that are enclosed by EPS and other organic fibers. This complex structure protects microorganisms from being lyzed and thus slows hydrolysis. In order to overcome this rate-limiting step, pre-treatment processes are often applied to solubilize the sludge for AD.

WAS is also less biodegradable than PS (Mao et al., 2004; Lafitte-Trouqué and Forster, 2002) because of the different compositions. PS consists of carbohydrates, proteins and organic particles, which can be more readily used by the anaerobic microbial consortium. However, WAS is primarily microorganisms with cell walls composed of macromolecules such as peptidoglycan. Kepp and Solheim (2000b) reported the production of 306 mL CH_4/g VS with PS against 146–217 mL CH_4/g VS for WAS. The need for cell lysis limits the rate of hydrolysis, which further limits the rate of the whole anaerobic process (Lafitte-Trouqué and Forster, 2002). The activated sludge process is dependent on microbial floc formation with structures enhanced by bacterial secretions.

In order to accelerate the AD of WAS, pre-treatments are carried out to disintegrate the microbial flocs and rupture microbial cell structures to release intracellular organics (Wang et al., 1999). Various pre-treatment methods have been reported with the number of publications on pre-treatment increasing over the last 20 years (Figure 1.4). Various combinations of pre-treatments have also attracted attention because of the potential for better performance over a single pre-treatment as a result of the deployment of several disruption mechanisms.

Pre-treatment of WAS has been proven to disrupt sludge structures, causing solubilization of organics and accelerating subsequent AD (Wang et al., 1999; Tiehm et al., 2001; Onyeche et al., 2002; Pérez-Elvira et al., 2009b). While individual pre-treatments have been reviewed (Odegaard, 2004; Carrère et al., 2010), there has been a steady increase in research papers on combined pre-treatment. This book intends to present the individual treatments strategies as well as combinations that are able to exploit synergistic effects and to discuss the resulting enhancements in performance.

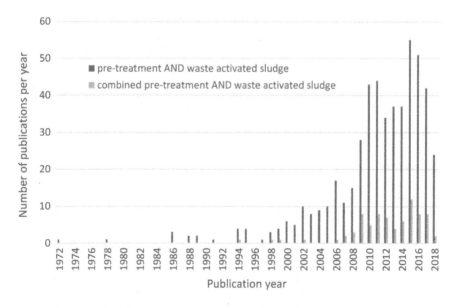

FIGURE 1.4 Number of publications per year with the words "pre-treatment" or "combined pre-treatment" and "waste activated sludge."

1.3   BASIC PRINCIPLES OF ANAEROBIC DIGESTION

AD is the result of the concerted action of several populations of bacteria and occurs in a series of steps. The degradable solids that make up the waste are firstly biologically hydrolyzed to smaller soluble molecules. Then, acid-forming bacteria use these soluble intermediates as substrates for energy and growth, resulting in the formation of fermentation products such as volatile fatty acids (VFAs). Finally, the methanogens or VFA-consuming bacteria produce methane and carbon dioxide. Several researchers (Gujer and Zehnder, 1983; Harper and Pohland, 1986) have described the anaerobic process in six recognizable steps (see Figure 1.5):

1. The hydrolysis of high molecular often insoluble organic polymers such as proteins, carbohydrates and lipids. Enzymes convert them to soluble fragments (monomers) such as sugars, amino acids and long-chain fatty acids.
2. The acidogenesis or fermentation of amino acids and sugars. The conversion of organic monomers to acetic, propionic and butyric acids (gathered under the "VFA" label), hydrogen, carbon dioxide and other organic products such as ethanol and lactic acid.
3. Anaerobic oxidation of long-chain fatty acids and alcohols to acetic acid, hydrogen and carbon dioxide by the obligate hydrogen-producing acetogenic (OHPA) bacteria.
4. Anaerobic oxidation of intermediate products such as volatile acids (with the exception of acetate).
5. Acetoclastic methane fermentation, i.e., conversion of acetate to methane.
6. Conversion of hydrogen to methane.

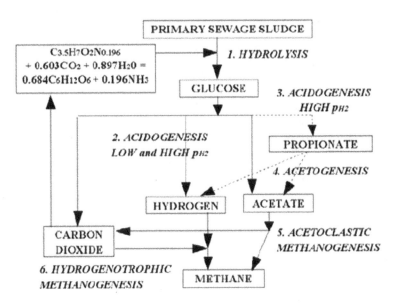

FIGURE 1.5   The main degradation pathways in anaerobic digestion. (Gujer and Zehnder, 1983.)

AD depends on a variety of different bacteria, i.e., a consortium. It is generally thought that a wider diversity of bacteria will more efficiently absorb stresses. The following subsections briefly describe the main characteristics of the different bacteria involved in AD.

1.3.1 HYDROLYTIC BACTERIA

Hydrolysis (Step 1 in Figure 1.5) is achieved by the release of extracellular enzymes and by cell-bound enzymes such as cellulase, lipase or protease; the microorganisms producing them can be either obligate or facultative anaerobes (Zehnder, 1988). Common hydrolytic microorganisms and their enzymes are listed in Table 1.2, along with the hydrolysis products. Key bacteria involved in the hydrolytic phase include *Clostridium, Cellulomonas, Bacteroides, Succinivibrio, Prevotella, Ruminococcus, Fibrobacter, Firmicutes, Erwinia, Acetovibrio* and *Microbispora*.

1.3.2 FERMENTATIVE BACTERIA

Acidogenic or fermentative bacteria are involved in Steps 2 and 3. They convert the hydrolysis products into VFAs (acetic, formic, etc.), alcohol (ethanol), carbon dioxide and hydrogen. This group of microorganisms include *Acetobacter* and *Pseudomonas*, which have a minimum doubling time of around 30 min. Other fermentative bacteria include *Peptoccus, Clostridium, Lactobacillus, Geobacter, Bacteroides, Eubacterium, Phodopseudomonas, Desulfovibrio, Desulfobacter* and *Sarcina*. Acetic acid is produced and excreted by the *Acetobacter* and *Clostridium*

TABLE 1.2

Hydrolysis Products of Biopolymers under Anaerobic Conditions

Biopolymer	Hydrolysis Products	Example of Microorganisms Involved	Enzymes
Carbohydrates:	Polysaccharides	*Bacteroides*	Cellulase
• Cellulose	Oligosaccharides	*Clostridia*	Hemicellulase
• Hemicellulose	Glucose	*Acetovibrio celluliticus*	Xylanase
	Polysaccharides	*Fungi*	
	Oligosaccharides	*Clostridia*	
	Hexoses		
	Pentoses		
Proteins	Polypeptides	*Proteus vulgaris*	Protease
	Oligopeptides	*Clostridia*	Peptidase
	Amino acids		
Lipids	Fatty acids	*Clostridia*	Lipase
	Glycerol		
	Alcohols		

Source: Gujer and Zehnder, 1983.

TABLE 1.3

Typical Reactions Involved in Anaerobic Digestion and Their Standard Free Energies (25°C)

Process Step	No.	Reaction	ΔG_0 (kJ/mol)
Fermentation	1	$C_6H_{12}O_6 + 2\ H_2O \rightarrow 2\ acetate^- + 2\ CO_2 + 2\ H^+ + 4\ H_2$	+9.6
Steps 2 and 3	2	$C_6H_{12}O_6 + 2\ H_2 \rightarrow 2\ propionate^- + 2\ H_2O + 2\ H^+$	+30.6
	3	$C_6H_{12}O_6 \rightarrow n\text{-}butyrate^- + CO_2 + H^+ + 2\ H_2$	+11.5
Step 4	4	$Propionate^- + 3\ H_2O \rightarrow acetate^- + HCO_3^- + H^+ + 3\ H_2$	+76.1
	5	$Propionate^- + 2\ HCO_3^- \rightarrow acetate^- + H^+ + 3\ HCOO^-$	+72.2
	6	$Butyrate^- + 2\ H_2O \rightarrow 2\ acetate^- + 2\ H_2$	+48.1
	7	$Ethanol + 2\ H_2O \rightarrow 2\ acetate^- + 2\ H_2$	+9.1
	8	$4H_2 + 2\ HCO_3^- + H^+ \rightarrow acetate^- + 4\ H_2O$	−104.6
Step 5	9	$Acetate^- + H_2O \rightarrow CH_4 + HCO^-$	−31.0
Step 6	10	$4\ H_2 + HCO_3^- + H+ \rightarrow CH_4 + 3\ H_2O$	−135.6
Sulfate reduction	11	$4\ H_2 + SO_4^- \rightarrow HS^- + 4\ H_2O$	−153
	12	$Acetate^- + SO_4^- \rightarrow HS^- + 2\ HCO_3^-$	−72

Source: Mosey, 1983; de Bok et al., 2004.

genus. Fermentation of glucose (Reactions 1–3 in Table 1.3) is accomplished by fermentative bacteria (Mosey, 1983).

1.3.3 ACETOGENIC BACTERIA

Acetogenic bacteria convert VFAs and alcohols into acetate, carbon dioxide and hydrogen or formate in Step 4. Typical bacteria are *Syntrophobacter wolinii* for propionate degradation (de Bok et al., 2004), and *Syntrophomonas wolfei* for the oxidation of butyrate and other VFAs (McInerney et al., 1979), but also *Syntrophus*, *Pelotomaculum*, *Syntrophothermus*, *Moorella* and *Desulfovibrio*. Acidogens are known to be fast-growing bacteria with a minimum doubling times of around 30 min (Mosey, 1983). There are two types of acetogenic bacteria:

1.3.3.1 Hydrogen-Producing Acetogenic Bacteria

These bacteria convert propionate (Reactions 4 and 5), butyrate (Reaction 6) and ethanol (Reaction 7) into acetate. As their free energies indicate, these reactions are readily stopped if dissolved hydrogen builds up. During the acetogenesis process, acetogens further decompose the higher organic acids (e.g., propionic and butyric acids) to form primarily acetic acid, and H_2 via β-oxidation; however, the conversions are only favorable thermodynamically under particularly low concentrations of the reaction products (acetate and H_2) (i.e., acetate concentration: 10^{-4}–10^{-1} mol/L; H_2 partial pressure required for propionate: 10^{-6}–10^{-4} atm; and for butyrate: $(1.0$–$7.0)\times10^{-3}$ atm). Kaspar and Wuhrmann (1978) reported a linear increase in propionic acid concentration when the hydrogen partial pressure was increased from 10^{-4} to 1.5×10^{-2} atm. Novak and Carlson (1970) observed that long-chain fatty acid degradation was inhibited by hydrogen. Barnes et al. (1983) observed an increase in VFAs immediately following an increase in the hydrogen partial pressure from 2×10^{-4} to 1.5×10^{-3} atm, as induced by a shock loading. As a result, it is recommended to keep the hydrogen concentration low in anaerobic digesters, which is achieved by maintaining a small distance between syntrophic bacteria (de Bok et al., 2004; Kim et al., 2002).

Later, it was discovered that hydrogen-producing bacteria can only survive when their products are removed. This is achieved by the hydrogen-consuming bacteria, i.e., the methanogenic ones. These OHPA bacteria have been estimated to provide the substrate for 54% of the total methane produced (Kaspar and Wuhrmann, 1978), but their doubling times are of the order of 2–6 days (Boone and Bryant, 1980; McInerney et al., 1979).

1.3.3.2 Hydrogen-Consuming Acetogenic Bacteria

Reaction 8 in Table 1.3 is achieved by hydrogen-consuming acetogenic bacteria such as *Clostridium aceticum*, and these bacteria will have to compete with the methanogenic bacteria involved in Reaction 10, which have a high affinity for hydrogen. This can result in the inhibition of acetate cleavage by methanogens when the level of hydrogen is high, as observed by van den Berg et al. (1976).

1.3.4 METHANOGENS

These microorganisms are involved in Steps 5 and 6 in the production of methane. They are obligate anaerobic microorganisms and are therefore very sensitive to oxygen. They have various morphologies (sarcinae, cocci, bacilli, filamentous) and are generally non-spore-forming and heterotrophic, though autotrophic methanogenesis may play an important role. They possess several unique features as opposed to bacteria: they have a distinctive 16S rRNA oligonucleotide sequence; they lack peptidoglycan in their cell walls; they have a unique lipid composition; and they contain cofactors (coenzyme M, F420, F430 and F432) not found in other organisms. For these reasons, these microorganisms form a separate phylogenetic and physiological group called Archaea. They require lower oxidation-reduction potential (ORP) than most anaerobic bacteria (usually below −330 mV). Typical substrates are H_2 and CO_2, formate, acetate, methanol and methylamines (some species). All H_2/CO_2-utilizing species require some unknown organic growth factor. All known methanogens utilize NH_3 as an N-source. Nickel is an essential component of cofactor F430, and Fe, Mo, Co are also important elements. All Archaea are capable of oxidizing hydrogen and reducing carbon dioxide, except *Methanothrix* sp., which can only utilize acetate. Acetic acid can only be utilized by two genera: *Methanosarcina* and *Methanothrix*. According to their ability to utilize the various substrates, methanogenic bacteria can be classified as follows (Dolfing and Bloemen, 1985):

1. Hydrogen-oxidizing acetotrophs (HOA) that can utilize both H_2 and CO_2 and acetate, i.e., *Methanosarcina*; Traore et al. (1983) showed that *Methanosarcina* spp. can use both substrates simultaneously at low substrate concentrations. However, at high concentrations, Dolfing and Bloemen

(1985) found that methane formation from acetate in *Methanosarcina* spp. is inhibited by high levels of hydrogen.

2. Non-hydrogen-oxidizing acetotrophs (NHOA) or aceticlastic methanogens that can utilize only acetate and not H_2 and CO_2, i.e., *Methanosaeta*.
3. Hydrogen-oxidizing methanogens (HOM) that do not cleave acetate, but use H_2 and CO_2 (or formate) as substrates, i.e., *Methanobacterium* or *Methanococcus*.

Some species can also grow on other substrates such as methanol and formate (Zehnder, 1988). Kaspar and Wuhrmann (1978) have reported that approximately 70% of the methane produced originates from acetate (Reaction 9), while about 30% is formed through hydrogen oxidation (Reaction 10). Doubling times of the methanogens are in the range of 0.2–2 days when hydrogen, formate or high levels of acetate are used as the substrate (Ghosh and Klass, 1978), but some species require more time, e.g., *Methanothrix soehngenii*: 9 days (Zehnder, 1988).

Methanogens have not only different substrates but also different kinetic parameters. Table 1.4 shows that NHOA such as *Methanothrix* sp. have much lower substrate utilization rates (U) and their maximum growth rate (μ_{max}) is only one-fifth that of HOA, although their affinity for acetate is much higher (smaller value of K_S). These observations corroborate the finding that *Methanosarcina* sp. develops preferentially throughout or in the inlet of reactors, wherever the acetate concentration is equal to or higher than 350 mg/L. When the acetate concentration falls below 350 mg/L, *Methanosaeta* becomes the prevailing species (Ehlinger et al., 1987).

Furthermore, nitrate-reducing bacteria and the sulfate-reducing bacteria (SRB) (*Desulfovibrio* sp.) are also found in anaerobic digesters (Harper and Pohland, 1986). They are known to compete with methanogens for the substrate (Griffin et al., 1998) and are involved in the oxidation of the intermediate products to acetic acid and carbon dioxide and the oxidation of acetic acid (Reaction 12), propionic acid and hydrogen (Reaction 11). Acidogenic bacteria play a key role in degrading macro-organics to hydrogen, ethanol and VFAs, which are then utilized by SRB to reduce sulfate. The symbiotic relationship between SRB and acidogenic bacteria has been examined, although no direct microbiological evidence has been presented previously (Zhao et al., 2008). SRB can utilize more than 100 different organic substances, although lactate has been shown to be the preferred electron donor.

TABLE 1.4
Biokinetics of Acetate Cleavage to Methane

References	Culture	μ_{max} (d)	K_S (mg COD/L)	Y (g VSS/g COD)	U (g COD/g VSS.d)
Smith and Mah (1978)	*Methanosarcina barkeri*	0.6	320	0.04	15
Zehnder (1988)	*Methanothrix soehngenii*	0.11	30	0.03	3.7

The good health of an anaerobic process is thus a matter of balance between the different families of bacteria involved. If the methanogens are not present in sufficient numbers to convert the VFAs into the final products, a buildup of these acids will occur, resulting in the acidification of the media. If acidification persists, the pH may decrease to under the acidity threshold of the methanogens, leading to irrecoverable failure. Several clues can indicate an imbalance: large amounts of propionic acid in the effluent can be a sign of stress and/or overloading depending on the reactions, leading to the formation of this acid (Pullammanappallil et al., 2001), while large amounts of butyric acid usually suggest failure (Asinari Di San Marzano et al., 1981). Products such as lactate and ethanol usually appear shortly after overloading and/or when some environmental stress has been applied to the reactor.

1.3.5 PROTEIN DEGRADATION

Ammonia is produced by the biological degradation of the nitrogenous matter present in the yard and food waste fractions of municipal solid waste (MSW), mostly in the form of proteins. A protein is a long, complex chain of alpha-amino acids, linked by peptide bonds. Complete hydrolysis of a protein reduces it to its constituent alpha-amino acids. During hydrolysis, intermediate compounds are produced and then further hydrolyzed. These compounds are composed of shorter chains of amino acids. As groups, they are called, in order of decreasing length and increasing water solubility, proteoses, peptones and polypeptides.

Besides proteins, other nitrogenous compounds present in the food and yard waste fractions of MSW include phospholipids, other nitrogenous lipids and nucleic acids. Each of these compounds is sequentially digested by the two groups of anaerobic bacteria (hydrolyzers and acetogens) and eventually their by-products are utilized by methanogens.

Purines and pyrimidines are produced in the digestion of a few amino acids. They also contain nitrogen and, therefore, also release ammonia when further degraded. Ammonia is released during hydrolysis, the first stage of bioconversion. There, hydrolyzing microorganisms deaminate nitrogenous compounds to produce ammonia (Kayhanian, 1999). Theoretically, the quantity of ammonia that will be generated from the anaerobic biodegradation of a biodegradable organic substrate can be estimated using the following stoichiometric relationship:

$$C_aH_bO_cN_d + \frac{4a-b-2c+3d}{4}H_2O \rightarrow \frac{4a-b-2c+3d}{8}CH_4 + \frac{4a-b+2c+3d}{8}CO_2$$

$$+ d\,NH_3$$

where

N_d is the amount of nitrogen present in the feedstock and
NH_3 is the amount of ammonia produced.

In other words, all of the organic nitrogen present in the feedstock will be converted to ammonia.

1.4 THEORETICAL ENERGY ASPECTS OF PRE-TREATMENT

The energy production (E) in an anaerobic process can be expressed in a simplified model as a function of the process efficiency (η_{AD}) and the sludge concentration (c) fed into the system (Cano et al., 2015). Some typical values for sewage sludge have been substituted here according to typical values from a municipal WWTP, and a calorific value of 11 kWh/N m^3 CH$_4$ for methane was set.

$$\text{Organic load (OL)} = (\text{VS/TS})[\text{kg VS/kg TS}] \cdot (c)[\text{kg TS/m}^3] \cdot (r_{COD})$$

$$[\text{kg COD/kg VS}]$$

$$= (0.7) \cdot (c) \cdot (1.4)$$

$$= (0.98c)[\text{kg COD/m}^3]$$

$$\text{Biogas produced (B)} = (\text{OL})[\text{kg COD/m}^3] \cdot (\eta_{AD})$$

$$[\text{kg COD}_{rem}/\text{kg COD}] \cdot (r_{CH4})[\text{Nm}^3 \text{ CH}_4/\text{kg COD}_{rem}]$$

$$= (0.98c) \cdot (\eta_{AD}) \cdot (0.35)$$

$$= 0.34c\, \eta_{AD}[\text{Nm}^3 \text{ CH}_4/\text{m}^3_{sludge}]$$

Total energy from biogas without pre-treatment (E)

$$= (B)[\text{Nm}^3 \text{ CH}_4/\text{m}^3] \cdot (\Delta H_c)_{CH4}[\text{kWh/Nm}^3 \text{ CH}_4]$$
$$= (0.34c\, \eta_{AD}) \cdot (11)$$

$$= (3.77c\, \eta_{AD})[\text{kWh/m}^3_{sludge}]$$

AD efficiency (η_{AD}) can be considered as the biodegradability extent in the digestion and acquires typical values between 40% and 50% in full-scale digesters. However, when pre-treatments are applied prior to digestion, the biogas production and the biodegradation extent can surpass a 40% enhancement, leading to efficiencies of over 60%. Substituting these figures in the foregoing equation, we obtain

Total energy from biogas without pre-treatment (E_F): $\eta_{AD} = 0.45$ (to be adjusted with real plant data)

$$E_F = (1.70c)[\text{kWh/m}^3_{fresh\ sludge}]$$

Total energy from biogas with pre-treatment (E_{PT}): $\eta_{AD} = 0.63$ (to be adjusted with real plant data)

$$E_{PT} = (2.38c)\left[\text{kWh/m}^3_{\text{pre-treated sludge}}\right]$$

Subtracting the last two equations gives the net amount of energy produced by the pre-treatment (ΔE):

$$\Delta E = E_{PT} - E_F = (0.68c)\left[\text{kWh/m}^3_{\text{sludge}}\right]$$

The energy contained in the biogas must be recovered and transformed in order to store it for selling or reusing it in the WWTP. A combined heat and power (CHP) system is an efficient way to produce electricity (electric energy [EE]) and recover heat in a gaseous stream at 400°C (exhaust gases [EG]) and in a liquid stream at low temperature (hot water [HW]) (see Figure 1.6).

According to typical values of commercial biogas engines, 15% of the biogas energy (EB) is lost and, from the rest, 35% is converted into EE and 65% into thermal energy (30% in EG and 35% in HW). Unfortunately, most of the time, just the EE is useful and generates profit, which only represents 30% of the total energy contained in the biogas. Therefore, when a pre-treatment is implemented prior to an anaerobic digester, its energy requirements (EPT) should be lower than the increase in EE that produces (ΔEE) in order to ensure an energy self-sufficient process. In other words, the condition for self-sufficient pre-treatment consuming electricity is

$$\text{EPT} \leq \Delta\text{EE} = 0.35(0.85\Delta E) = (0.20c)\left[\text{kWh/m}^3_{\text{sludge}}\right]$$

The maximum energy consumed is a linear function of the sludge concentration, which means that the energy invested in the pre-treatment increases proportionally as the solids content in the sludge rises (0.2 kWh for each kg TS). This

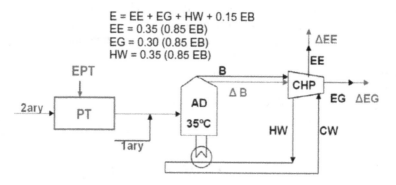

FIGURE 1.6 Energy recovery from biogas with a CHP system. (Reproduced from Cano et al., 2015, with permission from Elsevier.)

simple inequality enables a quick evaluation of different pre-treatment techniques, either applied at the laboratory scale or the industrial scale, to check if the energy balances are satisfied and the process is energetically self-sufficient. Using this approach, Cano et al. (2015) showed that all laboratory-scale experiments lead to energetically inefficient pre-treatments and full-scale experiments considerably reduce their energy consumption, leading in some, but not all, cases to energetically self-sufficient pre-treatments. Ultrasounds pre-treatment showed the most interesting behavior: laboratory-scale works have shown wide range energy consumption (27–118 kWh/m^3_{sludge}) for different sludge concentrations (5–60 g/L), but full-scale applications have demonstrated low energy consumption below 10 kWh/m^3_{sludge}. It can be seen from the equations that the sludge concentration is the key parameter for energy integration of pre-treatments and for energy integration of the whole WWTP. Assuming a sludge concentration of 150 g/L, the maximum energy consumption by the pre-treatment is 30 kWh/m^3_{sludge}. This value is the limit below which the pre-treatment will start to produce a net benefit for the process.

However, when talking about thermal pre-treatments, the recovery of extra heat that is produced in the CHP (EG mainly, ΔEG) for the pre-treatment step would lead to efficient energy integration and the amount of energy that could be recovered for the pre-treatment would be greater ($\Delta EG + \Delta EE$). This can be expressed as

$$EPT \leq \Delta EG + \Delta EE = 0.30(0.85\Delta E) + 0.35(0.85\Delta E)$$

$$= (0.37c)\left[kWh/m^3_{sludge}\right]$$

On the other hand, when a co-generation system (CHP) is considered to recover heat and electricity from biogas, only the hot gases thermal fraction (ΔEG) can be reused for the pre-treatment requirements, since the electric fraction (ΔEE) will represent a net profit or will be dedicated to satisfying the electric requirements of the process. This way, the inequality would be expressed by the following set of equations:

$$EPT_{thermal} \leq \Delta EG = (0.17c)\left[kWh/m^3_{sludge}\right]$$

$$EPT_{electric} \leq \Delta EE = (0.20c)\left[kWh/m^3_{sludge}\right]$$

However, taking into account that thermal pre-treatments require thermal energy as the energy source, it would be more appropriate to consider a biogas boiler that converts biogas energy into heat with an overall efficiency of 90%, which would be more efficiently reused in the pre-treatment process. This way, no EE is misused and maximum heat can be recovered. Then, the inequality becomes

$$EPT \leq (0.34c)\left[kWh/m^3_{sludge}\right]$$

This implies that the amount of thermal energy available for the pre-treatment is higher than the electric energy when considering a CHP system. This shows interesting perspectives concerning energy integration feasibility. In the case of thermal pre-treatments, the potential to be implemented with full energy integration in a WWTP is much higher, since they can recover heat from the biogas engine. This way, full energy integration can be achieved in thermal hydrolysis plants (Cambi, Exelys, CTH) and theoretical approaches set a minimum sludge concentration of 5% TS as the main key factor to ensure energy self-sufficiency.

2 Biological Treatment of Sludge

2.1 INTRODUCTION TO THE AB PROCESS FOR MUNICIPAL WASTEWATER

In recent years, there has been an increase in research efforts aimed at improving the energy efficiency of wastewater treatment processes at large centralized wastewater treatment plants (WWTPs). Concerns over global warming impacts, energy sustainability and biosolids generation are among several key drivers toward the establishment of energy-efficient WWTPs. WWTPs have been recognized as major contributors to greenhouse gas emissions as they are significant energy consumers in the industrialized world (Chai et al., 2015). In addition, the large quantity of biosolids – a by-product of wastewater treatment processes –poses solid waste disposal problems as a result of the limited capacity of landfill sites and air pollution problems from incineration sites. Furthermore, the biosolids management system is considered cost-intensive as it typically accounts for 25%–60% of the total operational costs of conventional activated sludge (CAS)-based WWTPs (Canales et al., 1994; Verstraete and Vlaeminck, 2011). Therefore, innovative design and treatment strategies are required to achieve cost-effective and energy self-sufficient WWTPs by minimizing energy consumption while increasing recovery. It has been reported that sewage contains up to 10 times more chemical energy (as organic pollution) than is needed for its treatment using CAS. However, it is difficult to utilize this energy due to the low concentration of sewage (Dolejs et al., 2016). Although primary sedimentation at WWTPs (or even chemically enhanced primary sedimentation) followed by mesophilic or thermophilic anaerobic digestion (AD) has already been recognized as a viable technology for energy-neutral wastewater treatment (Jenicek et al., 2013), it cannot utilize dissolved or colloidal organic compounds, which represent up to 30% of the organic matter present in sewage (Diamantis et al., 2014). High-rate activated sludge (HRAS) processes such as the AB process (Böhnke, 1977) and its modifications are based on very low hydraulic and solids retention times (HRT and SRT) of <1 h and 1 or 2 days, respectively (Figure 2.1).

One of the approaches toward an energy-neutral, if not energy-positive, wastewater treatment process is to recover the potential energy available in raw municipal wastewaters (Shizas and Bagley, 2004). A two-stage process, the so-called AB process, has been suggested for the recovery of caloric energy content from sewage organics (Versprille et al., 1985; Böhnke, 1977; Meerburg et al., 2015). The first stage is primary treatment in an extremely high loaded biosorption stage (A-stage) also called the high-rate CAS process, which is subsequently followed by secondary treatment in a low loaded biological stage (B-stage) to ensure the removal of dissolved organics

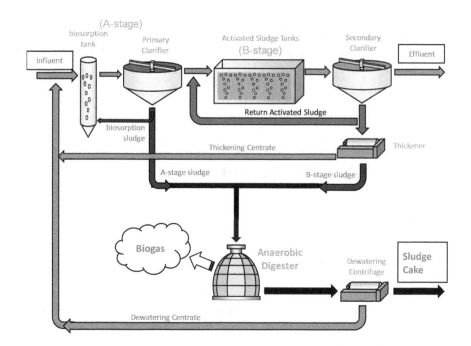

FIGURE 2.1 Application of the AB process as a pre-concentration strategy for raw munici-pal wastewater.

and ammonia. The A-stage treatment allows the biological concentration of sewage organics with minimum oxidation to CO_2, consequently producing a concentrated sludge stream to be channeled to the anaerobic digester. Typically, the A-stage con-sists of a contact tank or contact phase in which return sludge is mixed with influent under aerobic, complete-mix conditions, followed by a settling tank or settling phase to separate sludge from the effluent. A short HRT in the process (<30 min) is selected for the rapid removal of organics and typical removal efficiencies range from 50% to 70%. Furthermore, this system is operated at sludge-specific loading rates of 2–10 g biodegradable chemical oxygen demand (bCOD) per gram volatile suspended solids (VSS) per day and SRTs between several hours and 2 days.

The entrapped organics (chemical energy) could then be recovered through conversion to biogas without significant energy losses (Verstraete et al., 2009). A characteristic feature of the A-stage reactor is its operation with a high food to micro-organism (F/M) ratio, a short HRT and a short solids retention time (SRT), to achieve a high removal rate of sewage organics (Boehnke et al., 1997). Indeed, treatment with a short SRT has been demonstrated to significantly improve the biodegrad-ability of sludge in a downstream anaerobic digester (Ge et al., 2013). The separation of excess sludge in the A-stage can be achieved through an intermediate clarifier (henceforth referred to as an "A-stage clarifier") or a dynamic membrane filtration unit (Roest et al., 2012; Ersahin et al., 2012). The A-stage implements a combination of adsorption-sedimentation and the production of bioflocculants (extracellular poly-meric substances [EPS]), which is promoted at low SRT. Furthermore, it has been suggested that significant fractions of organic matter are removed by sorption (i.e., bio-flocculation) and storage (Meerburg et al., 2015).

Several carbon capture mechanisms during A-stage treatment have been suggested in the literature, namely organic removal by conversion to biomass by fast-growing microorganisms, sorption/bio-flocculation and microbial storage (Boehnke et al., 1998; Makinia et al., 2006; Haider et al., 2003b). Among these mechanisms, sorption/bio-flocculation processes have been commonly applied to enhance the A-stage primary treatment (Meerburg et al., 2015; Yu et al., 2009). Wett et al. (2007) presented a successful case of a net energy-positive municipal WWTP in Strass, Austria, with a two-stage AB process implementing sorption-based carbon entrapment in the primary step. During the biosorption process, the A-sludge retains particulates and colloidal organic substances within the biomass matrix, thereby leaving mainly dissolved organics in the effluent. Readily degradable dissolved organics are typically removed very rapidly and, depending on the SRT, the A-stage generally leaves behind inert or difficult to degrade dissolved organics (Haider et al., 2003b). This would mean a reduced aeration energy requirement and low sludge production in the following B-stage (Versprille et al., 1985), and therefore may lead to considerable energy savings and an overall reduction in biosolids generation. The A-stage configuration is currently applied at full scale in Dokhaven WWTP in Rotterdam, The Netherlands.

2.2 THE HIGH-RATE CONTACT STABILIZATION PROCESS FOR MUNICIPAL WASTEWATER

In recent years, HRAS processes have gained attention as pre-concentration technologies because of their capability to remove particulate, colloidal and dissolved organic matter, and produce high amounts of sludge. To advance toward energy-positive wastewater treatment, it is necessary to maximize the capture of organic matter, and thus increase the relative contribution of sorption and the storage of substrates rather than oxidation or extensive bioconversion processes. The contact stabilization (CS) process may therefore be better suited than the CAS process to achieving the recovery of chemical energy, because of the greater contribution of sorption and storage to the overall removal of substrates (Meerburg et al., 2015). The high-rate CS process is a variant of the AB process where the A-stage is configured differently, as shown in Figure 2.2.

In the CS system, return sludge is aerated in a stabilization phase or a separate stabilization tank for a period of several hours to regenerate the sludge. There, oxidation of adsorbed and stored products takes place as well as microbial growth. Afterward, the sludge is mixed with influent in a preferably anoxic contact phase or contact tank for a relatively short contact time – minutes to hours – and the substrate is adsorbed and stored within cells. After settling, effluent is discarded and return sludge is sent back to the stabilization phase. As such, a substrate gradient exists between the contact and stabilization phases. This subjects the sludge microorganisms to feast–famine conditions and a selective pressure to favor sorption and storage of the substrate. The high-rate CS process is defined as a high-rate system in a CS configuration, with a minimal sludge-specific loading rate of 2 g bCOD g^{-1} VSS d^{-1} and a maximal SRT

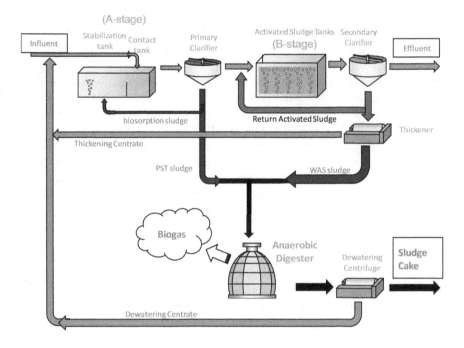

FIGURE 2.2   The high-rate contact stabilization process.

of 2 days. Such a process combines the advantages of high sludge-specific loading rates and a low SRT with a selective pressure toward sorption and storage (similar to the CS process).

In the stabilization tank, air is supplied to achieve a dissolved oxygen (DO) concentration of 0.5 mg/L. A low DO level is maintained to enhance the biosorption of organic matter and minimize the carbon oxidation. This is the main goal of HRAS systems. The concentration of 0.5 mg O_2/L was high enough to avoid filamentous bacteria, whose growth becomes a problem when the DO drops below 0.2 mg O_2/L. In comparison with other high-rate systems at lower solids retention time (SRT < 2 days), the CS system promotes well-settling sludge by selecting faster-growing flocculent microorganisms. The application of this process at the laboratory scale resulted in low substrate oxidation to CO_2 in the A-stage with a recovery of COD in the primary sludge in the range 30%–45% (Meerburg et al., 2015). However, COD is still escaping the A-stage in the form of particulates due to the difficulty to settle well in settling tanks. Further research is needed to improve the solid–liquid separation units in order to increase further the COD recovery.

2.3   PROCESS CONFIGURATION AND OPERATING CONDITIONS OF A PILOT-SCALE AB PROCESS

Thus far, the published literature has documented the biodegradability of the enhanced A-stage sludge at several plants treating municipal wastewaters, but there is virtually no report on the sludge characteristics and its comparison with the B-sludge, a more conventional type of sludge. Moreover, as both biological and physical processes play a crucial role in sewage and biosolids treatment, understanding the change in the bio-/physicochemical properties of the sludge before and after AD would also be of relevance. In this chapter, the abovementioned characteristics of the A- and B-sludge and its changes during the digestion process were examined. The ultimate aim of the effort was to achieve energy self-sufficiency without compromising effluent quality.

A pilot plant was operated with an AB process and was run in a continuous flow mode at ambient temperature (28°C–32°C) with an average wastewater flow of 1000 m³/day. The pilot plant comprised an equalization tank, a coarse (5 mm) screen, a high-rate A-stage, a primary/A-stage clarifier, a fine (2 mm) screen and an ultrafiltration membrane bioreactor (MBR) system that comprised five biological tanks (two anoxic tanks and three aerobic tanks), a membrane tank and a deoxygenation tank. A schematic diagram of the pilot plant is shown in Figure 2.3 and a photograph of the actual plant is given in Figure 2.4.

The raw influent was held in the equalization tank. It was drawn through submersible pumps operating in constant flow rate mode. Initial screening was subsequently performed through 5 mm slot-size screen units followed by a screw conveyor–type grit removal system. The A-stage was designed with an SRT of 0.5 days (calculated

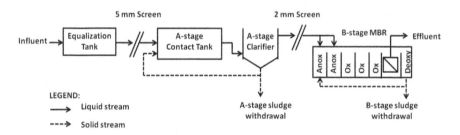

FIGURE 2.3　Schematic diagram of the pilot plant. (Reproduced from Trzcinski et al., 2016, with permission from IWA Publishing.)

FIGURE 2.4　Photograph of the AB actual plant.

over the entire contact tank and clarifier) and a total HRT of 2 h, consisting of 0.5 and 1.5 h for the contact tank and clarifier, respectively.

To protect the downstream MBR process, 2 mm fine screens were provided for the removal of smaller solid particles. The B-stage was operated with a 5-h HRT in the modified Ludzack–Ettinger (MLE) configuration with a step-feed of 50% influent to the first anoxic zone and the other 50% to the second anoxic zone. A target SRT of 5 days was set in order to maintain the slow-growing nitrifying organisms for N removal. The mixed liquor suspended solid (MLSS) in the B-stage was in the range of 0.54–6.1 g/L (average: 2.2 ± 0.9 g/L). The deoxygenation tank was installed after the membrane tank to deplete the DO concentration in the mixed liquor prior to recirculation to the first anoxic tank. DO concentrations were maintained at 0.3 and 1 mg O_2/L in the corresponding contact tank and aerobic tanks.

Biochemical methane potential (BMP) was determined in batch assays using an Automatic Methane Potential Test System (AMPTS II, Bioprocess Control, Sweden). The assay was performed to examine the biodegradability of the A-stage and B-stage sludges through the measurement of their cumulative methane production. The AMPTS bottles were seeded with anaerobic sludge, which was collected from a mesophilic anaerobic digester. The assay was conducted at 35°C for approximately 28 days. Prior to the assay, the inoculums were degassed at 35°C for 1 week to eliminate indigenous methane production. Biomedium containing nutrients and vitamins was prepared in accordance with Owen et al. (1979). Inoculum (200 mL), substrate (100 mL) and biomedium (50 mL) were added to each reactor, which was subsequently flushed with nitrogen gas for approximately 5 min. For comparison purposes, all bottles contained 100 mL of samples. The results were then normalized by dividing by the mass of volatile solids (VS) fed in each bottle and reported as mL CH_4/g VS_{fed}. A bottle without substrate addition was used as the negative control. Methane produced from that negative control was subtracted from the cumulative methane produced from the sludge samples. All assays were performed in duplicate. The biogas composition (CH_4, CO_2 and H_2) was determined using gas chromatography (GC; Agilent, USA) with a thermal conductivity detector after the assay was completed as previously reported (Tian et al., 2014a).

Sludge samples were taken weekly from the pilot plant during 62 consecutive weeks. Total solids (TS), VS, MLSS, mixed liquor volatile suspended solids (MLVSS) and COD concentrations were immediately analyzed in accordance with standard methods (APHA, 2012). The MLSS concentration was also used in sludge volume index (SVI) measurement (APHA, 2012) to determine the compatibility and settleability of the biomass. Calorific value was determined using an oxygen bomb calorimeter (IKA) to measure the energy content in the sludge. The calorimeter unit consisted of a stainless-steel bomb, a water jacket, an ignition unit, a thermometer and a mechanical stirrer. The internal volume of the stainless-steel bomb was approximately 350 mL and the volume of the water jacket surrounding the bomb was 2 L. The mechanical stirrer was used to keep the water jacket uniformly mixed. After centrifugation, the biomass pellet was frozen at –20°C and subsequently freeze-dried at 0.01 mbar vacuum and –45°C overnight. Next, the dried samples were crushed into powder, weighed and combusted using high-pressure oxygen (30 bar) in a bomb calorimeter. The temperature rise in the water jacket during combustion was used to calculate the energy content of the sludge samples. The heat capacity of the bomb was determined using benzoic acid as a standard (Shizas and Bagley, 2004).

Sludge dewaterability was determined using the capillary suction time (CST) test. The test was performed using a capillary suction timer (Part No. 294–50, OFI Testing Equipment, USA) as per the manufacturer's instruction. CST represents the rate in seconds at which water permeates through filter paper, which is a measure of the filterability of the sludge cake. Sludge dewaterability in seconds was normalized by dividing by the MLSS (g/L) to obtain the specific CST in s/L.g. The specific CST value was used in order to be able to compare various sludge samples having different solid contents.

Particle size distribution was analyzed by the laser light scattering technique with the Mastersizer 2000 Particle Size Analyzer (Malvern Instruments, UK). The zeta potential was determined using the Zetasizer Nano ZS (Malvern Instruments, UK) to measure the surface charge of the biomass. The measurement was based on the electrophoretic mobility of sludge particles in an electric field.

Adenosine tri-phosphate (ATP) concentration was measured on the same day as the sampling using QuenchGone21™ Wastewater Test Kit following the manufacturer's instructions (LuminUltra, USA). The assay was based on the conversion of chemical energy produced from ATP breakdown during luciferase reaction into light energy. The emitted light was quantified using a luminometer in relative light units (RLUs), which were converted to actual ATP concentrations (ng/mL) after calibration with 1 ng/mL ATP standard. Two independent ATP tests were conducted for each sample, i.e., total ATP and dissolved ATP tests. The total ATP includes ATP from living cells (cellular ATP) and extracellular ATP from dead biomass. The dissolved ATP represents the extracellular ATP – ATP released from cells as a stress response or dead cells. Hence, cellular ATP was determined by subtracting dissolved ATP from the total ATP. Additionally, the active biomass ratio was also examined by first converting cellular ATP to the active biomass equivalent. The biomass stress index (BSI) was calculated as the ratio of dead-cell ATP to total ATP.

2.4 PERFORMANCE OF THE AB PROCESS

Table 2.1 presents the performance data of the pilot plant. Incoming total raw wastewater COD varied within a wide range (290–1900 mg/L) and the COD removal by the A-stage varied generally between 20% and 90% (average: 44.6 ± 16%).

The MLSS decreased from 510 ± 250 mg/L on average in the influent to 203 ± 106 mg/L due to biosorption/microbial storage and settling in the A-stage. The MLSS removal by the A-stage was 58% ± 16% on average. These COD and MLSS removal rates are similar to those achieved by conventional primary sedimentation tanks (PST). However, up to 90% COD and 83% MLSS removal was observed in this study on certain days, which is not achievable by conventional PST. It has been reported that under optimal conditions, COD removal at the laboratory-scale A-stage can be 70%–80% (30% of it is soluble COD [SCOD]), while MLSS removals can be as high as 80%–95% (Zhao et al., 2000). Diamantis et al. (2014) reported 80% COD removal in a bench-scale A-stage treating municipal wastewater with a lower COD content (400–700 mg/L) than this study. However, previous studies in laboratories sometimes

TABLE 2.1

Influent and Effluent Characteristics of the A-Stage and B-Stage

COD Parameters	Units	N	Concentration		
			Min	Max	Average ± SD
Influent COD	mg/L	268	290	1900	775 ± 280
A-stage effluent COD	mg/L	267	180	1440	440 ± 145
A-stage COD removal	%	240	10	90.2	44.6 ± 16
B-stage permeate COD	mg/L	108	ND	108	20.5 ± 12
B-stage COD removal	%	98	79.2	99.3	95.1 ± 3
AB process COD removal	%	98	83.5	99.7	97.2 ± 2
BOD parameters					
Influent BOD	mg/L	77	166	1331	368 ± 180
B-stage permeate BOD	mg/L	63[a]	2	3.8	2.1 ± 0.4
AB process BOD removal	%	63	98.7	99.9	99.3 ± 0.3
MLSS parameters					
Influent MLSS	mg/L	95	116	1960	510 ± 250
A-stage effluent MLSS	mg/L	95	88	1050	203 ± 106
A-stage MLSS removal	%	95	8.3	83	58 ± 16

Source: Reproduced from Trzcinski et al., 2016, with permission from IWA Publishing.

Note: N, number of samples; ND, non-detected; SD, standard deviation.

[a] 14 samples had a non-detectable BOD value (<2 mg/L).

report only the best conditions and should therefore be interpreted with caution. Wett et al. (2014) reported 40%–85% COD removal from a full-scale A-stage unit, which is similar to the pilot-scale data in this study. Despite some occasional high removals, it seems therefore that the A-stage suffers from greater variability at larger scale.

Furthermore, the large variability obtained in this study may be due to the high oil and grease content in the influent, which is specific to municipal wastewater in South East Asia, and also from the sludge recycle from the A-stage clarifier to the contact tank. Oil and grease can inhibit the adsorption of SCOD and colloids because they will adsorb onto bioflocs preferentially due to hydrophobicity. This is consistent with several literature reports suggesting a poor settling performance of primary sludge thereby limiting the application of the AB process (Frijns and Uijterlinde, 2010; Jenkins et al., 2003).

Nevertheless, the A-stage acted as a buffer to remove organics and suspended solids and protect the B-stage against organic shocks from the incoming wastewater. Because of the membrane in the B-stage, there were no MLSS in the permeate and COD and biochemical oxygen demand (BOD) concentrations were in the non-detectable range up to 108 mg/L (average: 20.5 ± 12 mg/L) and 2–3.8 mg/L (average: 2.1 ± 0.4 mg/L), respectively. The total COD (TCOD) and BOD removals were 97.2 ± 2.2% and 99.3 ± 0.3% on average, demonstrating the high-quality effluent obtained from the membrane compartment of the B-stage despite the influent variability.

2.5   PHYSICOCHEMICAL CHARACTERISTICS OF A-STAGE AND B-STAGE SLUDGES

Figure 2.5 shows the process parameters monitored over 62 weeks including VS concentration, VS/TS ratio, soluble COD concentration and calorific energy content (kJ/g TS) in the sludge withdrawn.

A higher fluctuation of the VS content was observed in the A-sludge, ranging from 0.2 to 14.2 g VS/L, while the fluctuation was less pronounced for the B-sludge (1.5–4.8 g VS/L) (Figure 2.5a). Such variation was attributable to the suspended solids load in the influent originating from the bottom of the equalization tank at low water level, which was pumped into the A-stage contact tank. This was due to the diurnal pattern of municipal wastewater flow that contained high amounts of suspended solids every morning.

Two distinct solids spikes are observed in Figure 2.5a; the first spike occurred on Week 10 and was caused by influent bypassing the contact tank as a result of pump failure, whereas the second spike on Week 61 was due to an extreme solids shock load as previously elaborated.

A more dynamic VS/TS ratio (range: 36%–84%, average: 66% ± 15.6%) was also noted from the A-sludge in comparison with the B-sludge (range: 65%–81%, average: 75% ± 3.5%) (Figure 2.5b). The latter ratio is typical for CAS and has often been reported in full-scale WWTPs worldwide (Cao et al., 2013; WRC, 1984). The low VS/TS ratio in many A-stage sludge samples indicates a high proportion of inorganic material in the influent. These inert particles do not contribute to sorption or microbial storage mechanisms and this may have negatively affected the biosorption process, which would explain the variability in COD and MLSS removal rates by the A-stage.

There was difference in the SCOD concentration bandwidth for the two sludges. The SCOD was in the range 44–655 mg/L (average: 178 ± 93 mg/L) and nondetectable to 125 mg/L (average: 43.4 ± 27 mg/L) in the A-stage and B-stage sludge, respectively (Figure 2.5c). This highlights again the capacity of the A-stage to absorb organic shocks so that a reduced and relatively more stable load enters the B-stage. The results also show that low SCOD concentrations (<125 mg/L) were detected in the B-stage supernatant. This is relevant since membrane modules (ultrafiltration) are submerged in the B-stage membrane tank and SCOD represents soluble organics that will affect the fouling because they are the same size as the pore diameter (Mei et al., 2014). Therefore, a low SCOD concentration is highly desirable in the B-stage to prolong the membrane operation. This information is very relevant for plant operators and further research needs to be carried out to reduce these SCOD levels further in the B-stage, possibly with more baffled compartments and/or plastic biocarriers.

Furthermore, some variation in the calorific content was observed in the A-sludge due to the wide variation in the VS/TS composition of the sludge (Figure 2.5d). Zanoni and Mueller (1982) reported average calorific values of 15 and 13.5 kJ/g TS for primary and secondary sludge, respectively, compared with 15.9 and 12.4 kJ/g for Shizas and Bagley (2004) and 18.2 ± 2.3 and 16.8 ± 1.2 kJ/g TS in this study. The difference between this study and the two previous studies could be due to the

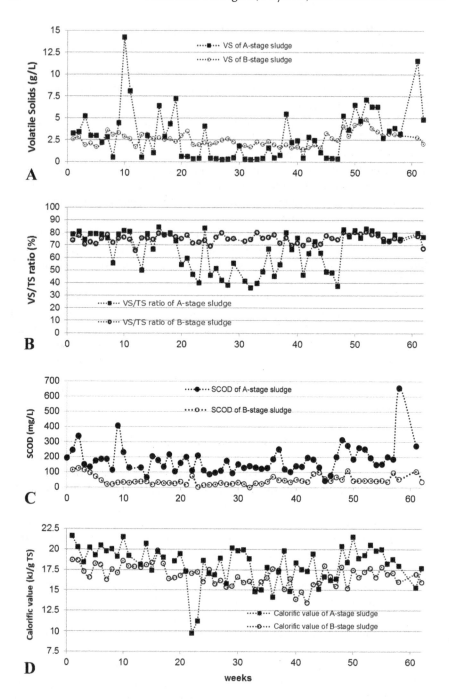

FIGURE 2.5 (a) Volatile solids concentrations in A-stage and B-stage sludges. (b) Volatile solids to total solids ratio in A-stage and B-stage sludges. (c) TCOD and SCOD concentrations in A-stage and B-stage sludges. (d) Calorific values in Joules per gram TS in A-stage and B-stage sludges over the 62-week study period. (Reproduced from Trzcinski et al., 2016, with permission from IWA Publishing.)

inherent characteristics of the wastewaters, i.e., high oil content due to Asian cuisine and a combination of municipal and a small amount of wastewater from small businesses for this study and completely municipal in nature for the previous studies. It could also be related to the very short HRT applied in this study, which preserves the easily biodegradable compounds. As Figure 2.5d demonstrates, the A-stage sludge generally had higher caloric energy content as compared to the B-stage, which is consistent with calorific values from conventional primary and secondary sludges reported elsewhere. This indicates that the A-stage sludge contained organic-rich substrates, and, on the other hand, a significant portion of the organic content in the B-stage sludge had already been consumed in biological processes.

The SVI is an important parameter used by plant operators to monitor sedimentation tanks in a WWTP. In this study, a relatively narrow SVI range (61–76 mL/g MLSS) was observed for the A-sludge during Weeks 57–62, whereas the B-sludge demonstrated a decreasing trend (from 244 to 102 mL/g MLSS) during the test period (Figure 2.6).

This finding, in general, indicated a good settling property of the A-stage sludge, which would be a relevant controlling parameter for the future implementation of the full-scale AB process. The reason for these different SVIs is thought to be due to the presence of some type of extra-polymeric substances (present in raw influent or

FIGURE 2.6 Photograph showing the A-stage sludge (left) with an SVI of 61 and a turbid supernatant versus the B-stage sludge (right) with an SVI of 102 and a clear supernatant.

produced at a short SRT) that can increase settleability by enmeshing larger particles together with the bioflocs making them denser. The excretion of microbial products is commonly thought of as a stress response of biomass upon environmental changes, high loading or exposure to undesirable toxic/inhibitory substance.

2.6   BIOCHEMICAL METHANE POTENTIAL OF A-STAGE AND B-STAGE SLUDGES

The methane yields were in the range 130–775 mL CH_4/g VS_{fed} for the A-stage sludge (average: 460 ± 152 mL CH_4/g VS_{fed}) and in the range 120–430 mL CH_4/g VS_{fed} for the B-stage sludge (average: 256 ± 70 mL CH_4/g VS_{fed}) over the 62 weeks study (Figure 2.7a). The data are plotted in comparison with the corrected calorific value, which was defined as the chemical energy content (as measured in kJ) per unit of organic mass (as measured in g VS) calculated using the VS/TS ratio (Figure 2.5b) and the measured calorific value (Figure 2.5d). It is clear from the data in Figure 2.7

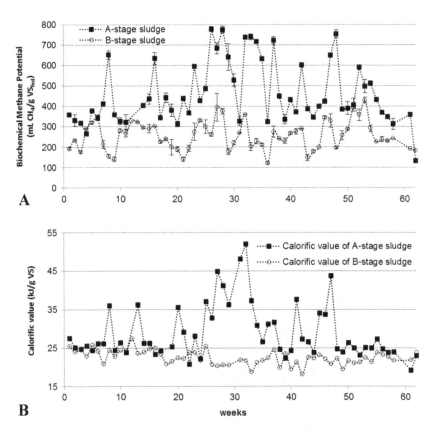

FIGURE 2.7 (a) Comparison between the biochemical methane potential of A-stage and B-stage sludges. (b) Comparison between corrected calorific value (the biomass-specific calorific value) of A-stage and B-stage sludges over the 62-week study. (Reproduced from Trzcinski et al., 2016, with permission from IWA Publishing.)

that the methane potential and corrected calorific value followed a similar trend and were consistently higher for the A-stage sludge.

From the data obtained in this study, between 14% and 493% more methane (average: $97 \pm 83\%$) was obtained from the A-stage sludge in comparison to the B-stage. This has not been reported in previous studies because only a single yield taken from a particular day is usually reported. Rapid transfer from aerobic condition in the A-stage to anaerobic conditions preserved the easily degradable organics entrapped in the A-sludge flocs, thereby leading to a higher methane yield and corrected calorific values of A-stage sludge. An exception, however, occurred in Week 61 when the methane potential of the A-sludge was at its lowest. In Week 61, an unusually high solids concentration was observed in the A-stage sludge (Figure 2.5a), which was coupled with a relatively low methane yield – the only measured values that were lower than the B-stage sludge (Figure 2.7). The BMP curves displayed no lag phase and the hydrolysis constant was generally in the range $0.19–0.23$ day^{-1} for both types of sludges, indicating no inhibitory compounds. However, methane production from the A-stage sludge was faster, as shown in Figure 2.8, where it can be seen that methane production plateaued after 2 weeks whereas the methane production rate from B-stage sludge was more sluggish and continued well after 20 days. Capturing carbon in the A-stage would therefore translate into a smaller anaerobic digester being built at a large scale to stabilize the A-stage sludge. This contrasts with a conventional anaerobic digester having typical residence times of 30 days.

Based on a COD mass balance, the biodegradability was found to be $64.2 \pm 15.3\%$ and $50.6 \pm 14.8\%$ for the A-stage and B-stage sludge, respectively. In conventional WWTPs, AD is generally applied to a mixture of primary and secondary (waste activated) sludge. But waste activated sludge is known to be more difficult to digest than primary sludge (Bougrier et al., 2007). For example, Kepp and Solheim (2000a) reported the production of methane of 306 mL CH_4/g VS for primary sludge against 146–217 mL CH_4/g VS for secondary sludge. The average methane yield of A-stage

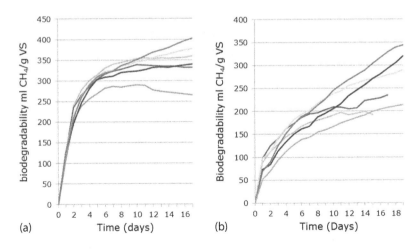

FIGURE 2.8 Comparison between anaerobic biodegradability of (a) A-stage sludge samples and (b) B-stage sludge samples taken at the same time.

sludge in this study was 460 ± 152 mL CH_4/g VS_{fed}, which is higher than reported values for conventional primary sludge. A closer look at Figure 2.7a reveals that 16 samples have a BMP close to or greater than 600 mL CH_4/g VS_{fed}, which is relatively high. This could be due to a high oil and grease content in the sampled domestic wastewaters, which is linked to the higher calorific values of the sludge and can explain the variability in COD and MLSS removal by the A-stage in this study. The occasional presence of oil and grease in the influent may have inhibited the adsorption of SCOD and the entrapment of colloids.

2.7 BIOMASS ACTIVITY

Table 2.2 presents a comparison of the biomass activity between the A- and B-sludge collected in Week 62 and their changes following AD.

The cellular ATP level reflects the quantity of living biomass and the value was used in the conversion to active VSS and subsequently in the calculation of the active biomass ratio. The results showed that the cellular ATP content of A-sludge was significantly lower than the B-sludge prior to AD: 7 ng/mL versus 255 ng/mL. This finding, coupled with the higher BSI of the A-sludge (95%), suggested that the biomass experienced a high level of stress in the A-stage. It could be hypothesized that due to the massive solids load, the supposed low-aerobic condition of the A-stage contact tank had turned to a localized anaerobic condition, leading to a decrease in

TABLE 2.2

Comparison and Changes in Biomass Activity of the A-Stage and B-Stage Sludge before and after Anaerobic Digestion (AD)

Parameter	A-Stage Sludge		B-Stage Sludge	
	Before AD	After AD	Before AD	After AD
Total ATP[a]	138 ng ATP/mL	62 ng ATP/mL	375 ng ATP/mL	53 ng ATP/mL
dATP[b]	131 ng ATP/mL	13 ng ATP/mL	120 ng ATP/mL	11 ng ATP/mL
cATP[c]	7 ng ATP/mL	49 ng ATP/mL	255 ng ATP/mL	42 ng ATP/mL
Active MLVSS[d]	3.4 mg biomass/L	24.5 mg biomass/L	127.3 mg biomass/L	20.8 mg biomass/L
Active biomass ratio[e]	0.025%	0.912%	4.271%	0.755%
BSI[f]	95%	20%	32%	21%

Source: Reproduced from Trzcinski et al., 2016, with permission from IWA Publishing.

[a] Includes ATP from living cells and ATP released from dead cells.

[b] ATP that was released from the dead cells (extracellular ATP).

[c] ATP content of the living cells (cellular ATP) – direct indication of total living biomass. Calculated as total ATP – dATP.

[d] Total mass of living microorganisms. Calculated as cATP*0.5, where 0.5 is an established factor to convert from ng ATP/mL to mg solids/L.

[e] Percentage of total MLSS that are living microorganisms. Calculated as Active MLVSS/MLTSS*100%.

[f] A measure of the stress level of the microbial community. Calculated as dATP/total ATP*100%.

the enzyme activities of the biomass. It can also be related to the very short SRT (0.5 days) and HRT (2 h) applied in the A-stage.

On the other hand, the A-stage sludge demonstrated a marginally higher cellular ATP level than the B-stage during post-AD measurement. This observation was also consistent with the results of the active biomass ratio wherein the A-sludge contained a significantly lower ratio before AD and a slightly higher post-AD ratio as compared to the B-stage sludge. This could be expected as the A-sludge was rich in carbonaceous materials that were highly biodegradable and hence resulted in a more active biomass in the digester and, as a consequence, higher methane production. This finding indicates that the foregoing ATP-based parameters could be useful in facilitating bioreactor operation, particularly in controlling and maintaining a stable living biomass population as well as in identifying reactor failure.

2.8 PHYSICAL PROPERTIES OF SLUDGE BEFORE AND AFTER ANAEROBIC DIGESTION

Table 2.3 presents the physical properties of both sludges, namely dewaterability, zeta potential and particle size distribution, and summarizes their changes after the AD process. The sludge dewaterability was measured using CST, which revealed the better dewatering capacity of the B-sludge as compared to the A-sludge before digestion.

The inferiority of the A-sludge dewatering property appeared to be due to the abundance of extracellular organic materials in the liquor, which were mostly present in colloidal forms. This finding is in contrast to the typical case of conventional wastewater treatment where primary sludge has been reported to have better dewaterability than secondary sludge. The post-digestion mixed liquor, nonetheless, showed a rather similar CST value that was within 4–5 s.L/g MLSS for both sludges. In general, post-AD sludge dewaterability (CST) was marginally deteriorated and this

TABLE 2.3
Comparison and Changes in Physicochemical Properties of the A-Stage and B-Stage Sludge before and after Anaerobic Digestion

Parameter	A-Stage		B-Stage	
	Before AD	After AD	Before AD	After AD
CST (s/L.g)	4.14 ± 0.27	4.57 ± 0.19	2.78 ± 0.24	4.11 ± 0.19
Zeta potential (mV)	-17.9 ± 1.4	-16.8 ± 0.5	-17.5 ± 1.4	-16.4 ± 0.3
Particle size distribution				
Modal value (μm)	33	79	99	79
D10	14	26	37	28
D50	58	75	96	79
D90	359	247	279	211

Source: Reproduced from Trzcinski et al., 2016, with permission from IWA Publishing.

could be expected as EPS were produced during the AD. EPS are highly hydrated and able to bind a large volume of water, thereby contributing to the decrease in the dewatering characteristic of the sludge.

The zeta potential was measured in Weeks 57, 58, 61 and 62. As shown in Table 2.3, no significant difference in zeta potential was observed from the A- and B-sludge of both pre- and post-AD. This occurred as (1) both sludges were in their original state due to the absence of any chemical treatment or surface charge manipulation; and (2) the surface active component was not degraded during AD.

Prior to AD, the A-sludge demonstrated a wider range of particle sizes than the B-sludge (2–1000 and 10–700 μm for A- and B-sludge, respectively). The post-AD mixed liquor for both sludges demonstrated a more uniform distribution in a similar range (10–500 μm) (data not shown). It was also observed that the modal value of the A-sludge was 33 μm, which was much smaller than that of the B-sludge (99 μm) (Table 2.3). This suggested fine colloidal particles dominating the A-sludge and contributing to its lower extent of dewaterability. Particles represent a dominant component (≤85%) of the total COD in sewage and their size is known to affect both the biological and physical processes of sewage treatment (Levine et al., 1985; Zeeman et al., 1997). Particle size can be converted via hydrolysis and this activity contributes to the changes in size distribution after AD.

In conclusion, the AB process is a two-sludge system designed to capture energy in the first step, the A-stage, such that minimum energy is required in the second step, the B-stage. Due to the membrane in the B-stage, the AB process achieved 97.2% COD removal, 99.3% BOD removal with permeate COD and BOD values lower than 108 and 3.8 mg/L, respectively. Throughout the 62 weeks study, results showed that the A-sludge was more biodegradable and delivered higher recovery of chemical energy from sewage organics as compared to the more conventional B-sludge. The ATP analysis revealed that the cellular ATP content of A-sludge was two orders of magnitude lower than the B-sludge. This translated to a BSI of 95% suggesting that A-stage biomass experienced a high level of stress due to the massive organic load, low HRT and low-aerobic condition.

2.9   HIGH-RATE ACTIVATED SLUDGE FOLLOWED BY AUTOTROPHIC NITROGEN REMOVAL

HRAS followed by autotrophic nitrogen removal (i.e., partial nitritation + anammox) has been proposed as a promising method for maximizing energy recovery from domestic wastewater (Laureni et al., 2016; Lotti et al., 2015; van Loosdrecht and Brdjanovic, 2014). Conventionally, the biological nitrification and denitrification (N&DN) process is used to remove nitrogen from wastewaters. In this process, first ammonium is biologically oxidized to nitrate via nitrite under aerobic conditions, and later nitrate is biologically converted to nitrogen gas under anoxic conditions in the presence of organic carbon. The discovery of the anaerobic ammonium oxidation (anammox) process has revolutionized the removal of nitrogen from NH_4^+-rich residual streams. Anammox is a biological process capable of the anaerobic transformation of NH_4^+ to dinitrogen (N_2) gas with NO_2 as an electron acceptor. The anammox process proved a better substitute for conventional N&DN due to a 60% decrease in oxygen demand (aeration), a 100% decrease in organic carbon source demand and less/no N_2O (global warming potential 310 times greater than CO_2) production. Above and beyond, 90% less sludge is generated in the anammox process as compared to the conventional N&DN process, which results in a reduced sludge treatment cost.

Nitrogen removal from wastewaters by the anammox process consists of two steps: partial nitrification (about half of NH_4^+ is oxidized to NO_2) and the subsequent anammox process (NH_4^+ is oxidized with NO_2 to N_2 gas) under anoxic conditions. This combination is known as partial nitrification and anammox (PN&A). In the beginning, the PN&A process was applied separately in two stages, the first stage the partial nitrification process, followed by the anammox process in the second stage. Later, partial nitrification and anammox processes were introduced in single-stage reactors (Ali and Okabe, 2015). Anammox bacteria are affiliated with the bacterial phylum Planctomycetes. Although extensive studies significantly advanced our understanding of Anammox bacteria, there are still some inherent challenges in the practical application of the anammox process. The biggest hurdle for the application of the anammox process is its slow growth rates (doubling time ranging between 7 and 14 days), causing a slow start-up to the process at full scale.

Wastewater is considered to consist of three forms of substrates: particulate, colloidal and soluble substrates. In the HRAS reactor, the particulate and colloidal substrates are removed by bioadsorption and the subsequent solids–liquid separation. The soluble substrate is removed by intracellular storage, biosynthesis and/or biological oxidation. The HRAS reactor can also be replaced with "A stage," which is mainly based on adsorption. Afterward, the nitrogen in the effluent of a HRAS reactor (or "A stage") is removed via partial nitritation in the anoxic/aerobic tank and anammox in the anaerobic tank, as shown in the schematic diagram in Figure 2.9. Since most organic carbon is biodegraded anaerobically (i.e., AD) in this process, the sludge production would be substantially reduced because of the low yield (<0.1 g COD/g COD) of anaerobic microbes. However, this process is still in its infancy and some challenges (e.g., unstable partial nitritation) still exist. Alternatively, nitrite can first be produced in a partial nitritation reactor followed by an anammox reactor. Therefore, further studies and full-scale tests are still needed (Wang et al., 2017). Figure 2.9 also shows the inclusion of a dewatering unit prior to the anaerobic digester, which will result in a significantly smaller digester. The mixture of A-stage sludge and sludge from the anaerobic tank is also expected to be highly biodegradable compared to a conventional B-stage sludge, resulting in significantly shorter residence times. A smaller digester operating at small residence times would translate into low capital and operating costs, but there is a lack of data for this type of process.

2.10 ANAEROBIC MEMBRANE BIOREACTOR FOLLOWED BY AUTOTROPHIC NITROGEN REMOVAL

An anaerobic membrane bioreactor (AnMBR) followed by autotrophic nitrogen removal (i.e., partial nitritation + anammox) is also a promising approach for maximizing energy recovery from domestic wastewater (Batstone et al., 2014;

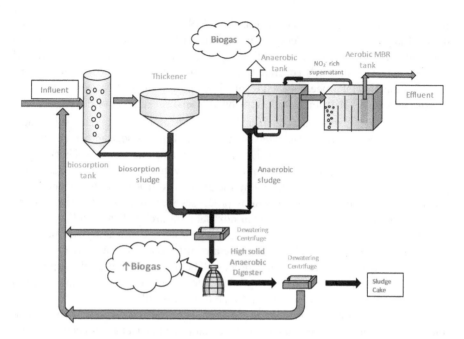

FIGURE 2.9 Flowsheet describing a high-rate activated sludge followed by partial nitrification to produce NO_2^- and recycle the NO_2^- supernatant to an anaerobic tank where nitrogen is removed by Anammox bacteria.

McCarty et al., 2011; Solley et al., 2015). In the AnMBR, wastewater is treated by anaerobic microbes to convert almost all the biodegradable organic carbon to biogas (Figure 2.10). The nitrogen is then removed via partial nitritation and anammox. Similar to the HRAS coupled with the autotrophic nitrogen removal process, the organic carbon is removed anaerobically in this process. Therefore, sludge production would be significantly reduced due to the low yield (<0.1 g COD/g COD) of anaerobic microbes (Metcalf and Eddy, 2014). Nevertheless, this process is still in its early stage and efforts are still needed to overcome the existing challenges (e.g., unstable partial nitritation). Another challenge is dissolved CH_4 in the AnMBR, which should be recovered because methane has a strong greenhouse gas potential. As a result, further studies and full-scale tests are required.

2.11   PREDATION OF PROTOZOA AND METAZOA

Protozoa and metazoa are small organisms that prey on bacteria in WWTPs. They are widely present in activated sludge. By preying on bacteria, they decrease the sludge production due to the dissipation of energy when energy transfers in the food chain (Wang et al., 2017). Sludge reduction via predation is generally achieved by adding protozoa and/or metazoa externally (Basim et al., 2016; Ratsak, 1994), which has been successfully applied at laboratory and full scale. *Tubifex tubifex*, *Lumbriculus variegates* and Tubificidae are the commonly used protozoa/metazoa,

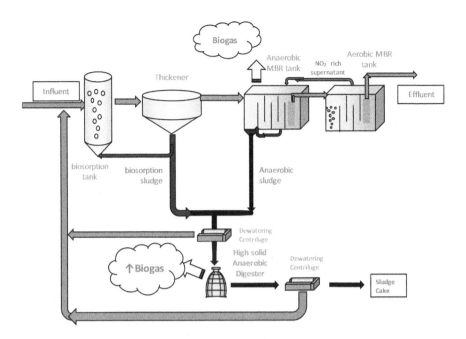

FIGURE 2.10 Flowsheet describing an anaerobic membrane bioreactor followed by auto-trophic nitrogen removal by Anammox bacteria.

through which a sludge reduction of 125%–75% has been reported. It should be pointed out that reducing sludge production via predation of protozoa and metazoa generally happens in WWTPs with a long SRT, under which the protozoa and meta-zoa would get sufficient time to grow.

2.12 CANNIBAL PROCESS

US Filter has developed a package of technologies they market as the Cannibal™ process. The Cannibal process uses cyclic environments to reduce the production of WAS from a secondary treatment system. The Cannibal system also differs from the previous two technologies in that it is a process tied in with the secondary treatment process to prevent production of WAS for digestion. Versions of the Cannibal pro-cess have been in operation since the late 1990s at municipal and industrial WWTPs. The Cannibal process consists of two primary processes: a physical separation step and a biological treatment step, as depicted in Figure 2.11.

The physical separation step, or solids separation module (SSM), consists of a very fine drum screen (250 μm perforations) that removes inert organic matter from the bioreactor. The screens treat approximately 50% of the return activated sludge (RAS) flow rate on a continuous basis. The second part of the SSM is a set of hydrocyclones that are intermittently operated on the RAS, generally on a monthly basis. The underflow from the hydrocyclones is discharged into a classifier similar to that used for grit washing in the headworks. These remove heavy organic material,

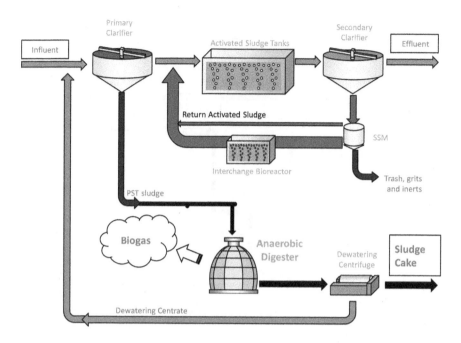

FIGURE 2.11   Flowsheet of the Cannibal process.

grit and dense inorganic particles that will accumulate without wasting sludge. The material produced by the SSM steps is a stabilized material that does not require digestion. The biological treatment part of the Cannibal process is the heart of the process and consists of an additional biological tank, called the "interchange reactor." The flow that would normally be considered WAS is wasted to this additional reactor on a daily basis, using a conventional calculation of the MLSS inventory in the main activated sludge process and a typical SRT of 8–15 days. The interchange reactor is sized for a 10–12 day SRT, with a solids concentration no greater than 1%. Air is intermittently applied to maintain the tank at the cusp between anoxic and anaerobic conditions and allow the growth of facultative microbes. These conditions are maintained by monitoring oxidation-reduction potential (ORP) and intermittently applying coarse bubble aeration to produce nitrates. In between aeration events, the solids are allowed to settle. Each day, a portion of the material in the interchange reactor is returned to the main bioreactor system. A mixer is activated prior to returning mixed liquor from this tank back to the activated sludge plant. The process continues in a closed loop cyclic interchange between oxic and facultative conditions, with no sludge wasting from the process. However, it is recommended that one inventory of solids be purged on an annual basis. The cyclic environments support the growth of different microbial communities, while microbes not adapted for that environment are stressed. This leads to changing the biodegradability of the cellular material and each community of microbes utilizing the other as a carbon source in their respective reactors. The end result of this process is that the amount of traditional WAS from the secondary process is significantly reduced. These savings are offset by the production of the screenings from the SSM, but the net benefit is a reduced solids production from the plant (Sheridan and Cutis, 2004).

There are a number of municipal and industrial installations with versions of the Cannibal process, with more currently in design or construction. The concept was first implemented in 1998 at a small 3.8 megaliters per day (MLD) sequencing batch reactor plant in Georgia. A 60.6 MLD facility was built in 2006 in Albany, Oregon. The technology was commercially introduced by US Filter in 2003. The SSM has been implemented at newer installations, including installations in southern California and Illinois that have been operating since late 2004. Installations with the fine screening and hydrocyclone steps show that this inert organic material is produced at a rate of approximately 0.2–0.3 kg total suspended solids (TSS)/kg BOD (±10%) applied to the bioreactor at a plant without primary clarification. The material is similar in form to paper pulp and dewaters fairly easily to 30% solids, is generally inert, 90% volatile and can be disposed of with conventional screenings.

Annual purging of the system produces a WAS quantity that is equivalent to <0.1 kg/kg BOD treated. US Filter provides a 2-year guarantee that the observed yield will be less than this, supported by performance at existing installations that have run in excess of 2 years with the Cannibal process and produced solids quantities significantly less than the guaranteed level. Currently, one disadvantage is that biological phosphorus removal cannot be integrated with the Cannibal system as there is no wasting. The continuous addition of metal salts to the secondary process would also require frequent wasting and is therefore not compatible with the process. Phosphorus can be removed chemically in tertiary treatment steps, such as filters. Odor generation from the interchange reactor is another potential disadvantage. At the start of the aeration cycle in the interchange reactor, a strong odor is normally discernible. It is recommended that the interchange reactor is covered and odor control provided at plants where odor is a sensitive issue (Roxburgh et al., 2006).

The economic advantages of the Cannibal process will be site specific and it is most likely to be favorable for small to medium plants, particularly those below the threshold for AD systems. The cost of constructing the interchange reactor to provide a 10 day SRT at ≤1% solids concentration can be a significant capital expense. Covers and treatment for odor control, if required, will add to the capital cost. Capital and operating and maintenance costs for pumps, interchange reactor aeration systems, screens, hydrocyclones and classifiers will also need to be included in the economic evaluation. Costs for the SSM portion are usually no lower than $1 million. However, savings in the construction, operation and maintenance of aerobic or anaerobic digestion systems, thickening and dewatering as well as significant reductions in solids disposal costs will provide offsets against the costs of the Cannibal system. For plants that have aerobic digester tanks that could be converted for use as interchange reactors for the Cannibal process, the economics would be particularly favorable, as this approach would use existing facilities, minimizing capital costs, while reducing energy costs associated with the high aeration demand for aerobic digesters. For larger plants that have existing digesters and co-generation equipment, the cost–benefit analysis and footprint requirements may not favor the Cannibal process over other available technologies, such as enhanced digestion processes.

3 Biological Treatment of Sludge

Application of the AB Process to Industrial Wastewater

3.1 INTRODUCTION

Conventional wastewater treatment plants include a primary sedimentation tank (PST) to remove suspended solids and organics. These units operate typically at 2–3 h hydraulic residence time (HRT) and suffer from low efficiencies. This results in high aeration costs in the subsequent biological tanks to degrade the organic matter. A possible alternative to PST is the adsorption/bio-oxidation (AB) process, which comprises a high-rate activated sludge (HRAS) process (A-stage), which removes chemical oxygen demand (COD) primarily by bio-flocculation, adsorption, bioaccumulation and settling, followed by biological nutrient removal (BNR) (B-stage) (Figure 2.1). In the A-stage, COD removal by activated sludge is preceded by rapid physicochemical adsorption of organic matter on active sites of bioflocs (biosorption) and intracellular storage depending on the pH, dissolved oxygen (DO), residence time, organic loading, type of organics (particulate, colloidal or soluble) and microorganisms (Lim et al., 2015). In some circumstances, up to 60% of incoming COD can be removed by an intracellular mechanism versus 40% by a surface sorption mechanism at 15 min contact time. Table 3.1 lists the typical performances of A-stage processes obtained with municipal wastewaters.

The advantages realized by operating the AB process is the overall reduction of the total biological reactor and clarifier volumes, the reduced aeration requirement and the redirection of more sewage carbon to anaerobic digestion for biogas generation. The main objective of an A-stage is to produce large amounts of raw waste sludge that can be converted to biogas by anaerobic digestion and reduce the organic load on the subsequent BNR process. As a result of reducing the organic load and providing a more stable influent to the BNR process, the aeration capacity and tank volume of the B-stage can be reduced. The A-stage can also be a buffer against shock loads and inhibitory industrial inputs to the B-stage biological process.

Biosorption has been reported to be suitable for wastewaters containing high suspended solids and colloids concentration. The main parameters to consider are the solid residence time (SRT), HRT, aeration control, velocity gradient inside the contact tank, settling time, concentration of mixed liquor suspended solids

TABLE 3.1
Typical Parameters of A-stage Processes

| Influent COD (mg/L) | Parameters (Biosorption/Settling) Tanks | | | | | | References |
	HRT (h)	SRT (day)	DO (mg/L)	Volume	COD removal	Chemical Aid	
400–700	0.9/1.8	0.3–0.5	2	3/6 L	80%	20 mg Fe^{2+}/L	Diamantis et al. (2014)
171 (COD) 107 (SS)	0.58/1.5	NR	1.9–3.2	1.2 m^3	68% 91%	Al_2O_3 Fe_2O_3	Zhang et al. (2007b)
450–800 (COD) 200–600(SS)	0.5/1.5	1.1–2	2–4	2.64/6.11 L	70%–80% 80%–95%	No	Zhao et al. (2000)
700	1/1	–	NR	1 L (batch)	70%	No	Yu et al. (2014)
NR	0.21–0.61/1.47–4.3	0.5–0.7	NR	644/4540 m^3	40%–85%	$FeCl_3$+polym.	Wett et al. (2014)

Source: Reproduced from Trzcinski et al., 2017, with permission from IWA Publishing.
Note: NR, not reported.

(MLSS), sludge recycle ratio and temperature. Under low SRT and HRT, the COD removal is due to adsorption and bio-flocculation and the degradation of organic compounds by metabolism is avoided. Biodegradation of organic matter typically represents <10% of the incoming COD load (Guellil et al., 2001; Haider et al., 2003a; Hernández Leal et al., 2010). Furthermore, coagulant dosage to the influent of the A-stage may enhance the removal of the carbon by precipitation onto bioflocs.

Biosorption does not exceed 10–15 min to reach equilibrium (Guellil et al., 2001). Yu et al. (2014) reported that biosorption of the colloidal fraction in batch tests reached equilibrium after 10 min, while 45 min were required for the soluble fraction. Under optimal conditions and with municipal wastewater, COD removal can reach 70%–80% (30% of it is soluble COD [SCOD]) and about 80%–95% of total suspended solids (TSS) can be removed (Zhao et al., 2000).

Biosorption sludge contains aggregates of microorganisms, adsorbed organic matter and extracellular polymeric substances (EPS). EPS are highly charged polymers (proteins, polysaccharides, lipids, glycolipids and glycoproteins) that are excreted by microorganisms or produced by cell lysis and hydrolysis. The main mechanisms include charge neutralization, hydrophobic interactions and bridging (Vogelaar et al., 2005). The effectiveness of bridging depends on the molecular weight of EPS, the charge of the polymer and the particle, the ionic strength and the mixing. Divalent cations may improve the biosorption efficiency of activated sludge due to the negative charge of EPS (Keiding and Nielsen, 1997). These authors showed that small particles in wastewater have a negative surface charge density and a change in the repulsive forces due to calcium concentration and ionic strength can cause floc disintegration.

Diamantis et al. (2014) operated a biosorption step as a pre-treatment to ultrafiltration at the laboratory scale with diluted (<300 mg/L COD) and concentrated (≈400–700 mg/L COD) municipal wastewater to study the removal of organics (particulate and SCOD) and the recovery of nutrients (total Kjeldahl nitrogen [TKN], ammonia and phosphorus). The HRT was 0.9 and 1.8 h in the biosorption and sedimentation tanks, respectively, and the SRT in the system was 0.3–0.5 days. They found that the removal of particulate COD was significantly higher when concentrated wastewater (400–700 mg/L COD, on average 524 mg/L) was used, while SCOD removal was improved with iron supplementation ($FeSO_4$ at 20 mg Fe^{2+}/L). The addition of a coagulant is known to enhance the biosorption capacity by co-precipitating iron phosphate and soluble carbon onto bioflocs.

In this chapter, the objective of the A-stage pilot plant with a capacity of 1 m^3/h was to evaluate the removal of particulate and soluble organics from a combined municipal–industrial wastewater and to study the effect of DO, SRT, organic loading rate (OLR) and ferric chloride dosage. Conventional PST operate at 2–4 h HRT and typically achieve 30%–35% biochemical oxygen demand (BOD) removal and 50%–60% MLSS removal (Metcalf and Eddy, 2014). Thus, there is a need to study other technologies capable of removing more organics at similar residence time. Compared with municipal sewage, the combined municipal–industrial wastewater has an unknown composition and a large fluctuation in water quality and there is currently a lack of data on the applicability of A-stage treatment for high-strength industrial wastewater, especially at the pilot scale. The objective was to determine the organics removals (MLSS, SCOD and total COD [TCOD]) of the novel A-stage process operating at a lower HRT than most conventional PST.

3.2 PROCESS CONFIGURATION AND OPERATING CONDITIONS

Sewage and industrial wastewaters were collected from various local industries including petrochemical, chemical, electroplating, food processing and pharmaceuticals industries. The wastewater parameters were determined from a composite sample collected by auto samplers over a period of 24 h.

The pilot plant comprised a sorption column ($V=0.5$ m^3; cylindrical; $D=0.4$ m; water depth$=4$ m) followed by a clarifier ($V=1.5$ m^3; circular; $D=1.1$ m; surface loading 1 m^3/m^2.h). A simplified schematic diagram is depicted in Figure 3.1 and the actual units are shown in Figure 3.2. The influent flow rate was fixed at 1 m^3/h and the overall HRT was 2 h.

A fraction of the settled solids was returned to the sorption column through the return activated sludge (RAS) line. The sludge volume of the RAS was estimated at 330 L including the conical part (320 L) and the RAS pipe (10 L). The sludge concentrations in the sorption column were controlled by adjusting the RAS and waste activated sludge rates. The influent and effluent samples were composite samples collected automatically every hour over a period of 24 h. The sorption column and RAS samples were grab samples from the sorption column and clarifier bottom, respectively. The excess sludge's total solids (TS), volatile solids (VS), TCOD, SCOD, calorific value and anaerobic biodegradability of the clarifier sludge samples were analyzed throughout the study.

The pilot plant was operated for about 200 days. During the first 144 days, various DO levels were tested in the sorption column during Tests 1–6. DO was controlled at 0.2, 0.4, 0.5, 0.7 and 1 mg/L through adjustment of the blower flow rate with proportional-integral-derivative (PID) control. The experimental plan is shown in Table 3.2.

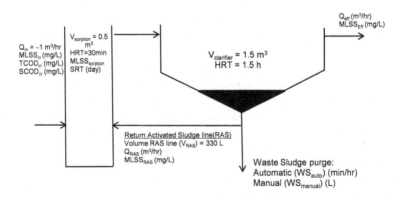

FIGURE 3.1 Schematic diagram of the A-stage pilot plant.)

FIGURE 3.2 Photograph of the biosorption column (left) and clarifier (right).

The SRT was controlled by withdrawing sludge in the recirculation line. To adjust the amount of sludge withdrawn, the opening time of the automatic sludge discharge valve was adjusted to 5–10 min every hour. The SRT in the A-stage process was calculated as follows:

$$
SRT = \frac{\text{Mass MLSS in the process} \left(g\ MLSS \right)}{\substack{\text{Daily MLSS removal in effluent} \\ \text{and WS automatic an manual removal}} \left(\dfrac{g\ MLSS}{day} \right)}
$$

$$
= \frac{V_{sorption} \cdot MLSS_{sorption} + V_{RAS} \cdot MLSS_{RAS}}{Q_{Eff} \cdot 24 \cdot MLSS_{Eff} + WS_{auto} \cdot Q_{RAS} \cdot 24 \cdot MLSS_{RAS} + WS_{manual} \cdot MLSS_{RAS}}
$$

From Day 145 to Day 210, the addition of a coagulant was considered to further enhance the sorption capacity by entrapment of the dissolved organic matter in iron phosphates precipitates. In this study, ferric chloride ($FeCl_3$, 38%) was used as the coagulant. The dosage rates were adjusted at 5 ppm (Test 7), 10 ppm (Test 8) and 20 ppm (Test 9) as Fe^{3+} in order to determine the optimum value while keeping a constant DO of 0.4 mg/L (Table 3.2). In Test 10, 20 ppm Fe^{3+} was tested together with a DO of 0.7 mg/L.

TABLE 3.2

Summary of Conditions Tested in the A-stage Process

Test	Days	Date	No. of Runs	Experimental Conditions						Influent			Reactor	RAS
				Influent	DO SP	RAS	WAS	SRT	FeCl₃	SS	TCOD	SCOD	MLSS	MLSS
				m³/h	mg/L	m³/h	kg/day	day	ppm	mg/L	mg/L	mg/L	mg/L	mg/L
1	14–71	4-9-13 to 18-10-13	6	0.95	**0.20**	0.21	10.8	0.52	0.0	1155	1553	527	5272	20,069
2	74–96	6-11-13 to 20-11-13	5	1.03	**0.40**	0.47	18.2	0.21	0.0	1338	1873	565	3733	10,533
3	110–112	4-12-13 to 6-12-13	2	1.00	**0.70**	0.45	18.2	0.20	0.0	3032	3026	262	4246	8,994
4	117–119	11-12-13 to 13-12-13	2	1.04	**1.00**	0.49	16.0	0.16	0.0	2883	3214	462	3276	7,866
5	124–126	18-12-13 to 20-12-13	2	0.95	**0.50**	0.29	9.8	0.31	0.0	846	1425	726	2092	21,872
6	132–144	26-12-13 to 8-1-14	3	1.06	**0.40**	0.50	14.1	0.24	0.0	687	1375	304	3355	8,873
7	151–160	15-1-14 to 24-1-14	4	1.04	0.40	0.50	12.7	0.24	**5.0**	3725	2266	712	3304	7,942
8	186–194	20-2-14 to 28-2-14	4	1.13	0.40	0.49	15.8	0.22	**10.0**	2559	2082	748	3800	10,057
9	201–203	5-3-14 to 7-3-14	3	1.13	0.40	0.50	13.0	0.21	**20.0**	2148	2649	619	3114	8,126
10	208–210	12-3-14 to 14-3-14	3	1.10	0.70	0.50	13.4	0.26	**20.0**	1876	2296	680	4559	8,112
		Average	34	1.04			14.20	0.26		2025	2176	561	3675	11,244

Source: Reproduced from Trzcinski et al., 2017, with permission from IWA Publishing. Tested variables are shown in bold.

AU: Table 3.2: Please explain the significance of the bold numbers in the table body.

3.3 PERFORMANCE OF THE A-STAGE

Influent MLSS is normally <700 mg/L in municipal wastewaters. The influent MLSS in this study ranged from 360 to 5370 mg/L and its average was 1690 mg/L, demonstrating the industrial nature of the influent. The exact ratio of the municipal and industrial wastewater was variable and unknown. The influent characteristics varied over a wide range, as shown in Table 3.3. Oil and grease and extremely high suspended solids were frequently found in the influent. Typical effluent characteristics are shown in Table 3.3 during a baseline test at 0.2 mg/L DO.

The incoming SCOD and TCOD were very high in the range 300–840 and 750–4120 mg/L, respectively. The SCOD to TCOD ratio was below 35% in the influent, and it increased to above 60% in the effluent, indicating that most of the effluent was soluble and particulate COD were effectively removed in the clarifier sludge despite the high fluctuations of raw wastewater. The A-stage could also remove some phosphorus as shown by a decrease in total phosphorus (Table 3.3), while the removal of TKN, NH_3-N and PO_4-P was not consistent. The BOD concentration decreased from 775 to 479 mg/L on average, showing the potential of the A-stage to remove

TABLE 3.3
Influent Characteristics, Mixed Liquor Properties in the Sorption Column and Effluent Parameters

Parameter (Influent)	# Samples	Units	Min	Max	Average ± Standard Deviation
pH	29	–	6.8	8.53	7.09±0.3
TCOD	29	mg/L	750	4,120	1790±830
SCOD	23	mg/L	300	840	490±130
BOD	35	mg/L	200	1,225	775±510
MLSS	29	mg/L	360	5,370	1690±1240
MLVSS	29	mg/L	190	3,115	770±590
TKN	35	mg/L	40.3	184	73.2±29.6
NH_3-N	35	mg/L	29.4	84	47.7±12.8
Total phosphorus (TP)	35	mg/L	9.2	75.7	20.8±12.3
PO_4-P	35	mg/L	1.7	12.6	6±2.7
Parameter (sorption column)					
MLSS	28	mg/L	760	17,930	5300±3420
MLVSS	28	mg/L	440	8,570	2640±1625
Parameter (clarifier effluent)					
TCOD	37	mg/L	290	1,715	814±310
BOD	35	mg/L	102	1,636	479±320
TKN	35	mg/L	33.1	117	65.5±18.4
NH_3-N	35	mg/L	27.4	89.4	49.2±13
Total phosphorus (TP)	35	mg/L	1.5	18.6	12.1±4.2
PO_4-P	35	mg/L	1	9.8	4.5±2.4

Source: Reproduced from Trzcinski et al., 2017, with permission from IWA Publishing.

organics at a high rate despite the fluctuations in the influent, which would considerably reduce the aeration costs in the subsequent biological stage. The BOD concentration in the clarifier effluent was in the range 102–1636 mg/L, which indicates that there would still be sufficient biodegradable matter for nutrient removal in the B-stage.

3.4 EFFECT OF DISSOLVED OXYGEN ON THE PERFORMANCE OF THE A-STAGE

A low DO environment can lead to the growth of filamentous bacteria, which would affect settling (Li et al., 2010). Yu et al. (2014) showed that the biosorption capacity of sludge decreased when it was mixed with anaerobic sludge. Air must therefore be provided during biosorption in order to reactivate the sludge and maintain its adsorption capacity. Higher DO would require a greater air supply, and therefore higher operating costs. It is thus important to investigate its impact on the process. Figure 3.3 shows the incoming MLSS, TCOD and SCOD, effluent MLSS, TCOD, SCOD and MLSS, TCOD and SCOD removal percentages at various DO tested during Tests 1–6.

Between 50% and 90% MLSS could be removed in the A-stage process, demonstrating the applicability of this compact treatment for industrial wastewater, which has not been shown before. Furthermore, at influent MLSS >2000 mg/L, 80%–90% of MLSS could be removed regardless of the DO level. Despite the high fluctuation and the presence of oil and grease, the A-stage process performs significantly better than conventional PST, which can only remove 50%–55% of MLSS from raw sewage. This significant improvement was expected because the adsorption properties of activated sludge floc are favorable for wastewater with high concentrations of MLSS and colloidal particles (Diamantis et al., 2014; Zhao et al., 2000), which was verified in this study with industrial wastewater. Entrapment of larger particles in the open structure of sludge flocs can also take place in carbon capture systems (Lim et al., 2015).

As can be seen from Figure 3.3 (bottom), the SCOD removal efficiency in the A-stage process was low and Figure 3.4 shows that it decreased as the DO in the sorption column was increased from 0.2 to 1 mg/L. When the DO was 0.5 mg/L or higher, the SCOD removal percentage was systematically negative. These results are very different from previous studies on municipal wastewater that reported 30% SCOD removal under optimum conditions (Zhao et al., 2000). The presence of oil and grease in the influent may have inhibited the adsorption of SCOD and the entrapment of colloids in this study. A high DO may also have caused hydrolysis of particulate COD to SCOD or the shear effect caused by vigorous aeration may have resulted in the breakage of bioflocs leading to the release of soluble components. From the results, a DO >0.4 mg/L is therefore not recommended.

3.5 EFFECT OF FE^{3+} DOSAGE ON THE PERFORMANCE OF THE A-STAGE

Raw data of MLSS, TCOD and SCOD from Tests 7 to 10 with a coagulant dosage are shown in Figure 3.5. The sorption column provided intimate contact between

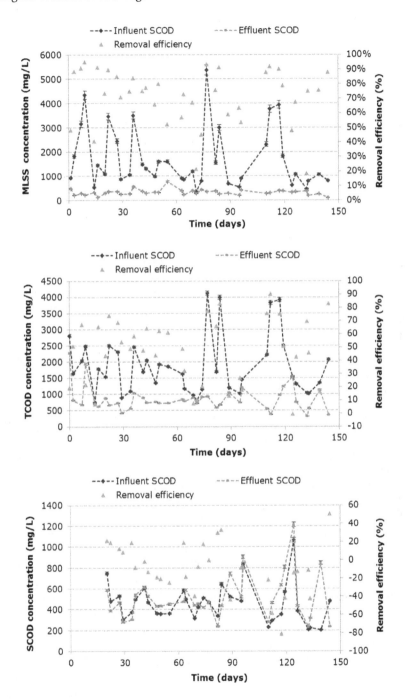

FIGURE 3.3 Evolution of MLSS (top), TCOD (middle) and SCOD (bottom) and the respective removal efficiency during continuous operation of the A-stage process (Tests 1–6). Error bars indicate standard deviation. The error bars were omitted when smaller than the marker. (Reproduced from Trzcinski et al., 2017, with permission from IWA Publishing.)

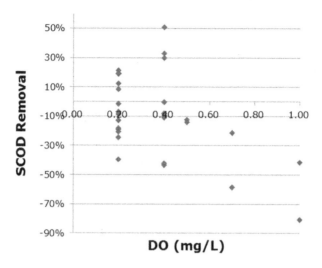

FIGURE 3.4 Effect of dissolved oxygen on SCOD removal percentage at DO in the range 0.2–1 mg/L (Tests 1–6). (Reproduced from Trzcinski et al., 2017, with permission from IWA Publishing.)

Fe^{3+}, influent MLSS and SCOD to investigate any benefit of Fe^{3+} dosing using real industrial wastewaters. DO was maintained at 0.4 (except for Test 10 at 0.7 mg DO/L) as it was observed that higher DO was detrimental for SCOD removal. The TCOD removal was in the range of 40%–76%.

However, TCOD and MLSS removals >90% were observed on certain days in this study, which are not achievable by conventional PST. It has been reported that under optimal conditions, TCOD removal in a laboratory-scale A-stage can be 70%–80% (30% of it is SCOD), while MLSS removal can be as high as 80%–95% (Zhao et al., 2000). Diamantis et al. (2014) reported 80% COD removal in a bench-scale A-stage treating municipal wastewater with a lower COD content (400–700 mg/L) than this study. However, previous studies in laboratories sometimes reported only the best conditions and should therefore be interpreted with caution. Wett et al. (2014) reported 40%–85% COD removal from a full-scale A-stage unit, which is similar to the pilot-scale data in this study. Despite some occasional high removals, it seems therefore that the A-stage suffers from greater variability at a larger scale in particular when treating industrial wastewater containing oils and grease. SCOD removal obtained by other researchers at the laboratory scale could not be replicated in this pilot-scale study with real municipal–industrial wastewaters.

The large variability obtained in this study may be due to the high oil and grease content from industrial and municipal wastewater in South East Asia, and also from the sludge recycle from the A-stage clarifier to the sorption column. Oil and grease can inhibit the adsorption of SCOD and colloids even when Fe^{3+} is added because it will adsorb onto bioflocs preferentially due to hydrophobicity. This is consistent with several literature reports suggesting a poor settling performance of primary sludge, thereby limiting the application of the AB process (Jenkins et al., 2003;

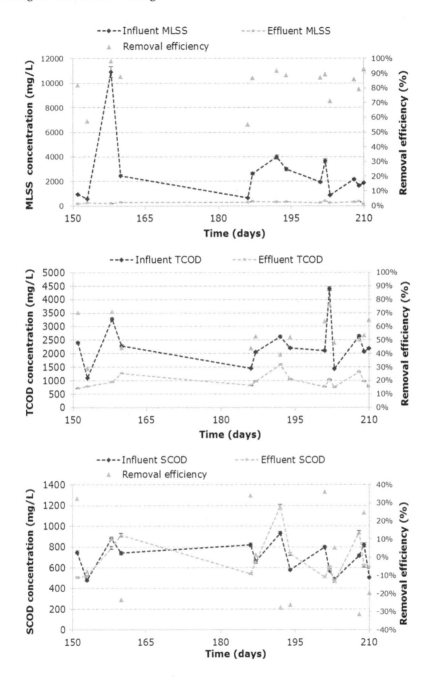

FIGURE 3.5 Evolution of MLSS (top), TCOD (middle) and SCOD (bottom) and the respective removal efficiency during continuous operation of the A-stage process with Fe^{3+} dosage (Tests 7–10). Error bars indicate standard deviation. The error bars were omitted when smaller than the marker.(Reproduced from Trzcinski et al., 2017, with permission from IWA Publishing.)

Frijns and Uijterlinde, 2010). Better results could be obtained with an oil and grease trap placed in front of the sorption column.

Figure 3.6 reveals that an Fe^{3+} dosage did not result in better removal. Increasing the DO to 0.7 mg/L in Test 10 did not have any significant effect. SCOD removal efficiencies highly fluctuated from negative values to a maximum of 50% regardless of the presence or not of a coagulant. This may be due to the particular nature of the industrial wastewaters used in this study. The high inorganic content of the spent sludge (VS/TS was in the range 24%–61%) can explain why it was not effective for SCOD adsorption. Alternatively, SCOD could not be removed by biosorption due to the nature of the organics, or due to the short SRT used in this study.

Since the HRT was maintained constant, the OLR varied according to the influent COD concentration. As shown in Figure 3.7, the OLR to the process had an impact on the MLSS removal. This is similar to Diamantis et al.'s (2014) results where the COD removal increased at OLRs in the range 5–20 kg COD/m³.day with municipal wastewater. It appeared from this study with industrial wastewater that high OLR (25–60 kg COD/m³.day) were favorable to remove MLSS. Up to 90% MLSS and 83% TCOD could be removed at an OLR of 36 kg COD/m³.day. This represents a significant advance compared to the conventional PST that can only achieve 50%–55% MLSS removal and 30%–40% BOD removal at 2–4 h HRT (Metcalf and Eddy, 2014).

When the OLR was below 20 kg/m³.day, the MLSS was generally >60%, which makes the A-stage an interesting competing technology for the treatment of industrial wastewater compared to conventional primary clarifiers. At low OLR, the activated sludge is prone to fragmentation due to an increase in water soluble EPS, while organic matter is desorbed from the floc and effluent quality is deteriorated (Guellil et al., 2001). Another reason is the lower strength wastewater (potentially diluted with rainfall). This is attributed to the ionic strength and divalent cation concentration of the raw wastewater.

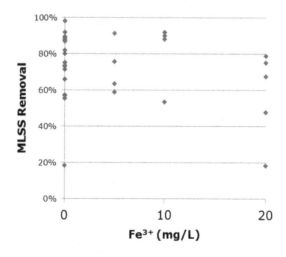

FIGURE 3.6 Effect of Fe^{3+} dosage on MLSS removal (Tests 1–10). (Reproduced from Trzcinski et al., 2017, with permission from IWA Publishing.)

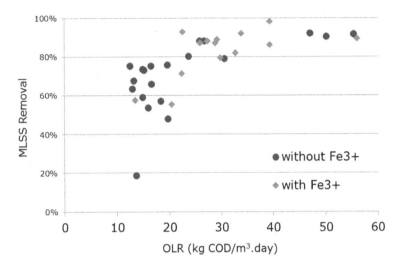

FIGURE 3.7 Effect of the organic loading rate (OLR) on MLSS removal with and without Fe^{3+} dosage (all data point in Tests 1–10). (Reproduced from Trzcinski et al., 2017, with permission from IWA Publishing.)

Considering the low HRT (2 h) and high OLR, it is evident that the removal mechanism is not biological degradation, but physical adsorption followed by settling. Indeed, biological degradation in HRAS systems normally takes place at 2–3 kg COD/m^{-3}.day (Tchobanoglous et al., 2003), which was not the case in this study. Furthermore, it was demonstrated that the pre-treatment of industrial wastewater can take place in a very compact A-stage system and can therefore be applied where land space is a constraint.

3.6 METHANE POTENTIAL AND CALORIFIC VALUE OF THE A-STAGE SLUDGE

In conventional wastewater treatment plants, anaerobic digestion is generally applied to a mixture of primary and secondary (waste activated) sludge. But waste activated sludge is known to be more difficult to digest than primary sludge (Bougrier et al., 2007). For example, Kepp and Solheim (2000a) reported the production of methane of 306 mL CH_4/g VS for primary sludge against 146–217 mL CH_4/g VS for secondary sludge. The proposed A-stage process configuration in this study considered the possibility of introducing sorption ahead of the aerobic process (B-stage) to concentrate organics from combined municipal–industrial wastewater and transfer carbon-rich biomass to the anaerobic digester, where high calorific value biomass can be recovered as methane. It is therefore important to confirm that the resulting sludge has a high calorific value and is indeed biodegradable and that the tested parameters, such as DO, did not affect the methane production rate in a negative way. It is also important to monitor its methane potential for designing future full-scale digesters. During all phases of the experimentation, the spent sludge (56 samples over

210 days) from the clarifier was analyzed for its methane potential (Figure 3.8) and calorific energy value (Figure 3.9). The methane yield was in the range 70–340 mL CH_4/g VS with an average of 205 ± 56 mL CH_4/g VS. Based on a COD balance, a biodegradability of $30\% \pm 10\%$ was found for this type of industrial sludge, which is not significantly different from that of the sludge (23.7%) taken from the full-scale PST on the same site.

Interestingly, in all the BMP tests methane production was completed in less than 2 weeks, which means that anaerobic digestion of this type of sludge could be carried out in a much smaller digester than conventional digesters treating thickened waste activated sludge at 30 days HRT. The process has, therefore, the potential to channel more carbon to the anaerobic digester due to better MLSS and TCOD removal than conventional PST and results in faster conversion to methane gas. At the end of the

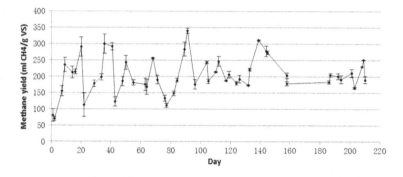

FIGURE 3.8 Methane yield of spent sludge. Error bars indicate standard deviation. (Reproduced from Trzcinski et al., 2017, with permission from IWA Publishing.)

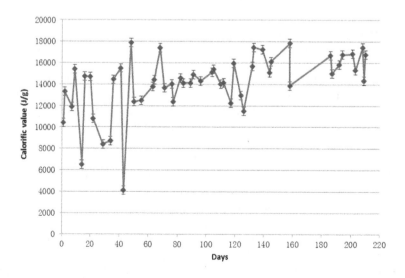

FIGURE 3.9 Calorific value of spent sludge.

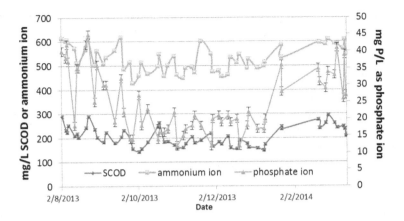

FIGURE 3.10 Evolution of SCOD and nitrogen (as ammonium ion) (left axis) and phosphorus (as phosphate ion) (right axis) in the supernatant of digested spent sludge.

BMP test, the supernatant of digested sludge was analyzed for SCOD, ammonia and phosphate (Figure 3.10). The average SCOD, NH_3-N and PO_4^{3-}-P were 210 mg/L, 530 mg N/L and 25 mg P/L. High ammonia and phosphorus concentrations in the centrate were expected due to their release under anaerobic conditions.

3.7 ELECTRICITY CONSUMPTION

Details of the electrical equipment used on the pilot plant are listed in Table 3.4. The average power consumption was 42.8 kWh/day and the average power consumption per treated water volume was 1.82 kWh/m³. Despite satisfactory solids removal, in this study the methane gas from the A-stage sludge could cover at best 16% of the electricity requirements (assuming an electricity yield of 2.2 kWh$_e$/m³ CH_4 found in

TABLE 3.4
List of Electrical Equipment used on the Pilot Plant and Their Specification

Equipment	Specifications
Blower	Roots-type blower 0.81 kW × 2, 10 Nm³/h
Diffuser	Fine bubble diffuser (ethylene propylene diene monomer), D 250 mm
Mixer	Submersible mixing pump 0.55 kW × 1, 50 L/min
Influent pump	Submergible pump 0.37 kW × 2 with strainer
Sludge pump	Magnetic drive pump 0.18 kW × 1, 0.75 m³/h
Chemical injection pump	Pulsing pump 0.02 kW × 2, ferric chloride

Source: Reproduced from Trzcinski et al., 2017, with permission from IWA Publishing.

Foladori et al., 2010). It is expected, however, that lower energy per m^3 would be possible as larger volume could be treated with the same equipment.

3.8 CONCLUSIONS

After 6-months operation of an A-stage pilot plant treating combined municipal–industrial wastewater, it was found that more than 60% MLSS removal could be achieved despite the high fluctuation in influent MLSS and COD concentrations, which is remarkable considering the high-strength wastewater. The process is operating at an overall HRT of 2 h and is therefore very compact in size, suitable for land-scarce countries or for decentralized applications. Typically, 60%–95% MLSS could be removed in the process at DO levels in the range 0.4–0.7 mg/L and when the influent MLSS was >2000 mg/L. TCOD removals >60% were demonstrated by the process, making it a promising alternative to conventional sedimentation tanks for the pre-treatment of industrial wastewaters. The spent sludge methane potential was on average 205 ± 56 mL CH$_4$/g VS.

4 Thermal/Biological Treatment of Sludge

4.1 THERMAL/BIOLOGICAL TREATMENT IN THE RETURN ACTIVATED SLUDGE (RAS) LINE

IDI through Ondeo-Degremont in France (now owned by Suez) has commercialized the Biolysis® process. It was developed as a process for improving sludge settleability while also reducing the quantities of waste activated sludge (WAS) produced by activated sludge treatment plants. Chemical and enzymatic stressing are used to make cellular material biodegradable, limit microbial growth and increase the energy requirements for the metabolism of bacteria. It is claimed that WAS production can be reduced by 30%–80%. Two versions of the process have been developed. The Biolysis 'O' process uses ozone for chemical oxidation to stress the bacteria (see Section 6.6.6). The Biolysis 'E' process uses a biological enzymatic process to achieve a similar effect. Mixed liquor from the activated sludge process is thickened and passed through a thermophilic enzymatic reactor operating at 50°C–60°C (Figure 4.1). This causes hydrolysis and enzyme release that prevent the reproduction of the bacteria. The treated sludge is sent through a heat exchanger for heat recovery and is returned to the activated sludge system. No external enzymes are used in this process. This process was originally developed in partnership with the Japanese company Shinko Pantec.

However, the return of this highly degradable stream may increase the oxygen requirement of the activated sludge by up to 40%. Lower sludge volume index (SVI) values have also been shown, although the level of improvement will depend on the baseline SVI and the presence of filamentous bacteria prior to the installation of the Biolysis® process. The capital costs for the Biolysis 'E' process include the enzymatic reactor, the heating and heat recovery system, as well as the mixed liquor suspended solid (MLSS) side stream pumping system. Operating and maintenance costs would include energy for heating the process side stream to thermophilic temperatures. The principal benefits of the process are the potential for improved performance and the capacity of the secondary clarifiers due to consistently low SVI values and benefits due to the reduced quantity of residuals that require conveyance, treatment and disposal. The Biolysis process may also have an important benefit over the Cannibal™ process (see Section 2.12), namely the lower footprint required. However, similar to the Cannibal system, savings in the construction, operation and maintenance of aerobic or anaerobic digestion (AD) systems, thickening and dewatering, as well as significant reductions in solids disposal costs will provide offsets against the Biolysis process costs. For larger plants that have existing digesters and co-generation equipment, the

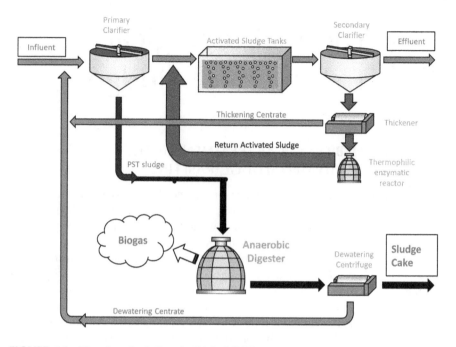

FIGURE 4.1 Flowsheet including the Biolysis® 'E' process.

cost–benefit analysis requirements may not favor the Biolysis process over other available technologies, such as enhanced digestion processes (Roxburgh et al., 2006).

4.2 THERMAL/BIOLOGICAL PRE-TREATMENT BELOW 100°C

One of the pre-treatment methods in the sludge line (Figure 4.2) is an anaerobic or aerobic biological method that requires either thermophilic (around 55°C) or hyper-thermophilic (between 60°C and 70°C) conditions, which typically result in an increase in hydrolysis activity, an increase in biodegradable chemical oxygen demand (COD) and pathogen destruction (Cabirol et al., 2002; Hartmann and Ahring, 2005; Carrère et al., 2010). This type of treatment is also known as "temperature-phased anaerobic digestion" (TPAD), which uses a pre-digester in either thermophilic (around 55°C) or hyper-thermophilic (between 60°C and 70°C) conditions, anaerobic or aerobic. Thermophilic processes, particularly the thermophilic hydrolytic activity of bacterial populations, were investigated 80 years ago, mainly at a temperature of 55°C (Carrere et al., 2012). The low retention time of the thermophilic or hyper-thermophilic anaerobic/microaerobic digester (i.e., pre-treatment digester) is followed by the long retention time of the mesophilic or thermophilic digester (mainly for methanogenesis) which increases the efficiency of AD in terms of biogas production and VS destruction (Bolzonella et al., 2012; Hasegawa et al., 2000). The TPAD allows the separation of the initial stages (i.e., hydrolysis, acidogenesis and acetogenesis) of the digestion process from methanogenesis, thereby encouraging

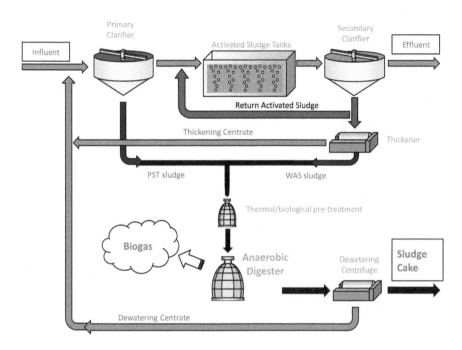

FIGURE 4.2 Flowsheet including a thermal/biological treatment in the sludge line.

the maintenance of the desirable operating conditions of the separate digesters and enhancing sludge reduction. The temperature in the pre-treatment digester is generally at 60°C–70°C with a retention time of 9–48 h. The enhanced volatile solids (VS) removal and improved biogas/CH_4 production are typically 7%–11% and 26%–50%, respectively. It was reported that TPAD was only able to improve the sludge degradation rate and had no impact on the sludge degradation extent (Ge et al., 2011). The TPAD has been successfully applied at the pilot scale.

Thermophilic treatment is usually characterized by accelerated biochemical reactions, higher growth rate of microorganisms and accelerated interspecies hydrogen transfer resulting in an increased methanogenic potential and lower hydraulic retention times (HRT) (Zabranska et al., 2000). Thermal pre-treatment has been reported to improve the stabilization and enhance the dewatering of sludge, reduce the number of pathogens and reduce foaming problems (Han et al., 1997), and can be realized at relatively low costs (Muller, 2001). Pre-treatments below 100°C have been shown to effectively increase biogas production in AD (Table 4.1). These pre-treatments are also safer, easier and cheaper to operate than thermal treatments above 100°C (Gavala et al., 2003b). Thermal treatment requires a long duration for obvious effects; normally tens of minutes are often needed (Bougrier et al., 2008). The treatment time appears to have little effect on AD compared to the temperature (Valo et al., 2004a; Bougrier et al., 2006).

Dhar et al. (2012) reported approximately five times higher soluble COD (SCOD) after treating WAS at 90°C for 30 min, while Eskicioglu et al. (2006a) used 96°C and reported ~3.6 times higher SCOD. Nges and Jing (2009) used temperatures of

TABLE 4.1
Thermal/Biological Pre-Treatment of Sludges

| Sludge | Thermal/Biological Pre-Treatment Conditions | Solubilization | | Anaerobic Digestion | | | | Scale, Mode, Temperature (HRT, days) | References |
		COD	DD	CH$_4$ Production	VS Destruction	TCOD Removal (%)	Dewaterability		
WAS	80°C, 1 h	55–850 mg/L	7.2%	+4.2%	31.1%	—	CST: 69.4–50.5 s	LS, B	Şahinkaya and Sevimli (2013a)
WAS	50°C, 30 min 70°C, 30 min 90°C, 30 min	SCOD/TCOD: from 6% to 18% from 6% to 20% from 6% to 36%	—	+14 +19 +13	25% 26% 27%	—	TTF: 81–48 s TTF: 81–56 s TTF: 81–54 s	LS, B	Dhar et al. (2012)
TWAS (4.7%–5.9%)	96°C, 80 min	SCOD/TCOD: from 6% to 27%	—	+475%	—	—	—	LS, B	Eskicioglu et al. (2006a)
PS+WAS (3:1)	50°C–70°C, 2 days	SCOD/TCOD: from 2% to 27%	—	+11%	+0%–9% (VS)	—	—	LS, B	Nges and Liu (2009)
WAS (TS: 20 g/L)	60°C–70°C, 1–2 days, microaerobic	40%	—	+26%–50%	+7%–11%	—	—	LS and PS, B and SC, M	Hasegawa et al. (2000)
TWAS (4.6%–5.5%)	50°C 75°C 96°C	SCOD/TCOD: 5%–15% 5%–19% 5%–24%	—	0% 2% 0%	—	—	—	LS, B	Eskicioglu et al. (2007c)
PS	70°C, 4 days	—	—	+16.2% (M) +80% (T)	—	—	—	LS, SC	Gavala et al. (2003b)

(Continued)

TABLE 4.1 (CONTINUED)

Thermal/Biological Pre-Treatment of Sludges

Sludge	Thermal/ Biological Pre-Treatment Conditions	Solubilization		Anaerobic Digestion				Scale, Mode, Temperature (HRT, days)	References
		COD	DD	CH$_4$ Production	VS Destruction	TCOD Removal (%)	Dewaterability		
SS	70°C, 2 days	—	—	+19.8% (M) +0% (T)	—	—	—	LS, SC	Gavala et al. (2003b)
WAS (TSS: 17.1 g/L)	70°C, 9 h	SCOD: 20–2200 mg/L	—	+41.8%	—	—	—	LS, B	Zhang et al. (2010)
WAS (SS: 7.7 g/L)	Microaerobic 65°C, 1 day	—	—	+0%	+17%	+14%	—	PS, C, M (44 days)	Dumas et al. (2010)
PS (TS: 26.9 g/L)	50–65°C, 2 days	—	—	+25%	+20%	—	—	LS, C, M (13–14 days)	Ge et al. (2010)
WAS (TS: 2%–3%)	50°C 60°C 65°C 70°C, 4 days	15% (7%) 27% 27% 27%	—	+42% +45% +50% +88%	34% (37%) 41% (37%) 48% (37%) 48% (37%)	—	—	LS, SM, M (14 days)	Ge et al. (2011)
PS+WAS (TS: 39 g/L)	70°C 9, 24, 48 h	—	—	+20%	—	—	—	LS, B&SC, T (10 days)	Ferrer et al. (2008) and Ferrer et al. (2009)
PS (TS: 15–17.9 g/L)	70°C 2 days	—	—	+48%	+12%	—	—	LS, C, T (13 days)	Lu et al. (2008b)
TWAS (TS: 58 g/L)	— — 65°C, 2 days	— — 19.6 g/L	—	Control +36% +48%	36% 48% 55%	35% 45% 55%	—	PS,M (20) T (20) T (18)	Bolzonella et al. (2012)

(Continued)

TABLE 4.1 (CONTINUED)
Thermal/Biological Pre-Treatment of Sludges

Sludge	Thermal/ Biological Pre-Treatment Conditions	Solubilization		Anaerobic Digestion				Scale, Mode, Temperature (HRT, days)	References
		COD	DD	CH₄ Production	VS Destruction	TCOD Removal (%)	Dewaterability		
PS+WAS (50:50) (TS: 3%–5%)	55°C, 4 days	–	–	+30%–100%	45% (32%)	–	–	LS, SM, M (20)	Han et al. (1997)
WAS (60 g/L)	50°C–55°C, 1–3 days	–	–	+16.5%	60% (48%)	–	–	FS, M (12–15)	Oles et al. (1997)
PS (TSS: 14.5 g/L)	70°C, 2 days	–	–	+11%	+28%	–	–	LS, C (13), T	Skiadas et al. (2005b)
WAS (TSS: 9 g/L)	70°C, 2 days			+38%	+616%			LS, C (13), T	

Note: Scale: full scale (FS), pilot scale (PS) and laboratory scale (LS). Mode: batch (B), semi-continuous (SC) or continuous (C). Temperature of anaerobic digestion: mesophilic (M) or thermophilic (T). TTF = Time to filter(s).

50°C–70°C and did not observe a difference between the two temperatures with the SCOD/total COD (TCOD) ratio in the range 2%–21%. Gavala et al. (2003b) applied a pre-treatment at 70°C for 7 days to obtain a 26% increase in methane production. Wang et al. (1997) investigated the thermal pre-treatment at 60°C–100°C and concluded that the methane yield can increase by 30%–52%. However, no significant difference was observed between the three temperatures, except that the rate was higher for the 60°C pre-treatment. Gavala (2003b) showed that the 70°C pre-treatment had a positive effect on the thermophilic AD of primary sludge, but the effect was minimal if the AD was carried out at mesophilic temperatures. The impact was also greater on primary sludge compared to secondary sludge. This is because primary sludge is mainly composed of carbohydrates, while secondary sludge mainly consists of bacterial cells, characterized by a higher protein content and much lower amounts of carbohydrates. Therefore, the authors argued that the temperature selection of the pre-treatment and AD step should depend on the composition of the sludge. The extreme thermophilic pre-treatment (65°C for 2 days HRT) was shown to increase anaerobic biodegradability due to a higher hydrolytic capability (Bolzonella et al., 2012). Barjenbruch and Kopplow (2003) noted an improvement in dewaterability when using a 90°C pre-treatment (–6% in capillary suction time [CST]) compared to the 121°C thermal pre-treatment (+10% in CST). This was confirmed at temperatures in the range of 50°C–90°C by other research groups (Dhar et al., 2012).

It has been reported that methane production can double as a result of thermal/biological pre-treatment (Han et al., 1997). However only a 14% methane increase was achieved by other researchers using different sludges (Val del Río et al., 2011). The solubilization of WAS by hyper-thermophilic heat treatment is induced by sludge lysis and further cryptic growth (lysis-cryptic growth) (Wei et al., 2003). In the lysis-cryptic growth, sludge reduction is achieved because some portions of lysates are consumed for catabolism and finally emitted as CO_2. For this reason, there may be a loss of carbon as CO_2 if the biological/thermal treatment is too long. It is not known what the thermal treatment time should be that minimizes this CO_2 production and optimizes the methane increase in the subsequent AD. The thermal pre-treatment of a few days retention time also helps to reduce the fecal coliform and odor of the final residue (Skiadas et al., 2005). Yan et al. (2008) used a simple heat treatment process (700 mL incubated at 60°C, 120 rpm for 24 h in a 1 L Erlenmeyer flask) and showed that there was a rapid increase in the population of thermophilic bacteria at the early stage of heat treatment and the emergence of protease-secreting bacteria. Hasegawa et al. (2000) showed that hyper-thermophilic aerobic microbes were identified as belonging to *Bacillus*. An increase in hydrolytic activity was also demonstrated and resulted in 40% volatile suspended solids (VSS) solubilization due to the pre-treatment. The production of biogas after AD of the microaerobically pre-treated sludge was increased by 1.5 times when compared with the sludge without pre-treatment. The destruction of 75% of organic solids from excess WAS was obtained at full scale, by combining a conventional municipal activated sludge process with a thermophilic aerobic sludge digester (65°C, HRT of 2.8 days) (Shiota et al., 2002). However, depending on the type of sludge (primary, secondary or a mixture of both), the residence time of this type of treatment is generally 2 days or longer. Therefore, the potential for increased performance is inherent in the sludge itself (Pilli et al., 2011) and thermal treatment at temperatures below 100°C is thus concurrent with biochemical reactions. However, it is not clear the extent to which the thermal and enzymatic reactions are responsible for the improvement. The increased thermophilic temperature may improve the production of extracellular enzymes to hydrolyze more complex or inert substrate materials, and the specialized microbial community may thrive under thermophilic environments (Ge et al., 2011). Recently, these biological/thermal pre-treatments have received particular attention for industrial application due to their efficiency and relatively low cost compared to chemical-physical methods that require greater capital investment and higher operating costs.

Thermal pre-treatment at temperatures below 100°C revealed an increase of more than 30% in gas production at lower temperatures such as 60°C and 80°C; however, the low temperature pre-treatment necessitated a longer contact time than the high temperature treatment (Hiraoka et al., 1985). During AD, the concentration of acetic and propionic acids increased with the reaction time, but there was little influence on butyric and organic acids with 5 carbon atoms or more (Hung-Wei et al., 2014). However, volatile fatty acid (VFA) release from primary hydrolyzed sludge was between four and seven times higher than from activated sludge due to the hydrolysis of unsaturated lipids (Wilson and Novak, 2009). Bi et al. (2013) reported that thermal pre-treatment was better suited to meso- rather than thermophilic digestion at thermal hydrolysis temperatures up to 120°C. Thermal hydrolysis pre-treatment enables an increase in the organic loading rate to the anaerobic digester by a factor of 2.3 compared to standard AD. The typical design loading rates with thermal pre-treatment are in the range of 5–6 kg VS/m^3.day (Oosterhuis et al., 2014).

The increase in anaerobic biodegradability may be linked to the particle size of bioflocs in sludge. However, some studies reported an increase in particle size to median size after thermal pre-treatment (Bougrier et al., 2006), whereas other studies reported a decrease in the average size from 70 to 35 mm (Barber, 2010) and a decrease in the average size from 107 to 66 mm (Neyens et al., 2004). This reduction in particle size is consistent with the fact that the hydrolysis rate has been reported to increase with decreasing particle size (Aldin, 2010).

Despite the advantages of thermal pre-treatment, it also has drawbacks for real application. For some sludge samples, the methane increase may not compensate for the energy input (Val del Río et al., 2011). Several authors have demonstrated that the yields obtained are sufficient to sustain the energy balance of the process (Bolzonella et al., 2007; Nges and Liu, 2010; Lu et al., 2008a; Han et al., 1997). Bougrier et al. (2007) stated that burning the additional methane would almost provide sufficient energy to heat the sludge. More data on the effects of thermal treatment below 100°C are also available in Chapter 9.

4.3   THERMAL TREATMENT ABOVE 100°C

4.3.1   Introduction

Historically, thermal pre-treatment has been shown to be effective (Stuckey and McCarty, 1984; Li and Noike, 1992; Bougrier et al., 2008). The heat disintegrates the floc structure, solubilizes organic particulates and even degrades some macromolecules into monomers and VFAs. Thermal treatment destroys cell walls thereby making intracellular materials available for the subsequent biodegradation (Camacho et al., 2005). Thermal pre-treatment is generally carried out at 165°C–180°C for 30 min, with a few cases performed at 121°C, although researchers have studied temperature ranges of 60°C–275°C for 10–180 min. The improved total solids (TS)/VS removal and CH_4 production were in the ranges of 7%–32% and 14%–90%, respectively. Thermal pre-treatment can improve both the sludge degradation rate and the sludge degradation extent (Batstone et al., 2009).

Thermal pre-treatment can be applied prior to both thermophilic and mesophilic AD (Skiadas et al., 2005; Qiao et al., 2011). Its effectiveness seems highly dependent on the sludge characteristics. An optimum in methane production (+60%) and COD solubilization (+40%) was found around 170°C–175°C (Haug et al., 1978). Temperatures above 200°C result in toxic compounds such as dioxins (Stuckey and McCarty, 1984; Bougrier et al., 2008). The increasing production of refractory material with increasing temperatures is clearly evident in the work of Stuckey and McCarty (1984) who noted a continual deterioration in gas production from thermally hydrolyzed activated sludge, which was 27% higher than a control at 175°C, similar at 250°C and lower than the control above that temperature. At thermophilic conditions, the impact of increasing temperature on reducing performance was further exacerbated. However, as well as increasing the quantity of refractory material, increasing the reaction temperature has also been linked to improved dewaterability (Everett, 1972; Higgins et al., 2015) and enhanced biogas production (Hung-Wei et al., 2014; Stuckey and McCarty, 1978a), implying different optimal operating conditions exist depending on the required project outcomes.

Thermal pre-treatment has been applied to full-scale wastewater treatment plants (WWTPs) and is the most widely used technology to achieve sludge reduction. The first plant with thermal pre-treatment (Cambi process) was operated in 1995 in Hamar, Norway (Kepp et al., 2000). Recently, the thermal pre-treatment technology has been commercialized by several other companies to achieve sludge reduction, including Veolia's Biothelys in 2006, Kruger-Veolia's Exelys in 2010, Sustec's Turbotec in 2011 with one pilot plant, Aqualogy's CTH in 2012 with an industrial prototype, Eliquo's Lysotherm in 2012 and Biorefinex in 2013 (Wang et al., 2017).

Full-scale application of thermal treatment such as the Cambi process at 170°C has resulted in a 47% increase in biogas and enhanced dewatering (Kepp et al., 2000). The sludge is dewatered to 12%–15% TS before the thermal treatment, which can save up to 50% of the anaerobic digester's volume. They reported that 24 kW of electricity was required for the hydrolysis step. The high degree of stabilization (60%) led to a 23% reduction in solids and the final sludge after dewatering contained 35% TS compared to 20% with the conventional process. This translated into a 20% increase in net electricity production. The thermal hydrolysis of sewage sludge involves the application of heat above autoclave temperature. The heat is typically provided by live steam injection at a design temperature and concomitant pressure, which is then rapidly released (exploded) in a flash unit, although some configurations use standard heat exchange (Barber, 2016).

Currently, there are 75 facilities of which 39 are operating and the remaining are in various stages of design (Barber, 2016). Barber (2016) summarized the following advantages of thermal treatment:

- Improves the biodegradability of activated sludge
- Improves the biodegradability of primary sludge
- Allows significantly higher loading rates resulting in smaller digestion plants
- Increases the rate of biogas production
- Reduces sludge viscosity
- Improves sludge dewaterability on all dewatering systems
- Sterilizes sludge providing pathogen-free biosolids
- Reduces odor and pathogen regrowth from dewatering
- Eliminates scum and foaming and produces conditions that do not encourage foaming
- Minimizes inhibition due to hydrogen sulfide
- Significantly reduces downstream requirements for drying and other thermal processes
- Numerous sites successfully operating at full scale

The disadvantage of the technology are

- Parasitic energy demand with some configurations (depends on process)
- Higher ammonia concentration than standard digestion
- Potential for the production of refractory material especially with food waste
- Potential increase in polymer demand for dewatering
- More complex than standard AD
- Requires boilers
- Sludge needs cooling prior to AD
- Requires centrifuge thickening to 16%–18% dry solids (DS)
- Higher release of nutrients with potential for salt crystallization and subsequent maintenance issues and deterioration of dewaterability

4.3.2 Effects of Thermal Treatment on Sludge Solubilization

Thermal treatment increases COD solubility in a linear-type fashion with both temperature (130°C and 170°C) and reaction time (10–60 min) (Hung-Wei et al., 2014). Everett (1972) showed that COD solubility increases between 150°C and 180°C and reaction time (0–90 min). However, at 170°C and above, no further solubilization occurred between 60 and 90 min. For temperatures lower than 200°C, COD solubilization was found to increase linearly with the treatment temperature for different sludge samples tested (Bougrier et al., 2007). Liu et al. (2012c) confirmed that COD solubility increases with temperatures between 125°C and 175°C.

Bougrier et al. (2008) reported that the solubility of carbohydrates also increased between 130°C and 210°C, but little influence was observed between 95°C and 130°C. In another study, the solubility of carbohydrates was reported to occur below 150°C, above which the solubility of proteins became more evident (Hung-Wei et al., 2014). According to Liu et al. (2012c), the solubility increased with temperatures between 130°C and 170°C with a strong linear increase between 10 and 30 min, after which further increase was negligible. Noike et al. (1985) reported an increase with temperatures between 125°C and 175°C. The reduction in the solubility of carbohydrates beyond 170°C is hypothesized to be due to conversion to refractory material (Bougrier et al., 2008). Regarding the solubilization of lipids, Bougrier et al. (2008) reported little influence from either temperature or reaction time, which was consistent with other researchers (Hung-Wei et al., 2014; Li and Noike, 1992).

Following thermal treatment at high pressure and temperature, a flash unit suddenly drops the pressure. This rapid decompression also contributes to cell lysis in WAS and consequently an increase in sludge solubilization. Some authors investigated the effect of the pressure drop in the flash unit on solubilization and found that the concentration of both carbohydrates and proteins increased in an approximately linear fashion with an increasing pressure drop between 3 and 6 bar. In contrast, the release of lipids was not influenced by the pressure drop (Perrault et al., 2015). The biogas production rate is also more rapid with an increasing pressure drop.

4.3.3 EFFECT OF THERMAL TREATMENT ON ANAEROBIC BIODEGRADABILITY

Thermal treatment significantly improves biogas production with an increase in temperature of between 130°C and 170°C (with constant reaction time of 30 min) (Hung-Wei et al., 2014). There is little difference in biogas production within the first 24 h; however, higher production was observed after 10 days. Concurrently, there are minor improvements in VS reduction during digestion with increasing thermal hydrolysis processing temperatures (Higgins et al., 2015). However, the methane yield (m^3 methane/kg VS added) was higher at 170°C compared to lower temperatures, which were all similar. WAS digestion significantly improved by 75%–80% by over-pressurizing to 21 bar, although there has been no discussion as to the additional energy requirements of over-pressurization (Phothilangka et al., 2008). Oosterhuis et al. (2014) reported an increase in gas production yield of over 100 L/kg VS fed at the laboratory scale with a 20:80 primary:activated sludge mixture. The VS reduction of WAS increased from 26% to 42% (relative increase of 62%), which is similar to the relative increase in WAS digestion of 55% at 140°C reported elsewhere (van Dijk and de Man, 2010). When pre-treating sludge by thermal hydrolysis before digestion at the large scale, Haug calculated a 25% increase in energy production compared to conventional digestion (Haug et al., 1978). In some cases, inhibition was observed during AD when a thermal pre-treatment above 175°C was applied (Stuckey and McCarty, 1984; Haug et al., 1978). In more recent studies, no inhibition was observed with a digester feed of 10.4% DS (Oosterhuis et al., 2014) and up to 13% DS (Chauzy et al., 2008) (Table 4.2).

TABLE 4.2
Thermal (above 100°C) Pre-Treatments of Sludges

Sludge	Thermal Pre-Treatment Conditions	Solubilization		Anaerobic Digestion			Dewaterability	Scale (HRT, days)	References
		COD	DD	CH₄ Production	VS Destruction	TCOD Removal (%)			
TWAS (TS: 20 g/L)	170°C, 90 min	+41% (+48% TS)	–	+51%	–	–	CST: 151–39 s	LS, B	Bougrier et al. (2006)
	190°C, 60 min	+48% (+38% TS)		+48%			CST: 151–29 s		
WAS (TS: 17.1 g/L)	130°C, 1 h	SCOD/TCOD: 2.7%–25.3%	–	+21%	–	–	–	LS, B	Valo et al. (2004a)
	150°C, 1 h	2.7%–43.9%		–					
	170°C, 1 h	2.7%–59.5%		+45%					
WAS (TS: 17.1 g/L)	170°C, 1 h	–	–	+54%	+92%	70.7% (44)	–	LS, C (M, 20 days HRT)	Valo et al. (2004a)
TWAS (TS: 14.5 g/L)	135°C, 30 min	5.9 g/L (5.6 g/L)	–	+12%	41% (39%)	58% (52%)	CST: 233 (481)	LS, SM, M (20)	Bougrier et al. (2007)
	190°C, 15 min	7.6 g/L (5.6 g/L)		+25%	57% (39%)	63% (52%)	105 (481)		
WAS (TS: 10%–12%)	130–180°C, 30 min	–	–	+47	+23	59 (40)	TS content: 35% (20)	FS, M (17)	Kepp et al. (2000)
AGS (TS: 106.1 g/L)	170–210°C, 20 min	57%	–	+14%–18%	–	–	–	LS, B	Val del Río (2011)
WAS (1% TS)	175°C, 30 min	68%	–	+60%–70%	36% (VSS)	–	–	LS, SC, M, 15	Haug et al. (1978)
WAS (7 g SS/L)	180°C, 1 h	30% (VSS)	–	+90%	–	–	–	LS, B, M	Tanaka et al. (1997)

Note: AGS, aerobic granular sludge. Scale: full scale (FS), pilot scale (PS) and laboratory scale (LS). Mode: batch (B), semi-continuous (SC) or continuous (C). Temperature of anaerobic digestion: mesophilic (M) or thermophilic (T).

4.3.4 EFFECT OF THERMAL TREATMENT ON VISCOSITY, DEWATERABILITY AND FOAM CONTROL

During thermal treatment, the viscosity falls significantly with increasing temperatures of between 130°C and 170°C (Higgins et al., 2015). Following thermal hydrolysis (175°C for 60 min), the viscosities of WAS, kitchen waste and vegetable and fruit residues reduced from 13,500 to 36,000; 6250– to 1625; and 1658 to 663 mPa.s, respectively (Liu et al., 2012c). The viscosity reduces further with thermal pre-treatment than it does with other technologies such as ozonation or ultrasonic treatment (Bougrier et al., 2006). The thermal treatment of sludge (170°C and 190°C) changes the rheological properties of WAS from non- to pseudo-Newtonian. Although thermal hydrolysis reduces apparent sludge viscosity by orders of magnitude, the viscosity of digested sludge is independent of the presence or absence of pre-treatment (Dawson and Ozgencil, 2009).

Besides an obvious increase in biogas production during AD, the pre-treatment has also been reported to be beneficial in terms of dewatering of sludge. In 1972, dewaterability was shown to improve with treatments between 180°C and 210°C (Everett, 1972). The heat treatment of sludge improved dewaterability whether or not the sludge was anaerobically digested (Haug et al., 1978). Dewaterability was improved in a linear fashion from approximately 27% to 32% DS with increasing temperatures between 130°C and 170°C (Higgins et al., 2015). Similarly, the DS content of the final dewatered sludge increased from 25.2% to 32.7% when a thermal pre-treatment of 180°C and 21 bar was applied (Phothilangka et al., 2008). Oosterhuis et al. (2014) reported an improvement from 26% to 33.9% DS.

The centrate from thermally hydrolyzed digested sludge is, however, one-third higher in COD and three times higher in ammonia compared to digested sludge (Barber, 2010). In terms of scum control, a pilot-scale plant was successful in treating scum comprised of *Gordonia*-type organisms when a thermal treatment was applied (Jolis and Marneri, 2006). Thermal hydrolysis at 170°C was successfully used to abate foaming issues in downstream pilot-scale digesters. A significant reduction in the abundance of the filamentous organisms *Microthrix parvicella* was noted by Alfaro et al. (2014). Heat treatment prior to AD removed foaming issues by destroying hydrophobic materials (Barjenbruch and Kopplow, 2003). A potential disadvantage is the increase in polymer demand during dewatering due to thermal hydrolysis, which was reported at the laboratory scale (Oosterhuis et al., 2014). The polymer dosage increased from 10.5 to 18 kg active polymer per ton dry solids (TDS). However, no difference was noted at full scale, with a dose of 8 kg active polymer/TDS for centrifuge dewatering (Lancaster, 2015). Moreover, odorous compounds normally associated with sewage sludge are significantly reduced during the digestion of thermally pre-treated sludge (Neyens and Baeyens, 2003). Significantly lower odor potential during dewatering for both belt presses and high-speed centrifuges was also reported (Chen et al., 2011).

4.3.5 MASS BALANCE OF THERMAL HYDROLYSIS

A key factor regarding thermal hydrolysis is the minimization of the energy requirement needed to reach the reaction temperature. Subsequently, it is important to optimize the quantity and temperature of the sludge being processed. As the sludge moiety has a lower specific heat capacity than water (Xu and Lancaster, 2009), increasing the DS of the sludge will intrinsically reduce energy requirements. For that reason, the sludge is typically thickened to approximately 15%–18% DS (Figure 4.3), but further thickening may incur heat transfer limitations as well as practical processing concerns.

It is not possible to extract data from full-scale mass balances due to numerous site-specific conditions. Therefore, Figure 4.3 shows a calculated water and sludge balance for a hypothetical plant processing all of a 60:40 blend of primary and activated sludge with an annual quantity of 10,000 tonnes, based on theoretical considerations and design criteria (Barber, 2016). After the initial wastewater has been thickened to reduce the energy requirements of hydrolysis, it is then diluted twice. Firstly, as a consequence of steam addition to reach the reaction temperature, and secondly, from the addition of treated water to control ammonia inhibition in downstream digestion.

4.3.6 Energy Balance of Thermal Hydrolysis

The steam demand is influenced by the influent sludge temperature, the temperature difference, the sludge thickness and the thermodynamic and physical properties of fluid. The energy required to provide steam can be supplied in numerous ways, including direct use of boilers running on either bio- or natural gas, co-generation using a reciprocating internal combustion engine with an auxiliary boiler running on bio- or natural gas or the use of a gas turbine in larger facilities, although the latter comes with a loss of power generation unless operated in a combined cycle (Fernandez-Polanco and Tatsumi, 2016). The quantity of energy required is typically described as a fraction of the biogas generated; however, comparing the literature data reveals

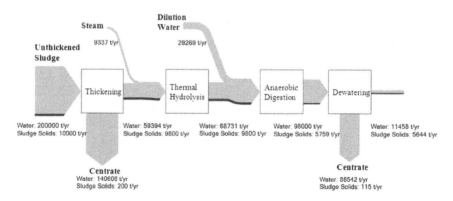

FIGURE 4.3 Typical water and sludge balance for standard thermal hydrolysis followed by anaerobic digestion. Blue line (water); brown line (sludge) for plant processing 10,000 tonnes dry solids of 60:40 primary:activated sludge mix. Based on live steam injection systems that make up over 99% of installed capacity. (Reproduced from Barber, 2016, with permission from Elsevier.)

that the energy required does not follow a predictable pattern (Lancaster, 2015; Pook et al., 2013; Merry and Oliver, 2015). This is due to several parameters including the type, efficiency, configuration and availability of the co-generation plant; the presence and quantity of gas storage; the gas production profiles; the configuration, operating temperature and retention time of AD plant; and, by far the most important parameter – albeit habitually overlooked – the sludge composition itself. In order to elucidate the wide-ranging energy demands observed in the literature, it is necessary to look closer at the energy balance for a theoretical situation that is independent of the variability of full-scale installations. The importance of thickening the sludge to exploit the lower specific heat capacity of the sludge fraction has been demonstrated (Barber, 2016).

Figure 4.4 shows the main energy flows for the hypothetical facility of 10,000 tonnes of sludge with both primary and activated sludge being thermally hydrolyzed. In the example, biogas yield and composition are determined independently for primary and activated sludge using elemental composition and stoichiometry (Barber, 2016). The expected VS destruction of both unthermally and thermally hydrolyzed sludges are also sludge dependent and based on expectations from the literature (Vesilind, 2003; van Haandel and van der Lubbe, 2007; Speece, 2008;

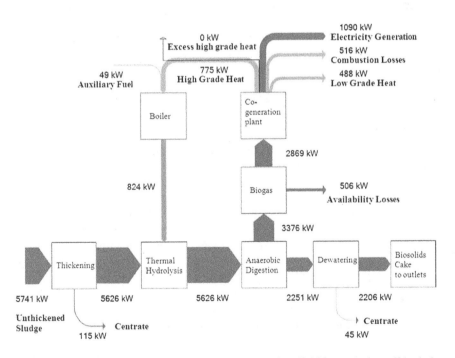

FIGURE 4.4 Typical energy balance for a plant processing 10,000 tonnes dry solids sludge with a primary:activated sludge ratio of 60:40. Both primary and activated sludge are thermally hydrolyzed. Energy balance based on the use of an internal combustion engine based on: 85% availability; 38% electrical efficiency; 27% high-grade heat; 17% low-grade heat. (Reproduced from Barber, 2016, with permission from Elsevier.)

Phothilangka et al., 2008; Winter and Pearce, 2010; van Dijk and de Man, 2010; Oosterhuis et al., 2014; Hung-Wei et al., 2014; Lancaster, 2015). In this instance, the high-grade heat can provide close to 90% of the energy required to raise the necessary steam. The power output is 950 kWhre/t DS processed, compared with an equivalent scenario with no thermal hydrolysis of 825 kWhre/t DS processed, typical of UK facilities (Mills et al., 2014).

Ironically, energy benefits improve with increasing quantities of primary sludge because it is more biodegradable irrespective of thermal hydrolysis (Haug et al., 1978; WEF, 1987) and it has an inherently higher energy content, both in combination, resulting in a higher biogas yield than an equivalent quantity of activated sludge (Barber, 2016). The greater biodegradability and subsequent gas production from primary sludge elevates the heat produced, concomitantly reducing the outstanding energy demand.

Although thermal hydrolysis enhances performance and biogas production, the auxiliary energy demand (49 kW in the configuration described in Figure 4.4) makes the energy benefit relatively modest when concentrating only on AD and co-generation.

However, where the energy benefit from deploying thermal hydrolysis is clearly evident, is downstream of AD. A combination of improved VS destruction combined with enhanced dewaterability characteristics results in significantly less digested sludge with, importantly, less water contained within it. The reduction in downstream energy requirements afforded by thermal hydrolysis has been exploited at several full-scale facilities. In Ringsend, Dublin, a planned expansion in drying capacity of 300% was reduced to 50% after thermal hydrolysis was installed (Pickworth et al., 2006; Panter and Kleiven, 2005). In the UK, a trend has been observed whereby the process has been used to downscale or remove sludge drying facilities altogether (Rawlinson et al., 2009; Merry and Oliver, 2015; Merry and Fountain, 2014).

In Europe, developments have been made to reduce the energy demand by altering the configuration of thermal hydrolysis. One configuration that is gaining traction is the concept of hydrolyzing only the activated sludge (as it is more amenable to treatment – Shana et al., 2013). The energy balance result in that case is 402 kW of excess high-grade heat versus 0 kW when both primary and activated sludge are thermally hydrolyzed. Although the primary treatment is not exposed to thermal hydrolysis, the biogas production falls only slightly in accordance with previous studies (Phothilangka et al., 2008; Oosterhuis et al., 2014; van Dijk and de Man, 2010; Wilson and Novak, 2009) with an equivalent power generation (based on the same assumptions) of approximately 900 kWhre/t DS processed. When comparing the thermal hydrolysis of all sludge against only activated sludge, Mills et al. (2013) observed a minor fall in biogas yield when processing only the activated sludge fraction (421 and 454 Nm^3/tDS for activated and all sludge processed, respectively). However, as only the activated sludge is processed, the hydrolysis unit is significantly reduced in size, and the steam demand is met with no auxiliary fuel. Another approach showing promise is that of positioning the thermal hydrolysis unit downstream of AD to concentrate on the inherently less biodegradable effluent prior to a second stage of digestion (Mills et al., 2013; Chauzy et al., 2014). The first recorded application of this approach was in Hillerød (Gurieff et al., 2011). Subsequently, pilot-scale studies showed an increase in biogas yield of over 10% to 503 Nm^3/tDS when this configuration was compared with standard pre-digestion positioning (Mills et al., 2013).

5 Mechanical Treatment of Sludge

5.1 INTRODUCTION

Compared to other pre-treatment methods, mechanical pre-treatments have the following advantages: no chemicals, easy operation, low investment and operational costs and high lysis efficiency (Zhang et al., 2012a). Mechanical sludge disintegration is categorized into different types, such as stirred ball milling (SBM), lysis-thickening centrifuge (LTC), high-pressure homogenizer (HPH), ultrasonication, high-pressure jet and collision and rotor-stator disintegration (Foladori et al., 2010). These treatments are applied on secondary sludge or mixtures of primary and secondary sludges as depicted in Figure 5.1. Some commonly applied mechanical pre-treatment technologies will be introduced here.

5.2 STIRRED BALL MILLING

SBM occurs in a cylindrical grinding chamber equipped with a central rotating shaft. Beads are put into the chamber to provide more shear force for sludge disintegration. A picture of a stirred ball mill is shown in Figure 5.2.

The shear force caused by the beads helps to disintegrate the sludge and increases biogas production. Baier and Schmidheiny (1997) reported the efficiency of SBM at treating sludge. A slight increase in biogas production of 10% is observed for activated sludge with a sludge age of 7 days. However, a 62% increase in biogas production was detected for thickened anaerobic digested sludge.

5.3 LYSIS-THICKENING CENTRIFUGE

LTC is designed based on a thickening centrifuge machine. Additional cutting blades are installed in the rotating shaft for sludge disintegration. Special disintegration gear is incorporated at the end of the centrifuge machine to give out the lysate (Figure 5.3).

An increase in sludge disintegration and methane production can be achieved. An average biogas increase of 31.8% for treated excess sludge than untreated excess sludge was observed. However, it should be noted that this average value was obtained from 8.1% to 86.4% with the changing sludge characteristics (Dohányos et al., 1997). Moreover, LTC seems to have a lower efficiency at treating a mixture of primary and excess sludge. For the same treatment condition, an average biogas increase of only 13.6% (range from 0% to 24%) was observed (Dohányos et al., 1997). Full-scale applications of LTC were reported by Zábranská et al. (2006). The increase in biogas production varied from 14.5% to 26%.

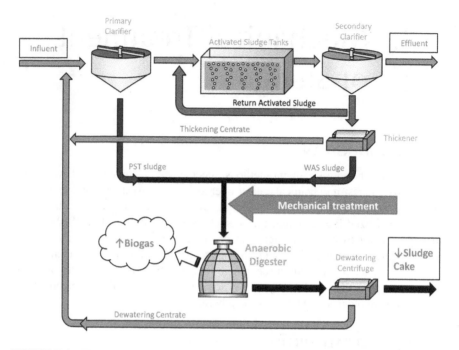

FIGURE 5.1 Flowsheet including a mechanical pre-treatment in the sludge line.

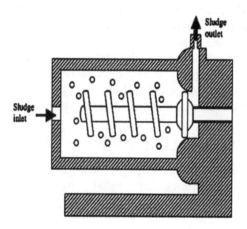

FIGURE 5.2 Sketch of stirred ball milling.

5.4 HIGH-PRESSURE HOMOGENIZER

HPH is an alternative mechanical treatment technology. The major principle of HPH relies on abrupt pressure gradient, high turbulence, cavitation as well as strong shearing forces, which are aroused under the strong depressurization of highly compressed sludge suspensions (up to 900 bar) While the sludge enters the HPH, its speed increases up to 50 times its original speed (Figure 5.4), which results in cavitation

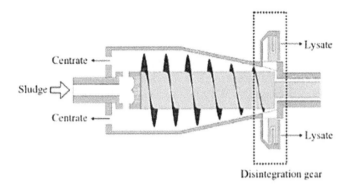

FIGURE 5.3 Sketch of lysis-thickening centrifuge.

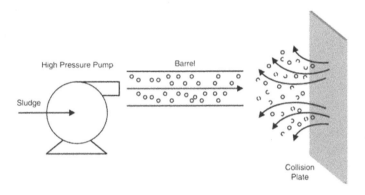

FIGURE 5.4 Jetting and colliding pre-treatment.

and collisions among sludge particles (Rai and Rao, 2009). These will subsequently induce sludge disintegration and cell destruction. A sludge reduction of 20%–94% has been reported at the pilot scale (Camacho et al., 2002).

Rai and Rao (2009) observed that sludge DD_{COD} increased to 4.5%, 10.7% and 12.5% at 200, 300 and 400 bar treatments, respectively; however, there was no further improvement for higher homogenization pressure. In a laboratory-scale semi-continuous experiment, Wahidunnabi and Eskicioglu (2014) compared two-phased anaerobic digestion (2PAD) with HPH pre-treatment (HPH++2PAD) to conventional anaerobic digestion (i.e., single stage [control] and 2PAD) of municipal sludge. The homogenizing pressure was found to be the most significant factor ($P < 0.05$) affecting the solubilization of particulate chemical oxygen demand (COD) and biopolymers in sludge. At 12,000 psi and 0.009 g NaOH/g total solids (TS), the HPH++2PAD system achieved maximum methane production (0.61–1.32 L CH_4/L digester-d) and volatile solids (VS) removal (43%–64%), as well as significant pathogen removal and a positive energy balance. HPH is characterized by easy operation, low investment and high-energy efficiency and, accordingly, has been in high demand for large-scale implementations over the past years. In a demonstration project described by

Onyeche (2007), a modified homogenizer at 150 bar with a flow rate of 2.7 m³/h was employed to pre-treat concentrated sludge; the results indicated about 23% sludge reduction with more than 30% increased biogas production, leading to enormous savings in sludge disposal costs. In addition, Rabinowitz and Stephenson (2005) applied a patented MicroSludge™ unit in a demonstration project in Los Angeles wastewater treatment plants (WWTPs) in October 2005. The process used chemical pre-treatment to weaken cell membranes and a HPH to burst the cells.

Nah et al. (2000) used a pilot-scale (2000 L) anaerobic digester with a jetting and colliding pre-treatment at 30 bars as shown in Figure 5.4. The soluble COD (SCOD) increased from 250 to 800 mg/L and the suspended solids decreased from 20.7 to 19.4 g/L, indicating that solubilization took place in the mechanical pre-treatment. The gas production and VS removal efficiency were 790–850 L/kg VS and 30%, respectively, at 6 days residence time, which was a significant improvement compared to 610 L/kg VS and 24% without pre-treatment at 20–30 days HRT.

Zhang et al. (2012a) used a laboratory-scale HPH and reported an optimal homogenization pressure of 50 MPa for one homogenization cycle and 40 MPa for two homogenization cycles (Figure 5.5). The mechanism of the HPH sludge pre-treatment is sludge disintegration by pressure gradient, turbulence, cavitation and shear stresses acting on solid surfaces. The sludge disintegration involves disruption of the sludge flocs and microbial cell walls and the release of extracellular polymeric substances (EPS) and cytoplasm, which leads to an increase in the protein concentration and SCOD (350–2167 mg/L) in the sludge supernatant. The VS removal and total COD (TCOD) removal, respectively, increased by about 24% and 45% when the anaerobic digestion was combined with the HPH pre-treatment at a homogenization pressure of 50 MPa with two homogenization cycles, showing that HPH pre-treatment of the sewage sludge effectively improved the anaerobic biodegradation of the organic matters. Biogas production increased by 64% and 115% with one and two homogenization cycles, respectively, and the methane

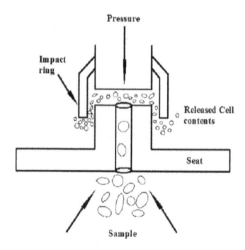

FIGURE 5.5   A HPH used as pre-treatment of WAS (Zhang et al., 2012a).

percentage in the biogas was 47% in the control digester and 64% in the digester with pre-treatment. The energy consumption was 3380 kJ/kg TS at a homogenization pressure of 40 MPa with two homogenization cycles.

5.5 ELECTROKINETIC DISINTEGRATION

Electrochemical technology may be used in some situations for the treatment of industrial effluents that contain bio-refractory organic pollutants such as landfill leachate, phenol, cyanides, cigarette industry, textile (dye) and tannery wastewaters (Song et al., 2010). Song et al. (2010) pre-treated waste activated sludge (WAS) with a pair of RuO_2/Ti mesh plate electrodes prior to aerobic digestion. At the power input of 5 W, the VS and volatile suspended solids (VSS) removals were 2.3% and 4.8% after 30 min electrolysis time, respectively, while the SCOD increased from 50 to 900 mg/L after 240 min indicating that electrochemical pre-treatment was suitable for solubilization. At the optimum pH of 10, the VS and VSS removals were 2.9% and 8.4%, respectively. In the subsequent aerobic digestion, the sludge reductions for VS and VSS after a solids retention time (SRT) of 17.5 days were 34.25% and 39.6%, respectively. Without the pre-treatment, an SRT of 23.5 days was required to achieve a similar reduction. Treatment costs were estimated at 5 kWh/m³ and 7–11 kWh/m³ for the pre-treatment and conventional aerobic digestion, respectively. Electrochemical pre-treatment could reduce digestion time, and could therefore reduce reactor volume and space requirement. However, no anaerobic digestion was examined. The literature on electrochemical treatment of WAS remains relatively rare.

High electrokinetic disintegration (or pulsed electric field) is one of the high-voltage electric field methods (Zhen et al., 2017). During the disintegration process, the charges created by the high-voltage field induce a sudden disruption of rigid sludge flocs and cellular membranes, thereby making the nutrients easily available to the fermenting bacteria. Lee and Rittmann (2011) achieved a 110%–460% increase in soluble compounds (NH_3-N, volatile fatty acids [VFAs], sugar and protein) after electrokinetic treatment at a specific energy input of ~34 kWh/m³, which provoked a 33% and 18% increase in CH_4 production and TCOD removal, respectively, in an anaerobic continuous stirred-tank reactor (CSTR) with an SRT of 20 days. In particular, the electrokinetic treatment saved 40% digester size with a CH_4 conversion of 0.23 g CH_4-COD/g COD. According to Choi et al. (2006), electrokinetic disintegration of activated sludge resulted in a 4.5 times increase in the SCOD/TCOD ratio and a 2.5 times higher biogas production at 19 kV, 110 Hz for 1.5 s.

Electrokinetic disintegration, as a newly developed sludge pre-treatment technology, has been extensively implemented in industry. For example, a full-scale installation in a WWTP sludge digestion has been described by Rittmann et al. (2008). The electrokinetic pre-treatment of 63% of the input waste sludge increased biogas production by over 40% and reduced the biosolids requiring disposal by 30%. They further estimated that for a plant treating 76,000 m³/day of wastewater (380 m³ sludge/day), electrokinetic treatment generated an annual economic benefit of approximately $540,000 net of electricity and other operating and maintenance costs. Most recently, Chiavola et al. (2015) applied electrokinetic disintegration in a full-scale

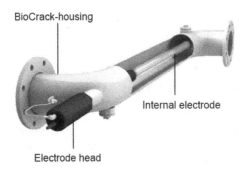

BioCrack-housing

Internal electrode

Electrode head

FIGURE 5.6 Sectional drawing of the BioCrack module (http://www.engineered-to-work. com/web/infomaterialien/biocrack_bga_ka_en.pdf).

WWTP for sludge reduction. The electrokinetic disintegration was able to drastically reduce the amount of biological sludge produced by the plant, without affecting its treatment efficiency, which gave rise to a considerable net cost saving for the company. In another full-scale implementation at the Northwest Water Reclamation Plant (NWWRP) in Mesa, Arizona, it similarly confirmed the net positive economic benefit as a result of the energy offsets from the increase in biogas (60%) and the reduction in biosolids disposal (40%) (Long and Bullard, 2014).

German Vogelsang is one of the representative electrokinetic disintegration device producers (called a "BioCrack module," as shown in Figure 5.6). The BioCrack module is composed of a system of pipes with electrodes alongside that apply a voltage of around 30–100 kV. The company claims that the application of the BioCrack module to pre-treat sludge increases biogas yields by up to 20% at a poised power 35 W per module while offering roughly 30% downstream energy savings (pumping, mixing, etc.).

6 Chemical Treatment of Sludge

6.1 INTRODUCTION

Chemical pre-treatment includes the addition of alkaline (ALK) agents (e.g., NaOH), acidic agents (e.g., H_2SO_4) and a strong oxidant (e.g., O_3). These chemicals can disintegrate the complex extracellular polymeric substances (EPS) matrix in waste activated sludge (WAS) as well as promote cell lysis. These treatments are applied to secondary sludge or mixtures of primary and secondary sludges, as depicted in Figure 6.1. A short introduction of the different chemical pre-treatments is given as follows.

6.2 ALKALINE PRE-TREATMENT

Alkali pre-treatment has been widely used for lignin breakdown in biomass. The basic principles of the alkali hydrolysis are based on solvation and saponification, which induce depolymerization and cleavage of lignin–carbohydrate linkages, accordingly rendering the uneasily biodegradable substances more accessible to the extracellular enzyme. In addition, it solubilizes the xylan hemicellulose by saponifying the intermolecular ester bonds (e.g., acetyl and uronic acid substitutions), though to a lesser degree than acidic pre-treatment. The alkali method gains high popularity in sludge disintegration before being sent to the digesters when considering its unique benefits in providing additional alkalinity that increases the buffer capacity of systems, its specific methanogenic activity and process stability. Table 6.1 lists the pre-treatment conditions and performance improvements in methane production, and the results are relatively encouraging. Among the ALK reagents, NaOH is the most effective in sludge solubilization and enhancing biogas production, with its ranking of efficacy being (NaOH > KOH > Mg(OH)$_2$ and Ca(OH)$_2$).

Alkali are often used to control the pH during the anaerobic digestion step. It has therefore also been used as a pre-treatment chemical. It has been reported as a sludge pre-treatment process since the late 1970s (Stuckey and McCarty, 1978a). NaOH was reported to induce higher sludge solubilization followed by potassium hydroxide (KOH) and calcium hydroxide (Ca(OH)$_2$) sequentially at the same pH value (Kim et al., 2003). ALK was suggested to work as follows. The addition of alkali provides a large quantity of hydroxyl ions and increases the sludge pH. This creates a hypotonic environment in which microorganism cells cannot withhold a proper turgor pressure and lyze (Neyens et al., 2003a). Additionally, the chemical oxygen demand (COD) solubilization has also been reported to be the consequence of protein degradation and various reactions such as the saponification of uronic

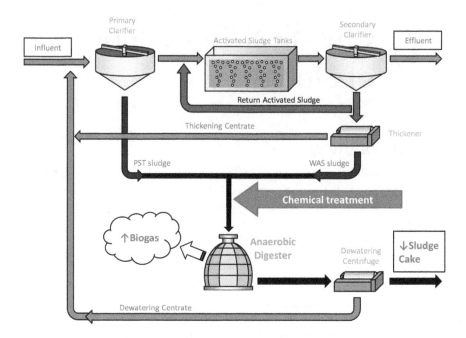

FIGURE 6.1   Flow sheet including a chemical pre-treatment in the sludge line.

acids and acetyl esters, reactions occurring with free carboxylic groups and reactions of neutralization of various acids formed from material degradation. Previous studies quantified ALK pre-treatment differently. The most commonly used terms are NaOH concentration, pH and NaOH dosage.

ALK treatment time was normally longer than 30 min in past results (Kim et al., 2010; Li et al., 2012b). However, as shown in Figure 6.2, the reaction between NaOH and sludge is a fast process. For an NaOH dosage higher than 0.02 M, significant solubilization of the COD was observed after the first minute of NaOH addition. A slight increase in soluble COD (SCOD) followed in the second minute and thereafter the SCOD concentration became relatively stable. However, for an NaOH dosage of 0.02 M, only a slight increase in COD was observed and for an NaOH dosage of 0.01 M, almost no increase in SCOD was notified. This suggested that even though NaOH is a highly reactive and corrosive reagent, 0.02 M is recommended as the lowest effective dosage threshold.

The protein and carbohydrate concentration was also tested. As no obvious increase in SCOD could be determined after the second minute, the 5-min treated sludge was selected as representative and the protein and carbohydrate concentration was measured. After 5 min, the increases in soluble protein were 2120, 1490, 320 and <50 mg/L at an NaOH concentration of 0.1, 0.05, 0.02 and 0.01 M, respectively. The solubilized protein concentration represented around 90% of the solubilized COD when these protein concentrations were roughly converted to an equivalent COD concentration by a factor of 1.5. Protein solubilization dominates sludge solubilization during the NaOH treatment.

TABLE 6.1
Alkaline Pre-Treatments of Sludge

Sludge	Alkaline Conditions	Solubilization		Anaerobic Digestion				Scale	References
		COD	DD (%)	CH₄ Production	VS Destruction	COD rem. % (Control)	Dewaterability (s)		
WAS (TS: 3.9 g/L)	0.01 N NaOH 0.03 N NaOH 0.05 N NaOH 0.08 N NaOH 0.10 N NaOH	—	3.9 7.5 14.4 15.3 16.1	—	—	—	—	LS, B	Şahinkaya and Sevimli (2013b)
WAS (TS: 17.1 g/L)	1.68 g KOH/L, 1 h 3.65 g KOH/L, 1 h	SCOD/TCOD: 2.7% to 9.3% 2.7% to 30.7%		+0%	—	—	—	LS, B	Valo et al. (2004a)
WAS (TS: 30.6 g/L)	4.6 g/L NaOH 26.1 g/L NaOH	SCOD/TCOD: 20%–63% 20%–75%		+150% –75%	—	—	—	LS, B	Penaud et al. (1999)
WAS (TSS: 17.5 g/L)	NaOH pH 10, 8 days	20–11,800 mg/L		+340%	—	—	—	LS, B	Zhang et al. (2010)
WAS TS: 1% TS: 1% TS: 2%	20 meq/L NaOH 40 meq/L NaOH 20 meq/L NaOH	251–2,962 mg/L 251–4,783 mg/L 251–3,368 mg/L		+33 +30 +163	+10 +12 +29	46 (38) 51 (38) 52 (38)	148–389 (309–735)	LS, SC, M (10)	Lin et al. (1997)
WAS (TS: 1%)	40 meq/L NaOH, 24 h	230–4,500 mg/L	—	—	—	—	—	LS, B	Chiu et al. (1997)

(Continued)

TABLE 6.1 (CONTINUED)
Alkaline Pre-Treatments of Sludge

Sludge	Alkaline Conditions	Solubilization		Anaerobic Digestion				Scale	References
		COD	DD (%)	CH₄ Production	VS Destruction	COD rem. % (Control)	Dewaterability (s)		
WAS (TS: 1%)	30 meq/L NaOH	45%	—	+112%	—	—	—	LS, B, M	Rajan et al. (1989)
WAS (7 g SS/L)	0.6 g NaOH/g VSS, 1 h	15% (VSS)	—	+40–50%	—	—	—	LS, B, M	Tanaka et al. (1997)
WAS (TS: 38 g/L)	7 g NaOH/L	SCOD/TCOD: 17.6%–85.4%	—	Biogas: +13.4%	20.5%–29.8%	—	Remark: Methane percentage was around 70% in all the biogas samples	LS, B	Kim et al. (2003)
Pulp and paper sludge	8 g NaOH/100 g TS	SCOD: +83%	—	+83%	—	—	+56–192% (settling velocity)	LS, B, M	Lin et al. (2009)
WAS 10 g TS/L	pH 9–11 (4 mol/L NaOH), 24 h	—	—	+7.2–15.4%	+10.7–13.1% (TSS)	—	Improved	LS, B, M	Shao et al. (2012)
WAS	0.1 mol NaOH/L	DD$_{COD}$: from 22.3%–26.9%	—	+1.5%	—	+26.4%	—	LS, B, M	Li et al. (2012a)
WAS 11.7 g TS/L	8 g NaOH/m³ wet sludge (pH 8)	SCOD/TCOD: 2%	—	−7.1%	+9.7% (TS), +11.5% (VS)	+18.1%	—	LS, C, T (21)	Wonglertarak and Wichitsathian (2014)

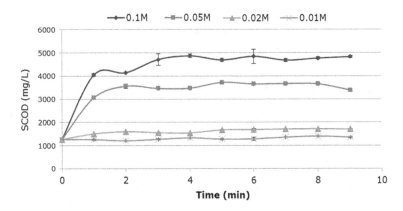

FIGURE 6.2 Example of sludge solubilization caused by alkaline addition at various concentrations.

In addition, the ALK treatment of organic material has been reported to induce swelling of particulate organics, making the substrate more susceptible to enzymatic attack (Vlyssides and Karlis, 2004). As a result, biodegradability may be significantly enhanced. ALK pre-treatment was also capable of increasing WAS biodegradability (Table 6.1). Şahinkaya and Sevimli (2013b) found that 0.05 N was the optimum NaOH dosage. Penaud et al. (1999) used 125 meq/L (0.125 N) NaOH and obtained a 40% increase in biodegradability. NaOH (20 meq/L) was observed to increase methane production by 33% by Lin et al. (1997). However, only a 13% increase in biogas production with 175 meq/L NaOH was observed by Kim et al. (2003). These results indicate that the ALK pre-treatment performance is strongly related to the sludge characteristics and operating conditions. The addition of chemicals not only means additional cost but also the potential for more sludge production (Foladori et al., 2010). Moreover, too high concentrations of sodium or potassium ion may inhibit anaerobic digestion (Appels et al., 2008). Zhang et al. (2010) employed a long period sludge alkalinization with a duration of 8 days at a pH of 10. This is different from the conventional pre-treatment with the treatment period less than 1 day. In their work, methane production was 330% higher compared to untreated sludge and 130% higher compared to thermal-ALK pre-treated sludge. Besides, anaerobic digestion was accelerated and only took 9 days to reach a plateau. The total digestion period was 17 (8 + 9) days, which was no longer than the untreated sludge (around 20 days).

These results are in opposition to those of Valo et al. (2004a) who did not observe any improvement using 3.65 g KOH/L (pH 12) at room temperature. This could be due to the inhibition of methanogens by refractory compounds solubilized by the base or insufficient organics solubilization. Lin et al. (1997) reported an improvement in dewaterability in terms of capillary suction time (CST) from 309 to 148 s after anaerobic digestion of sludges pre-treated with NaOH.

Lin et al. (2009) studied the NaOH pre-treatment of pulp and paper sludge (8 g NaOH/100 g total solids [TS]) and reported an 83% increase in methane productivity. However, the results tend to underline that the alkali pre-treated sludge with overloading NaOH (>16 g/100 g TS) is not compatible with anaerobic biological digestion. A too high dose of Na^+ might interfere with the metabolic pathway of anaerobic microflora and deteriorate methane output.

6.3   ALKALINE/MECHANICAL PRE-TREATMENT OF SLUDGE

Apart from alkali treatments, the combination of mechanical treatment with alkali was found to result in better effects. Cho et al. (2014a) showed the feasibility of the combined ALK treatment with a mechanical treatment crushing device with cutting blades. Such combined treatment resulted in 64% solubilization and an 8.3 times higher methane production than the control. In addition, a 50:50 ratio of such pre-treated WAS and untreated WAS was found to have the maximum economic benefits and induced synergistic methane production.

Zhang et al. (2012b) found that the performance of a high-pressure homogenizer (HPH) pre-treatment was significantly improved by adding NaOH. An NaOH concentration ≤0.04 mol/L and a homogenization pressure of ≤60 MPa were found to be suitable conditions for the combination. Fang et al. (2014) further investigated such combined treatment on sewage sludge anaerobic digestion. The optimal ALK dosage was found to be 0.04 mol/L. The maximal SCOD, total COD (TCOD) and volatile solids (VS) removals were 73.5%, 61.3% and 43.5%, respectively, from batch anaerobic digestion results. In addition, the biogas and methane yield from each gram of VS added was found to be linearly related to the disintegration degree (DD) as follows: biogas = 4.51 DD + 153.76 and methane = 4.66 DD − 9.69.

Stephenson et al. (2003) have patented a HPH that combines ALK addition under the trade name MicroSludge™. The process uses the ALK pre-treatment to weaken cell membranes and reduce viscosity, and a HPH to provide an enormous and sudden pressure change to burst the cells. The liquefied WAS is more readily converted to biogas in the digestion process.

A WAS at 4% TS is soaked at pH 8.5–10 for 1 h retention time. Chopper pumps are used to break up large particles and a fine 800 μm screen is used to remove particles that could damage the homogenizer. A gas/liquid separator and pulsation dampener are also required prior to the homogenizer to prevent gas locks occurring. The homogenizer itself is essentially a valve with a narrow passageway of <1 mm, through which the sludge is pumped at 830 bar. This rapid change in velocity results in high shear and cavitation, which solubilize the sludge. The principal costs of the process are chemicals and power. The principal benefits of the process are the increased biogas production associated with the increased VS reduction (VSr), a reduction in the digester hydraulic retention time (HRT) and a reduced quantity of digested biosolids that requires dewatering and disposal.

The first full-scale demonstration of the process was carried out in 2004 at the Chilliwack wastewater treatment plant (WWTP) in Canada (Rabinowitz and Stephenson, 2005). The Chilliwack plant had two anaerobic digesters operating in sequence, treating primary sludge and trickling filter/solids contact sludge in a ratio of 65:35, with total digester HRT of 13 days. As one digester was approximately twice the size of the other, the setup was not amendable to the use of one digester as a test digester, with a parallel control digester. The test was therefore conducted with MicroSludge testing from May through July 2004, followed by monitoring of the performance without MicroSludge from August through October 2004. As part of the test, the WAS was thickened in a rotary drum thickener to 4% TS. Thickening and homogenization was conducted 10 h per day. The homogenate was stored in a holding tank and fed to the digesters over each 24-h period. The Chilliwack test had some shortcomings. The control test could not be conducted in parallel with the MicroSludge test. Digester gas production was not measured as the gas meters were not reliable, and the total biosolids hauled from the plant were not measured, which did not allow for checks on the digester mass balance calculations used to measure digestion performance.

Since late 2005, a full-scale demonstration of the MicroSludge process has been operating at the 1,500,000 m^3/day Joint Water Pollution Control Plant in Los Angeles County, California. The plant has 24 mesophilic anaerobic digesters, each with a volume of 14,500 m^3 and an HRT of approximately 19 days. The digesters are fed with a 68:32 mix by mass of primary sludge and WAS from a pure oxygen-activated sludge process. As part of the full-scale demonstration, two 4000 L/h homogenizer modules have been continuously operated for 24 h/day to process approximately 192 m^3/day of thickened WAS at a nominal solids content of 6%. Primary sludge and pretreated WAS have been fed to a dedicated experimental digester. A parallel control digester has also been monitored to provide side-by-side performance comparison. It was anticipated that this test would address the shortcomings of the Chilliwack test. One critical difference between the setup of the Los Angeles test compared with the Chilliwack test is the absence of the homogenate holding tank, as the MicroSludge units were sized to operate and feed the digester continuously. Other differences inherent between the two installations are the type of secondary process, the HRT in the digesters and the difference in scale.

The Chilliwack test showed significant digester performance improvement, based on the solids mass balances calculated across the digesters. The MicroSludge pretreatment of the thickened WAS resulted in an average VSr of 78% despite an HRT of only 13 days. An extrapolation of data from the demonstration suggests that the VSr in the WAS alone was over 90%. This compares with a typical WAS VSr of 35%–40% in typical mesophilic digestion systems. Further, the process appears to liquefy both the volatile and the fixed solids fractions of the sludge. During the control test, the VSr at the plant was 60%. This is higher than the VSr performance typically seen across North America (low to mid 50%), particularly given the very short HRT. As noted above, the mass balance calculations could not be verified by digester gas production or total dewatered biosolids volumes.

The systems include the caustic system, chopper pump, screen and homogenizer, housed within a shipping container. Costs quoted in 2005 ranged from around $2 million for an 8,000 L/h unit (190 m^3/day) to $4.4 million for the largest module of 24,000 L/h and future costs will change to reflect changing commodity and labor prices. The installation of a MicroSludge system would entail other costs, including storage of thickened WAS (TWAS) or homogenate, piping, pumping to or from the system, electrical, instrumentation and control costs, site preparation, utilities, screenings handling, construction costs and costs for an alternative building if a shipping container is not the preferred means of housing. The life cycle cost–benefits

will be more favorable for thicker sludges. Electricity, natural gas and biosolids handling costs will also affect the economic balance and payback period. The heat produced through the homogenization process raises the temperature of the sludge by
approximately 20°C, which reduces the heat required for digestion.

The power cost for solubilization was estimated to be $45/ton TS and the total
cost of sludge management using this technology was estimated to be $152/ton TS.
All these results confirmed the supplement effects of ALK treatment and mechanical treatment. Actual economics and payback periods will be site specific. The economic benefits of MicroSludge pre-treatment are greater for plants facing higher
unit power and biosolids disposal costs. Other site-specific aspects that may impact
the maintenance costs for the MicroSludge equipment include the presence of abra-
sives and chlorides in the sludge, both of which will reduce the life of metal components. The economics for the installation of a MicroSludge system on a return
activated sludge (RAS) recycle for the reduction of WAS production from the activated sludge process would be considerably different. Gas production from the
digesters may be reduced due to the reduction in the quantity of WAS produced.
The heat produced through the homogenizer would not have a direct natural gas
heating offset, but may have some benefits for reaction rates within the activated
sludge plant, particularly in colder climates. The addition of caustic may also be
a benefit in nitrification in plants that may be alkalinity or pH limited. Aeration
requirements in the activated sludge process may also increase, with a corresponding increase in electricity consumption. However, there would be potential savings
in operating costs and future expansion capital costs for digester and dewatering
facilities and biosolids management.

6.4   ACID PRE-TREATMENT

Hydrochloride acid (HCl) and sulfuric acid (H_2SO_4) are two acidic pre-treatment
reagents. H_2SO_4 is reported to have a better performance in cell lysis than HCl
(Rocher et al., 1999). The effectiveness of acidic or alkali pre-treatment may vary
with the types and characteristics of the studied substrates because of their distinct
affinity to organic components. Acidic pre-treatment is indicated to be more effective for lignocellulosic biomass. The main reaction that occurs in this process is
the hydrolysis of hemicellulose, which releases the monomeric sugars and soluble
oligomers from the cell wall matrix into the hydrolysate, thereby improving the
enzymatic digestibility (Zhen et al., 2017). The method offers good performance in
hemicellulose removal but has hardly any impact on lignin hydrolysis, and the lignin
condensates and precipitates. Additionally, it may induce the formation of toxic by-
products, such as furfural and hydroxymethyl furfural (HMF), strong inhibitors of
microbial fermentation. Other drawbacks associated with the acidic method include
greater toxicity and stronger corrosivity because of extremely low levels of pH;
therefore, special materials are required for the reactor construction. For instance,
for a 14.3% increase in the methane yield of WAS, a pH as low as 2 was required
(Devlin et al., 2011). The acidic pre-treatment led to a four- and six-fold increase in
soluble carbohydrates and proteins, respectively, while reducing the polymer dosage
for dewatering by 40%.

Chen et al. (2007) tested the influence of an acidic environment, a basic environment and a neutral environment on the COD solubilization and methane production from sludge. A pH value of 4 led to a slight increase in SCOD, but the increase was negligible compared to a pH value of 11. Despite the high capability of COD solubilization in an ALK environment, a higher methane production is observed at a pH value of 6 than both neutral pH and a pH value of 8. In practice, acid pre-treatment is not recommended. A harsh acidic environment requires high-cost construction materials and may also bring more adverse effects on subsequent biological treatment than ALK pre-treatment. The integration of acidic-thermal pre-treatment has been presented in the literature, e.g., Neyens et al. (2003c) with thickened sewage sludge, Nielsen et al. (2011) with WAS, as well as Takashima and Tanaka (2014a) with anaerobically digested sludge. In a semi-continuous mesophilic reactor (20 days HRT) treating anaerobically digested sludge using H_2SO_4 (pH 5–6) at 170°C for 1 h, Takashima and Tanaka (2014a) increased volatile suspended solids (VSS) removal by a factor of 2–2.5 and methane production by 14%–21% while improving dewaterability by 22%–23%.

Peracetic acid (PAA) is sold as an aqueous quaternary equilibrium mixture of acetic acid, PAA and hydrogen peroxide, according to the following reversible reaction:

$$CH_3COOH + H_2O_2 \leftrightarrow CH_3COOOH + H_2O$$

It has an oxidation potential of 1.81 eV and has many applications in the medical and food industries as a disinfectant and in water purification as an oxidizing agent. The main advantage is that no toxic degradation products are formed. The oxidation of organic compounds takes place through the formation of hydroxyl radicals. Eventually, PAA degrades itself into acetic acid and water (Appels et al., 2011). A linear relationship was obtained between SCOD and the PAA dosage with a maximum of 6000 mg/L SCOD at 100 g PAA/kg TS. Interestingly, this dosage of PAA also generated volatile fatty acids (VFAs) (8.4 g/L) due to the oxidation of organic matter. However, the optimum biogas production (+21%) was obtained at a dosage of 25 g PAA/kg TS. For dosages in excess of 40 g PAA/kg TS, less biogas was produced compared to the untreated sample due to the inhibition by VFAs.

6.5 FENTON OXIDATION PRE-TREATMENT

Another well-established oxidation technology is the Fenton process, which involves reactions of hydrogen peroxide (H_2O_2) with catalyst iron ions (Fe^{2+}) to produce highly active hydroxyl radicals (OH•) (Zhen et al., 2017). Hydroxyl radicals have a higher oxidation potential of +2.80 V (in acidic conditions) than hydrogen peroxide (+1.36 V) and ozone (+2.07 V), and are particularly effective in the disintegration of sludge EPS and the cell lysis of microorganisms, resulting in the release of both intracellular materials and bound water.

Fenton oxidation is considered an attractive technology for the oxidation of a variety of organic pollutants. The process requires a Fenton's reagent, which is composed of a solution of ferrous iron and hydrogen peroxide. Fenton oxidation is highly accepted, due to iron being a non-toxic element and (due to its abundance) a cheap metal, while the use of hydrogen peroxide rapidly generates oxidants. The Fenton oxidation process is primarily completed in four steps: (1) the adjustment of the pH to a low acidic condition; (2) oxidation; (3) neutralization; and (4) by-products separated by coagulation. The pH factor generally hinders the Fenton process due to low acidic conditions being important for optimum operation. However, the efficacy of the Fenton process is more closely linked to the oxidation step, with its mechanism highly dependent on the generation of a hydroxyl radical (OH•) under low pH values by the catalytic dissociation of hydrogen peroxide (H_2O_2) in the presence of ferrous iron (Fe^{2+}) (Sahinkaya et al., 2015). The oxidizing radicals produced by this process contain redox potential (+2.33 V), in comparison to that produced by hydrogen peroxide alone (+1.36 V) (Erden and Filibeli, 2010a). Thus, due to the high redox potential, the solids have been reduced by up to 60% and have transformed the sludge into Class A biosolids (Pilli et al., 2015).

The application of Fenton oxidation as a sludge pre-treatment has a high capability to degrade EPS and break the microbial cell walls, thus releasing the intracellular organic content. The sludge may contain refractory organic compounds, which can be transformed into two available forms (e.g., readily and soluble organics) for subsequent biological treatment (Anjum et al., 2016). The effectiveness of this process depends on several variables, namely reagents concentrations (H_2O_2 and Fe^{2+}), Fe^{2+}/H_2O_2 ratio, treatment time as well as the initial pH and temperature. Dewil et al. (2007) have observed a 75% increase in biogas production from the anaerobic digestion of Fenton pre-treated sludge (50 g H_2O_2/kg DS and 0.07 g Fe^{2+}/g H_2O_2 at pH 3). Similarly, a further study has found a solubilization and biodegradation efficiency of up to 70% after applying the Fenton pre-treatment at the optimal conditions of temperature (i.e., 25°C) and duration (i.e., 60 min) (Pham et al., 2010). Pilli et al. (2015) treated secondary sludge with 60 g H_2O_2/kg TS, 0.07 g Fe^{2+}/g H_2O_2 and pH 3, and they noticed a 15% increase in methane yield (from 430 to 496 m^3 CH_4/Mg $VS_{degraded}$), a 3.1 times increase in net energy as well as considerably reduced greenhouse gas (GHG) emissions.

The optimum conditions for the Fenton pre-treatment of sludge may vary according to the type of sludge and the biological treatment process. Furthermore, Sahinkaya et al. (2015) investigated and compared the efficiency of conventional Fenton (Fe^{2+}/g H_2O_2, CFP) and Fenton-type ($Fe°$/g H_2O_2, FTP) processes in sludge disintegration and the enhancement of anaerobic biodegradability and stated that CFP was a more effective process because of using catalyst ferrous iron (Fe^{2+}) in its dissolved form. Under the optimal conditions of 4 g Fe^{2+}/kg TS, 40 g H_2O_2/kg TS and pH 3 within 1 h oxidation period, CFP and FTP enhanced methane production by 26.9% and 38%, respectively, relative to the untreated reactor. Also, a 10% improvement in methane production was found by Zhou et al. (2015) with 50 mg H_2O_2/g TS, 7 mg Fe/g TS and pH 2.0 for 30 min; in this case, the researchers applied indigenous iron in the sludge as the catalyst. A major drawback for the Fenton reagent is the necessity for low pH values (pH < 4.0) to avoid hydrolysis and the precipitation of Fe^{3+}; moreover, treated sludge needs neutralization before digestion. Apart from this, excess H_2O_2 or Fe^{2+} doses may also scavenge hydroxyl radicals, lowering the concentration of OH• radicals.

Preliminary laboratory-scale investigations on the use of Fenton oxidation (H_2O_2/Fe^{2+}) on sewage sludge were presented by Mustranta and Viikari (1993) and

Neyens et al. (2003b). The optimum conditions obtained in these laboratory tests were the addition of 25 g H_2O_2/kg TS in the presence of 1.67 g Fe^{2+}-ions/kg TS at pH 3 (using H_2SO_4) at ambient temperature and pressure. However, these studies did not report the effect on anaerobic digestion. Neyens et al. (2003b) investigated the peroxide pre-treatment of WAS at the pilot scale. As a result, the amount of TS per equivalent inhabitant per day was reduced from 60 to 33.1 g TS/IE.day and the percentage TS of the sludge cake was 47%, which is high compared with the 20%–25% achieved in a traditional sludge dewatering facility. An economic assessment for a WWTP of 300,000 IE confirms the benefits. Considering the fixed and variable costs and the savings obtained when the sludge is incinerated after dewatering, a net saving of approximately 140 Eur/ton TS can be expected.

Eskicioglu et al. (2008) treated TWAS (6.4% TS) with H_2O_2 (1 g/g TS) and found that the SCOD/TCOD ratio increased from 4% in the control to 17%. Despite the solubilization, the methane potential dropped from 308.4 mL CH_4/g VS_{added} in the control to 304.3 mL CH_4/g VS_{added} in the treated sample. This was due to the decrease in the methane potential of WAS via advanced oxidation and a lower dosage was recommended for future studies. The dewaterability was, however, improved with a CST that decreased from 180.7 to 160.7 s. Dewil (2007) studied several peroxidation techniques such as Fenton peroxidation and novel reactions involving peroxymonosulfate (POMS) and dimethyldioxirane (DMDO). The results of the treatments show a considerable increase in COD and biochemical oxygen demand (BOD) in the sludge water, and an increase in the BOD/COD ratio. A maximum increase in biogas production of 75% was measured with Fenton, while the POMS and DMDO treatment increased the biogas production by a factor of nearly 2 and 2.5, respectively.

6.6 OZONE PRE-TREATMENT

6.6.1 INTRODUCTION

Ozonation is a widely reported chemical sludge treatment technology. Ozone chemically reacts with the sludge and destroys microorganism cell components (Chu et al., 2009; Yan et al., 2009). It is also capable of solubilizing the EPS and breaking down solubilized complex macromolecules (Yan et al., 2009). In addition, ozone is able to convert refractory organic matters into biodegradable forms (Volk et al., 1993; Nishijima et al., 2003). Ozonation is a fast process without resulting in extra sludge production compared to other chemical pre-treatments. However, ozonation may induce mineralization, which converts organic carbon into inorganic carbon dioxide (Ahn et al., 2002; Lee et al., 2005). Therefore, a proper ozone dosage needs to be selected.

Sludge ozonation is the combined chemical effect of direct and indirect ozonation. Ozone can directly react with the dissolved/particulate substances or form a secondary oxidant like hydroxyl radicals to oxidize the target substances (Staehelin and Hoigne, 1985). Scheminski et al. (2000) suggested that the direct ozonation rate is lower and reactant dependent, while the indirect hydroxyl radicals reaction is not reactant dependent. Yan et al. (2009) found that hydroxyl radical concentrations decreased with increasing ozonation time and suggested that the indirect hydroxyl oxidation was suppressed with increasing ozonation time.

Ozone is able to lyze microorganism cells and oxidize both soluble and insoluble macromolecules into smaller monomers (Scheminski et al., 2000). Cesbron et al. (2003b) investigated the competition of particulate matter and soluble matter for ozone access and showed the importance of mass transfer. At the beginning, ozone reacts simultaneously with soluble organics and small particles, the diameter of which is smaller than the thickness of effective film (Cesbron et al., 2003b). After easily oxidized reactants are used up, the thickness of effective film will increase, which makes the oxidation of bigger particulates possible (Cesbron et al., 2003b). They also pointed out that a screen effect will be caused on particulate matter when the reaction between the soluble fraction and ozone is high (Cesbron et al., 2003b). Zhang et al. (2009) supported this conclusion by showing ozone's better performance in cell lysis than sludge matrix solubilization. Ozone could penetrate through cell walls, damaging the cell membrane structure, leading to cell lysis and the release of intracellular substances. However, this mechanism is favored at higher ozone dosage because low ozone dosage may not provide enough driving force for this penetration (Zhang et al., 2009).

Due to its strong oxidative properties, ozone has been used for water and wastewater treatment. Ozone has been used since 1970 to reduce the COD in the final effluent and to degrade WAS (Moerman et al., 1994). It was reported to reduce excess sludge production and that ozonation could be toxic to non-acclimatized sludge (Sakai et al., 1997). Ozone decomposes itself into radicals and reacts with the soluble and particulate fractions, organic and mineral fraction.

6.6.2 Quantification Units

Many papers quantified the ozonation treatment in terms of ozonation time (s) (Zhang et al., 2009; Xu et al., 2010a). Use of the ozonation time allows the calculation of energy input when the power of the generator is given. However, the ozone production efficiency varied among different devices and the applied ozone dosage was not available. In addition, sludge concentration also has an impact on ozonation performance. Therefore, the more reasonable ozone quantification unit would be the ozone dosage, which is the ozone mass received by each gram of TS (g O_3/g TS) (Goel et al., 2003b; Erden and Filibeli, 2011). A number of papers also quantified the ozone as ozone received per gram TSS (g O_3/g TSS) (Yan et al., 2009). However, since ozone reacts with both suspended and soluble solids, the use of ozone received per gram TS is more reasonable.

6.6.3 Laboratory Set-Up and Conditions

Ozone is produced from the reaction between oxygen atoms and oxygen molecules. Oxygen atoms are from oxygen molecules split using electrical energy, as depicted in Figure 6.3.

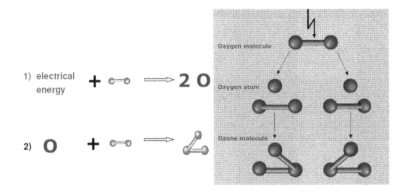

FIGURE 6.3   Chemical reactions involved in the production of ozone.

Ozone pre-treatment of sludge can be performed with an ozone generator such as the Wedeco ozone generator (GSO 30) shown in Figure 6.4. Pure oxygen, as in Figure 6.4, or air can be used as the feed gas and converted to ozone with a high-voltage converter. Due to the heat released during the reaction, cold water is recirculated using a water chiller. The power input of the ozone generator was 180 W. The applied ozone dosage was quantified according to the potassium iodide method (Konsowa, 2003).

For the study of ozone pre-treatment, an example of a possible set up is depicted in Figure 6.5. The oxygen pressure was 0.15 bar and the power was set at 10% of the maximum power. The power consumption was stable at around 170 W during operation.

FIGURE 6.4   Photograph of an ozone generator (left), water chiller (middle) in a fume cupboard and oxygen gas cylinder (right).

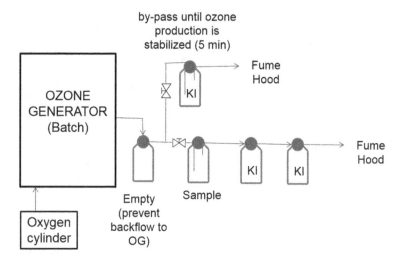

FIGURE 6.5 Example of ozone set-up using KI to destroy residual ozone. OG, ozone generator.

6.6.3.1 Effective Ozone Dosage

Before starting any experiment with WAS, a calibration of the ozone generator should be carried out in order to determine how much ozone is produced by the generator (actual dosage) and how much ozone can be dissolved into water (effective dosage). For the effective dosage, 500 mL of distilled water is sparged with ozone gas and the dissolved ozone concentration is measured using the Indigo colorimetric method (APHA, 2005). In that method, ozone reacts with KI to produce iodine I_2 according to the following reaction:

$$2KI + O_3 + H_2O \rightarrow 2KOH + O_2 + I_2$$

The iodine concentration is then measured using Leuco violet crystal that produces a violet color when in contact with iodine. The iodine concentration is then analyzed on a spectrophotometer. From Figure 6.6, it can be predicted that bubbling ozone gas through 500 mL of water for 1 min results in an ozone concentration of 3.32 mg/L.

6.6.3.2 Actual Ozone Dosage

For more consistency with the literature data, the actual ozone dosage can be determined. This is the total amount generated by the instrument regardless of how much is solubilized into the medium. An alternative to the colometric method is to use an ozone sensor that provides the ozone concentration in real time in water samples. From Figure 6.7, it can be predicted that 1 min of ozone generation will produce 13.7 mg of ozone. This is significantly more than the effective ozone dosage previously determined due to the difficulty of dissolving ozone in water. The solubility of ozone gas is 109 mg/L at 25°C, which is much more than oxygen solubility in water. However, the concentration of ozone in the gas is very low (typically 1%–3%), which

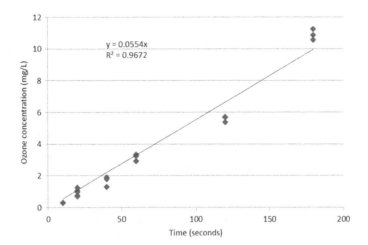

FIGURE 6.6 Ozone concentration in distilled water as a function of time.

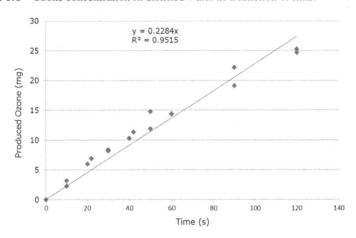

FIGURE 6.7 Ozone mass produced by the ozone generator as a function of time.

is why it is difficult to dissolve it in water. Also, ozone reacts very quickly with contaminants, itself and water. This linear relationship can then be used to derive the ozone dosage according to the duration of the operation. This dosage can then be normalized to the amount of solids in the sludge sample to obtain g O_3/g TS.

6.6.4 EFFECT OF OZONE PRE-TREATMENT ON SLUDGE SOLUBILIZATION

It can be seen from Figure 6.8 that the color of sludge changes from black to light gray as the treatment time increases. In this experiment, the specific ozone dosage ranged from 4.1 mg O_3/g TS (for 1 min treatment time using 200 mL) to 123 mg O_3/g TS (30 min treatment time). This range of ozone dosage gave a more realistic dosage compared to data published in the literature that typically ranges from 0.01 to 0.2 g O_3/g TS.

FIGURE 6.8   Ozone treatment of WAS. From left to right, treatment time: 30, 20, 15, 10, 5, 3 and 1 min.

When the same volume of sludge was centrifuged, it was observed that the amount of solids that settled down was smaller with increasing times, indicating a solubilization of particulate solids (Figure 6.9).

The COD results are shown in Figure 6.10 and the SCOD increased from 1200 to almost 5000 mg/L. There is a phenomenon of mineralization at high ozone dosage as illustrated, where the TCOD can be seen to decrease slightly over time due to the conversion of organics to carbon dioxide.

The TS, VS and TSS, VSS results are shown in Figures 6.11 and 6.12, respectively. There was a slight decrease in TS and VS, while the decrease in TSS and VSS was more significant, indicating a solubilization of solids.

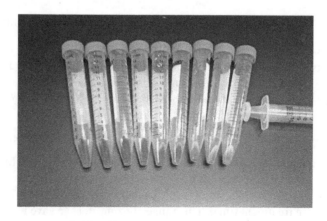

FIGURE 6.9   Centrifuged samples after ozone treatment of WAS. From left to right, treatment time: 1, 3, 5, 10, 15, 20 and 30 min.

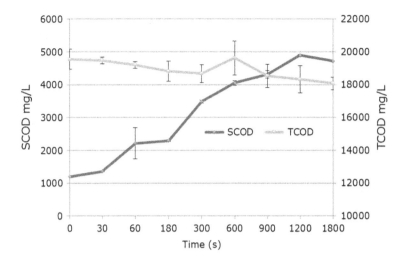

FIGURE 6.10 Evolution of COD during the ozone treatment of WAS.

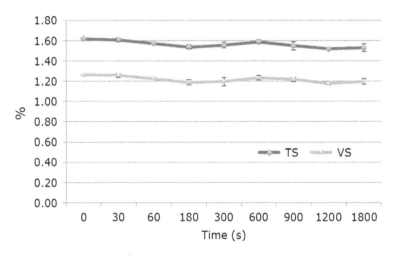

FIGURE 6.11 Evolution of TS and VS during the ozone treatment of WAS.

The mean particle diameter remained stable around 40–45 μm, as shown in Figure 6.13. Ozone solubilizes EPS and destroys the floc structure, but the particle size change is not very obvious. Bougrier et al. (2006) observed a decrease in the median diameter from 36.3 to 33.2 and 32.6 μm with an ozone dosage of 0.1 and 0.16 g O_3/g TS, respectively. This change was negligible compared to the median diameter of raw sludge. Erden and Filibeli (2011) reported similar results that the median diameter of raw sludge decreased from 84 to 75 and 70 μm, respectively, with an ozone dosage of 0.1 and 0.25 g O_3/g TS. In both cases, the particle size reduction was not obvious after ozonation.

Although ozonation does not show good particle size reduction, a significant decrease in suspended solids (SS) was reported. This could be explained by the fact

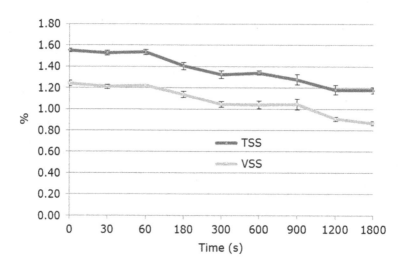

FIGURE 6.12   Evolution of TSS and VSs during the ozone treatment of WAS.

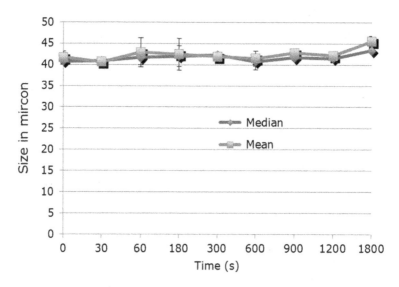

FIGURE 6.13   Evolution of mean and median particle size during the ozone treatment of WAS.

that ozone was efficient in attacking soluble or small particulate matters rather than large particulate matters (Cesbron et al., 2003a). Weemaes et al. (2000) observed that the TSS concentration decreased from 10.8 to 7.04 g/L, 6.11 and 3.83 g/L and the VSS concentration decreased from 6.4 to 3.78 g/L, 2.86 and 1.74 g/L after ozone dosages of 0.05, 0.1 and 0.2 g O_3/g COD were applied, respectively. Unlike mechanical sludge disintegration, ozonation sludge disintegration not only physically solubilizes SS but also chemically causes mineralization of solubilized organics (Ahn et al., 2002). After a certain dosage of ozone, the concentration of solubilized organics is a combined effect of organics mineralization and the solubilization of particulate (Ahn et al., 2002). Goel et al. (2003a) reported 37% VSS solubilization but only 5% VSr at an ozone dosage of 0.05 g O_3/g TS. This indicates that the degree of solids solubilization is much higher than mineralization solubilized substances at low ozone dosage. This also means the accumulation of solubilized organics in sludge, which is beneficial for subsequent biological treatment.

As expected, the concentrations of soluble proteins and soluble carbohydrates (Figures 6.14 and 6.15) also increased during the treatment following the same trend as observed for SCOD. Erden and Filibeli (2011) suggested that high ozone dosage caused soluble protein hydrolysis. They observed an increase in soluble protein and SCOD until a dosage of 0.1 g O_3/g TS with the soluble protein concentration dropping significantly thereafter. However, similar results were not reported elsewhere, which means further support for this phenomenon is needed. Sludge ozonation can cause cell lysis, splitting macromolecules and solubilizing organic particulate. This results in readily degradable organics available in the supernatant phase and enhances the anaerobic digestion. However, when ozone is overdosed, solubilized organic can be mineralized and converted to inorganic carbon, which is not desirable for the subsequent anaerobic digestion (Weemaes et al., 2000; Ahn et al., 2002). Therefore, the optimization of ozonation is crucial for a successful pre-treatment process.

Optimum ozone dosages have been reported in the range 0.05–0.5 g O_3/g TS, but Yeom et al. (2002) showed that the rate of mineralization was 5.1%, 20.1%, 29.2% and 49.2% at ozone dosages of 0.1, 0.5, 1 and 2 g O_3/g SS. Ozone causes damage to cells, the sludge matrix as well as other micro particulates in sludge. Extracellular and intracellular organics are solubilized by ozone and the organic fraction in the liquid phase increases (Scheminski et al., 2000). Weemaes et al. (2000) observed

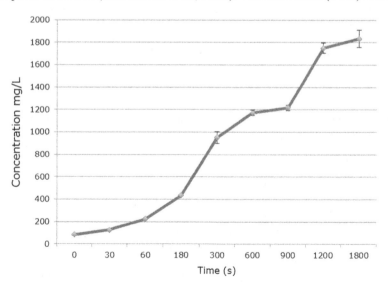

FIGURE 6.14 Evolution of soluble protein concentration during the ozone treatment of WAS.

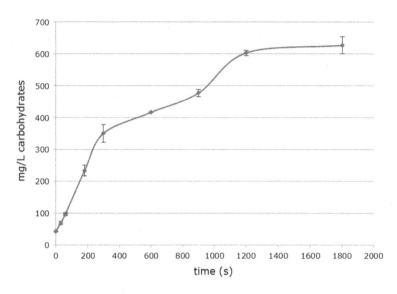

FIGURE 6.15 Evolution of soluble carbohydrates concentration during the ozone treatment of WAS.

a SCOD increase from 100 to 2660 mg/L for an ozone dosage <0.1 g O_3/g COD. When the ozone dosage increased to 0.2 g O_3/g COD, 200 mg/L less SCOD was observed than 0.1 g O_3/g COD. This decrease is generally recognized as the mineralization of solubilized organics. Ahn et al. (2002) investigated organic carbon change during ozonation. They found that around 5% of mineralization took place at an ozone dosage of 0.2 g O_3/g TS, 50% of the organics was in the soluble fraction and 45% of the organics carbon remained in the particulate phase. When the ozone dosage increased to 0.5 g O_3/g TS, organics in the soluble fraction decreased to 46% accompanied by an increase in mineralization to 20%. Yeom et al. (2002) indicated that after a dosage of 0.05 g O_3/g TS, mineralization became obvious and increased all the way with an increasing ozone dosage. Whereas, there is an optimum dosage (in their case: 0.5 g O_3/g TS) for an increase in organics in the soluble fraction, beyond that an ozone dosage increase causes adverse effects on SCOD concentrations. They found that biodegradability increased with an ozone dosage up to 0.2 g O_3/g TSS. Further increases in ozone treatment did not improve biodegradability.

Zhang et al. (2009) investigated the change in SCOD together with other biopolymers concentrations like DNA and protein with increasing ozone dosages. They observed rapid protein and DNA concentration increases in the first 15 min ozonation. As protein and DNA exist more in intracellular substances than extracellular substances, cell lysis was believed to occur in the first phase of ozonation (Zhang et al., 2009). Sludge matrix solubilization was believed to occur after 45 min, because the solubilization of polysaccharide, an important part of EPS, became obvious (Zhang et al., 2009). Zhang et al. (2009) also suggested that ozonation is more effective in lyzing cells than solubilizing sludge matrix.

6.6.5 EFFECT OF OZONE PRE-TREATMENT ON ANAEROBIC BIODEGRADABILITY

An example of the enhancement of the biodegradability of sludge following ozone pre-treatment is shown in Figure 6.16. In this biochemical methane potential (BMP), 10 mL of treated WAS and 40 mL of anaerobic inoculum were used and the cumulative methane production is shown in Figure 6.16. The 15 and 30 min treatment times corresponded to an ozone dosage of 0.06 and 0.123 g O_3/g TS, respectively. These two ozone dosages resulted in a 25% and 32% increase in methane production, respectively.

Weemaes et al. (2000) reported that ozone pre-treatment increased WAS biodegradability (Table 6.2). An ozone dosage of 0.1 and 0.05 g O_3/g COD caused a 70% and 50% increase in biodegradability, respectively. However, for an ozone dosage of 0.2 g O_3/g COD, a 3-day lag phase was observed and only a 30% biodegradability increase was noticed. Yeom et al. (2002) also observed the lag phase of anaerobic digestion when the ozone dosage was higher than 0.1 g O_3/g TSS, but total methane production was not influenced. Goel et al. (2003b) showed that ozone pre-treatment increased both VS degradation and methane production in a laboratory-scale continuous reactor. An ozone dosage of 0.015 g O_3/g TS only showed a slight increase in VS degradation and methane generation. When the dosage was increased to 0.05 g O_3/g TS, VS degradation and methane production were almost doubled. Saktaywin et al. (2005) found 60% of ozone-induced SCOD at the early stage was biodegradable; however, with the increase in ozone dosage, the remaining solubilized organics were relatively refractory. Bougrier et al. (2006) compared the biodegradability of solubilized COD from different pre-treatments and indicated that the SCOD caused by ozonation was less biodegradable than that caused by ultrasonication. They also suggested that refractory compounds may form when ozone dosage is high (Bougrier et al., 2006). Braguglia et al. (2012b) compared the ozonation and ultrasonication pre-treatment from an economic view. They indicated that for the same low energy input, ozonation was less effective than ultrasonication. Ultrasound pre-treatment (~2500 kJ/kg TS) led to a 19% VSr increase and a 25% biogas production increase,

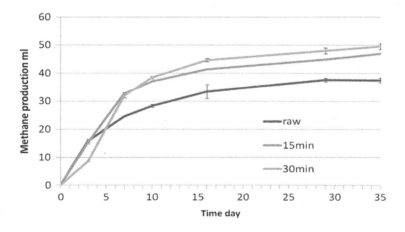

FIGURE 6.16 Cumulative methane production from ozone-treated WAS.

TABLE 6.2
Ozone Pre-treatments of Sludge

Sludge	Ozone Conditions (g O₃/g TS, TSS or COD)	Pre-treatment COD Solubilization	DD	Anaerobic Digestion Methane Prod. (%)	VS rem. %	Dewaterability (s)	Scale and Process Conditions (HRT, days)	Cost or Energy Consumption	References
TWAS (TS: 20 g/L)	0.1 g O₃/g TS	+18% (+28% TS)	—	+11%	—	151–382 (CST)	LS, B	—	Bougrier et al. (2006)
	0.16 g O₃/g TS	+22% (+26% TS)		+24%		N.R.			
P+WAS (TSS: 9.5 g/L)	0.05 g O₃/g COD	0.06–1.6 g/L	—	+50	—	716–86 (80)	LS, B, M	0.375 Eur/kg TS	Weemaes et al. (2000)
	0.1 g O₃/g COD	0.06–2 g/L		+70		862–115 (80)			
	0.2 g O₃/g COD	0.06–2.3 g/L		+30		372–74 (80)			
WAS (TS: 1.2%)	0.02 g O₃/g TSS	0.8%–9.1%	—	+26	—	—	LS, B, M	—	Yeom et al. (2002)
	0.05 g O₃/g TSS	0.8%–19.6%		+75					
	0.1 g O₃/g TSS	0.8%–23.9%		+114					
	0.2 g O₃/g TSS	—		+130					
	0.5 g O₃/g TSS	0.8%–32.7%		+135					
	1 g O₃/g TSS	0.8%–31.5%		—					
	2 g O₃/g TSS	0.8%–27.7%							
WAS	0.015 g O₃/g TS	+19% (VSS)	—	+17	+29	—	LS, SC, M, 14	20 kWh/kg O₃	Goel et al. (2003b)
	0.015 g O₃/g TS	—		+5	+11		28		
	0.05 g O₃/g TS	+37% (VSS)		+109	+86		14		
	0.05 g O₃/g TS	—		+82	+100		28		
BS (TS: 1.42%)	0.1 g O₃/g TS	240–960 mg/L	51.1	+25%	—	—	LS, B, M	—	Erden and Filibeli (2011)

(Continued)

TABLE 6.2 (CONTINUED)
Ozone Pre-treatments of Sludge

Sludge	Ozone Conditions (g O₃/g TS, TSS or COD)	Pre-treatment COD Solubilization	Pre-treatment DD	Anaerobic Digestion Methane Prod. (%)	Anaerobic Digestion VS rem. %	Anaerobic Digestion Dewaterability (s)	Scale and Process Conditions (HRT, days)	Cost or Energy Consumption	References
WAS	0.05 g O₃/g TS	–	2.5	−5.7%	+7.4%	1–14.8 (9.5)[a]	LS, SC, M, 10	–	Braguglia et al. (2012b)
	0.07 g O₃/g TS		6	+17% (biogas)	+25.9%	1.4–11.9 (8.9)			
Maize canning sludge	0.18 g O₃/g DS	BOD5/COD: 26%–93%	–	+8.2 times (biogas)	–	–	LS, B, M	–	Beszédes et al. (2009)
WAS	10 mg O₃/g TSS, 20 cycles, 30 s/ cycle	DD$_{COD}$: 18%, −18% in VSS	–	+800% (biogas)	+1.6 fold	–	LS, B, M	–	Cheng and Hong (2013)
WAS	0.09 g O₃/g MLSS, pH 11	40% COD: −30% TS	–	–	37%	–	LS ASMBR	–	Kumar et al. (2015)

Note: ASMBR, aerobic submerged membrane bioreactor; B, batch; BS, biological sludge; LS, laboratory scale; M, mesophilic anaerobic process; P, primary sludge; SC, semi-continuous anaerobic process; TWAS, thickened waste activated sludge; WAS, waste activated sludge.

[a] Normalized CST (s/L.g TS).

whereas ozone pre-treatment (0.05 g O_3/g TS, equals around 2000 kJ/kg TS) only led to a 7% VSr but no noticeable biogas production increase. When the ozone dosage was increased to 0.07 g O_3/g TS, VSr and biogas production increased rapidly by 27% and 17%, respectively.

Weemaes et al. (2000) and Liu et al. (2001) studied the effect of ozonation on the characteristics of WAS. Ozone pre-treatment removes or solubilizes two-thirds of the organic matter in sludge, thereby enhancing the subsequent anaerobic digestion. The CST of the untreated and the treated sludge was comparable after anaerobic digestion. The disintegration of the sludge cells was also reflected in decreasing SS and VSS contents of sludge. Erden and Filibeli (2011) showed that the SS and VSS content of sludge decreased by 34% and 12.3%, respectively, with 0.25 g O_3/g DS. These solubilization rates were comparable with the results of Chu et al. (2008), but lower than the results of Weemaes et al. (2000).

The improvement in solid degradation efficiencies observed after ozone pre-treatment (up to 65% solids degradation at 0.05 g O_3/g TS) (Goel et al., 2003b) was significantly higher than improvements reported for other pre-treatment options like mechanical disintegration (Kopp et al., 1997), ultrasound treatment (Tiehm et al., 1997) and mechanical jet smashing (Choi et al., 1997). A comparison with other technologies is difficult because it depends on the sludge source, energy inputs during pre-treatment and other operational parameters. The level of improvement, however, seems to be better than thermal/thermochemical treatment at low pre-treatment temperatures and comparable at higher temperatures of 175°C.

Goel et al. (2003b) showed that ozone could reduce sludge treatment and disposal cost by 11% assuming an ozone dosage of 0.05 g O_3/g TS and an ozone production cost of 20 kWh/kg O_3. However, their long-term data also suggested that biomass acclimation to ozonated sludge was necessary before higher degradation efficiencies could be achieved. Even though the ozone pre-treatment shows its cost-effectiveness, the real decisions regarding the application of such a system will depend on capital investments, energy costs and sludge disposal costs. The proposed systems seem to be interesting for larger treatment plants and for sludges that require higher disposal costs.

Ozone was also applied in the RAS line. In a nitrifying sequencing batch reactor operated in alternating anoxic/aerobic conditions, Dytczak et al. (2007) recorded a 14.7% decrease in sludge production and up to a 60% improvement in the denitrification rate when a pre-ozonation of 0.08 mg O_3/mg TSS was applied. Nie et al. (2014) even achieved a zero sludge production system with an ozone dose of 0.1 g O_3/g SS.

6.6.6 FULL-SCALE APPLICATION OF OZONE FOR SLUDGE TREATMENT

IDI through Ondeo-Degremont in France (now owned by Suez) has commercialized the Biolysis® process. It was developed as a process for improving sludge settleability while also reducing the quantities of WAS produced by activated sludge treatment plants. Chemical and enzymatic stressing are used to make cellular material biodegradable, limit microbial growth and increase the energy requirements for the metabolism of bacteria. It is claimed that WAS production can be reduced by 30%–80%. Two versions of the process have been developed.

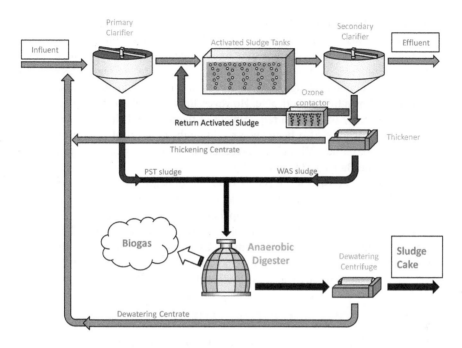

FIGURE 6.17 Implementation of ozone treatment in the sludge recycle line.

The Biolysis "O" process uses ozone for chemical oxidation to stress the bacteria. The process consists of bringing mixed liquor from the activated sludge process, injecting ozone and passing it through a specially designed contact tower (Figure 6.17). The ozonated stream is then returned to the activated sludge process. The ozone stresses and oxidizes the biological material, making it more readily biodegradable when it is returned to the activated sludge process, thereby reducing the reproducibility of a portion of the bacteria. The ozone is produced in a standard on-site generator.

Full-scale demonstration testing of the Biolysis "O" process has been carried out in the United Kingdom and Lebanon, Oregon. Testing in 2004 at the Broomhaugh STW (Northumbrian Water Ltd.) in the United Kingdom showed that Biolysis "O" reduced biological sludge by 78% in one phase of the test and up to 100% in a second phase. Percentage reduction was measured relative to a reference train of the plant. The representative ozone dosage for the test was established at 0.13 kg O_3/kg dry solids removed. It was also shown that the oxygen requirement of the activated sludge increased as the returned mixed liquor was oxidized with a mean increase of 41% as compared to the reference train. Sludge settleability was also measured during the test with a sludge volume index (SVI) improvement from an average of 200–250 mL/g before the trial to 70 mL/g following it.

A full-scale demonstration of the Biolysis "O" process was also conducted at the Lebanon WWTP, Lebanon, Oregon. The Lebanon facility is sized to treat 50,000 PE (18.9 MLD). Pretest average annual sludge production was 1533 kg dry solids/day. The sludge reduction objective for the demonstration was 80% on a TS basis at an ozone dose of 0.13 kg O_3/kg dry solids removed. The sludge SVI improvement objective for the test was to maintain a value of <100 mL/g. Data from test installations of the Biolysis "O" process support the target reduction of up to 80% in WAS TS, without the removal of inert solids. The design for a facility in France was based on a target 70% reduction in sludge. An ozone dose of 0.13 kg O_3/kg dry solids removed has also been shown to be practicable. However, the return of this highly degradable stream may increase the oxygen requirement of the activated sludge by up to 40%. Lower SVI values have also been shown, although the level of improvement will depend on the baseline SVI and the presence of filamentous bacteria prior to the installation of the Biolysis "O" process.

The main capital costs for the Biolysis "O" process are the ozone generator, the mixed liquor suspended solid (MLSS) sidestream pumping system and the injection system and contact tower. The principal operation and maintenance costs of the process are the energy demand for pumping, increased aeration demand and ozone generation. Initial reviews for a facility in the United States indicate that the costs of the ozone system are significant and may not, depending on site-specific factors, be as cost-effective as alternative options (Roxburgh et al., 2006).

6.7 FREE NITROUS ACID PRE-TREATMENT

Free nitrous acid (FNA, i.e., HNO_2) treatment is a recent sludge treatment technology to reduce sludge production, which has only been demonstrated at the laboratory scale with synthetic wastewater, although large-scale trials are ongoing (Wang et al., 2017). Pijuan et al. (2012) showed that FNA, at parts per million (i.e., mg/L) levels, had a strong biocidal effect on bacteria in WAS. It has been demonstrated that 50%–80% of the bacterial cells in WAS could be damaged by FNA at 1–2 mg N/L for 24–48 h. Based on the biocidal effect of FNA, Wang et al. (2014b, 2013b) proposed the FNA-based sludge reduction technology whereby a sludge reduction of 11%–28% can be achieved while part of the RAS is treated by FNA at 1.35–2.0 mg N/L. FNA treatment can also be applied in the sludge recycle line or as a pre-treatment prior to the anaerobic digester.

It has also been reported that 1.0–2.5 mg HNO_2-N/L for 5–24 h resulted in an increased CH_4 production of 15%–56%. It was reported that FNA pre-treatment was able to improve both the sludge degradation rate and the sludge degradation extent (Wang et al., 2014a, 2013a). It should be highlighted that FNA is a renewable chemical that is able to be produced as a by-product of wastewater treatment via nitritation of anaerobic sludge digestion liquor (Law et al., 2015). After pre-treatment, FNA contained in pre-treated WAS is diluted and rapidly removed in the anaerobic digester, thereby forming a closed loop pre-treatment technology. FNA has also been combined with heat (55°C) to further achieve sludge reduction (Wang et al., 2014a). It was reported that the sludge degradation (VS basis) increased from 35% (without pre-treatment) to 41% after implementing FNA pre-treatment of ~1.0 mg HNO_2-N/L. Moreover, combined FNA and heat pre-treatment achieved a sludge degradation of 44% (Wang et al., 2014a). Zhang et al. (2015b) also applied FNA + H_2O_2 to further reduce sludge amounts. The sludge degradations (VS basis) were found to be

30%, 49% and 59%, respectively, for the WAS without pre-treatment, WAS pre-treated with FNA (1.5 mg HNO_2-N/L) and WAS pre-treated with FNA + H_2O_2 (1.5 mg HNO_2-N/L + 50 mg H_2O_2/g TS).

6.8 ENZYMATIC PRE-TREATMENT

Hydrolytic enzymes can be added to WAS anaerobic digesters to solubilize macromolecules and increase the hydrolysis rate (Yang et al., 2010). A specific enzyme only works on its corresponding substrate. Different enzymes have been tested to disintegrate WAS, such as protease, amylase and their mixtures. It was reported that a mixture of enzymes has a better performance in sludge disintegration than single enzyme addition (Yang et al., 2010). This may be attributed to the complex composition of WAS. EPS mainly consists of carbohydrates and proteins; proteins and nucleic acid are main components of intracellular substances. Additionally, bacteria cell walls also contribute to carbohydrates.

Figure 6.18 illustrates the effect of the addition of a commercial enzyme targeting protein in WAS. Proteinase K (EC3.4.21.64) was obtained from Sigma-Aldrich. Proteinase K exhibits broad substrate specificity. It degrades many proteins in their native state even in the presence of detergents. Proteinase K was isolated from a fungus, *Engyodontium album* (formerly *Tritirachium album*), which is able to grow on keratin. Consequently, Proteinase K is able to digest native keratin (hair), hence the name "Proteinase K." The predominant site of cleavage is the peptide bond adjacent to the carboxyl group of aliphatic and aromatic amino acids with blocked alpha-amino groups. It is commonly used for its broad specificity. One unit of Proteinase K

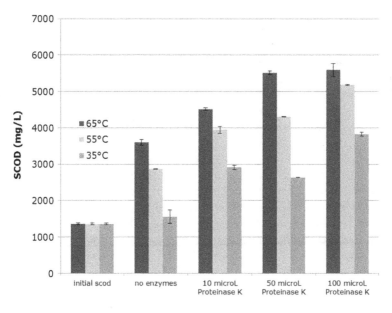

FIGURE 6.18 Solubilization of sludge due to the addition of a commercial protease (Proteinase K) at 35°C, 55°C and 65°C.

will hydrolyze urea-denatured hemoglobin to produce color equivalent to 1.0 mmole (181 mg) of tyrosine per minute at pH 7.5 at 37°C. Various volumes (0, 10, 50 and 100 μL) of a Proteinase K solution (600 U/mL) was added to 5 mL of sludge and incubated at three different temperatures for 24 h. After that, the SCOD was measured to determine the efficiency of solubilization due to enzymatic treatment.

It can be seen from Figure 6.18 that when no enzymes were added, the SCOD increased with increasing temperature due to thermal treatment and the activation of indigenous proteolytic enzymes in sludge. However, when Proteinase K was added, the solubilization significantly improved with greater temperatures, in particular at 55°C and 65°C, indicating that the commercial enzymes were more active at thermophilic temperatures. Adding 100 μL of enzymes resulted in more solubilization except at 65°C. It was also noticed that the thermal treatment at 65°C was equivalent to the enzymatic treatment with 100 μL at 35°C. These data support the fact that indigenous enzymes are already present in WAS and that thermophilic and hyperthermophilic temperatures can activate them. The indigenous enzymes are not very active at 35°C, in contrast with the commercial enzyme Proteinase K, which is very active even at 35°C. Adding commercial enzymes can therefore improve the solubilization of WAS, but they are expensive and their effect is limited due to the complexity of WAS and the multiple types of substrates that would require different specific enzymes, each having an optimum pH and temperature range.

Nevertheless, enzyme pre-treatment of WAS can increase anaerobic gas production. A mixture of two glycosidic enzymes was reported to increase the daily biogas production by 10%–20% (Yang et al., 2010). Ayol et al. (2008) achieved more than 50% daily biogas increment by adding either α-amylase or β-glucanase into anaerobic digesters. Enzymatic pre-treatment has its limitations. Due to the complexity of WAS, the addition of enzymes may not always achieve the expected performance (Barjenbruch and Kopplow, 2003). The optimum temperature for hydrolytic enzymes is around 50°C and the optimum performance of enzymes may not be obtained under mesophilic conditions.

Jung et al. (2002) have investigated the disruption of excess sludge using a mill and found that among protease, amylase, glucosidase, lipase and dehydrogenase, protease was the one with the highest activity. Almost 69% of the protease activity was recovered by protein precipitation using ammonium sulfate, while the recovered enzyme solution lost its activity by 32% after preservation at −20°C for 1 month. After the recovery of protease, the authors studied the applicability of the protease using milk as a model. A maximum protease activity of 2210±308 U/g MLSS was reported in a municipal wastewater sludge at a temperature of 50°C, while for a laboratory-cultivated sludge the maximum activity was 3450±124 U/g MLSS at 75°C. This suggested that each sludge has different microbial populations, therefore different protease activity and optimum temperatures. Gessesse et al. (2003) also found a protease activity of about 4000 U/g MLSS from enzymes extracted from WAS using Triton (0.5%).

In another study, proteases and lipases were extracted from WAS using ultrasound (24 kHz, 3.9 W/cm^2, 30 min, 5°C using water-ice bath) combined with a non-ionic detergent (Triton X100) (Nabarlatz et al., 2012). It was observed that the concentration of Triton has a strong influence on the extraction of protease, but not on lipase.

Proteases were purified by precipitation with ammonium sulfate and dialysis. The authors mentioned that the recovery of valuable products from sludge could be used in the sludge degradation itself, but they did not attempt it. The authors also found that the activity of extracellular protease in an activated sludge tank was much lower than that of intracellular protease.

Barjenbruch and Kopplow (2003) investigated the pre-treatment of a mixture of primary and secondary sludge (40:60) using a HPH, enzymes (carbohydrase) and thermal (80°C, 90°C and 121°C for 60 min) treatment. The improvement in the gas production of thermal at 80°C, 90°C and 121°C, enzymatic and HPH was 16%, 21%, 20%, 12.5% and 17.5%, respectively. The VSS degradations were 4%, 5%, 5.5%, 1% and 6%, respectively. The dewaterability improved particularly with the enzymes (–25% in CST) compared to 90°C (–6% in CST), 121°C (+10% in CST) and HPH (+10% in CST). This was due to the better delivery of water through the degradation of microbiological slimes by the enzymes. Parmar et al. (2001) used a mixture of industrial cellulase, protease and lipase to hydrolyze an anaerobic sludge and observed a TSS reduction of 30%–50% and better settleability. Due to the high cost of US\$3/m^3 or US\$100/dry ton, optimization on the small scale is recommended before large-scale application.

Thomas et al. (1993) have used a mixture of carbohydrases, proteases and lipases on anaerobic sludge and found that concentrations in the range of 2.5–5 ppm could reduce the CST by 50%. However, a long incubation time of 16 h and mixing were required.

7 Thermal/Chemical Pre-Treatment of Sludge

Some authors have combined chemical addition and thermal treatment (Figure 7.1 and Table 7.1). As explained in Chapter 6, alkalis have been used to solubilize various substrates such as lignocellulosic materials and waste activated sludge (WAS) (Pavlostathis and Gossett, 1985; Ray et al., 1990). Alkalis allows significant solubilization and improves biodegradability performances. When combined with thermal treatment (Li and Noike, 1992; Stuckey and McCarty, 1978b; Tanaka et al., 1997), it also leads to important solubilization. However, extreme pH conditions are not compatible with anaerobic biological treatment. Penaud et al. (1999) combined NaOH addition (26.1 g/L) at 140°C for 30 min, which led to 85% chemical oxygen demand (COD) solubilization instead of 53.2% at ambient temperature. This emphasized that heating improved the pH effect. However, at high pH the biodegradability performance was limited due to the formation of refractory compounds and not due to sodium toxicity. At 4.6 g/L NaOH and 140°C, an increase of 163% in biodegradability was obtained (Penaud et al., 1999), but beyond 5 g/L the biodegradability performance decreased.

Vlyssides and Karlis (2004) treated WAS (10% w/v) at 50°C, 60°C, 70°C, 80°C and 90°C with pH at 8, 9, 10 and 11. At pH 11 and 90°C and after 10 h of hydrolysis, the soluble COD (SCOD) reached 69,000 mg/L and the volatile suspended solids (VSS) solubilization was around 40%. The methane yield was 0.28 L CH_4/g VSS compared to 0.11 L CH_4/g VSS without pre-treatment. However, this contradicts other studies where it was demonstrated that thermo-alkaline pre-treatment did not generate any variations in biodegradability (Haug et al., 1983). Liu et al. (2008) found that alkaline/thermal treatment was slightly superior to ultrasonic/acid treatment in terms of solids and COD solubilization.

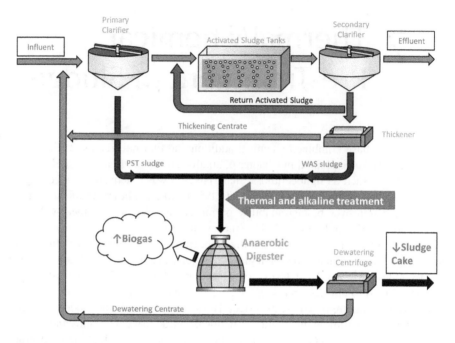

FIGURE 7.1   Flowsheet including a thermal and alkaline pre-treatment of sludge.

TABLE 7.1
Combined Thermal and Chemical Pre-Treatments of Sludges

Substrate	Thermal Conditions	Alkaline Conditions	Solubilization		Anaerobic Digestion			Scale, Mode, Temperature (HRT)	References
			COD	DD (%)	CH$_4$	VS Removal	TCOD Removal		
Activated sludge (TS: 17.1 g/L)	130°C, 1 h 170°C, 1 h	3.65 g KOH/L 3.65 g KOH/L	SCOD/TCOD: 2.7%–63.1% 2.7%–83%	–	+28% +45%	–	–	LS, B	Valo et al. (2004a)
WAS (TS: 17.1 g/L)	90°C, 1 h 90°C, 1 h 130°C, 1 h	150 mM H$_2$O$_2$ 300 mM H$_2$O$_2$ 150 mM H$_2$O$_2$	SCOD/TCOD: 2.7%–30.7% 2.7%–23.6% 2.7%–35%		+0% – –	– 	–	LS, B	Valo et al. (2004a)
WAS (TS: 17.1 g/L)	130°C, 1 h 170°C, 1 h	3.65 g KOH/L –	DD$_{COD}$: ~60%		+74% +54%	+72% +92%	60.2% (44%) 70.7% (44%)	LS, C, M (20)	Valo et al. (2004a)
WAS (TS: 30.6 g/L)	140°C, 30 min	4.6 g/L NaOH 26.1 g/L NaOH KOH (pH 12) Mg (OH)$_2$ (pH 12) Ca (OH)$_2$ (pH 12)	SCOD/TCOD: 20%–80% 20%–85.1% 20%–83.7% 20%–55.6% 20%–51.1%		+163% –77%	–	–	LS, B	Penaud et al. (1999)
WAS (TSS: 17.5 g/L)	90°C, 10 h	NaOH pH 11 (10 h)	20–3500 mg/L		+91.1%	–	–	LS, B	Zhang et al. (2010)
WAS (7 g/L SS)	130°C, 5 min	0.3 g NaOH/g VSS (=34 mM)	40%–50%		+200%–220%	–	–	LS, B, M	Tanaka et al. (1997)
WAS (4.3% TS)	175°C, 1 h	300 meq/L NaOH (1 h)	54%–55%		+15%	–	–	LS, B, M	Stuckey and McCarty (1984)

Note: Scale: full scale (FS), pilot scale (PS) or laboratory scale (LS). Mode: batch (B), semi-continuous (SC) or continuous (C). Temperature of anaerobic digestion: mesophilic (M) or thermophilic (T).

8 Physical Pre-Treatment of Sludge
Ultrasound

8.1 INTRODUCTION AND MECHANISMS OF ULTRASOUND

Ultrasonic waves are acoustic waves with a frequency between 20 kHz and 10 MHz (Tiehm et al., 1997). Ultrasound causes cavitation effects in treated samples. During cavitation, an abundance of cavitation bubbles are produced. These bubbles are classified into stable bubbles and transient bubbles. Stable cavitation bubbles last for hundreds of acoustic cycles (Tronson et al., 2002) and have almost equilibrium size during cavitation. Transient bubbles usually persist for only a few cycles (Tronson et al., 2002) and their bubble size may grow to at least two times their original size. These two types of bubbles are not mutually exclusive, as stable cavitation bubbles can be converted to transient cavitation bubbles via rectified diffusion.

The collapse of transient bubbles tends to cause more changes to the ultrasonicated liquid. The life span of a transient bubble can be concluded as bubble nucleation, bubble growth and bubble collapse. Ultrasonic longitudinal waves propagating through a liquid cause a strong oscillation within the liquid medium, which will lead to a periodical pressure change within the liquid. The total external pressure, P_e, has been proved to be a function of time:

$$P_e(t) = P_0 + P_A \sin \omega t$$

where
P_0 is the hydrostatic pressure (normally equal to atmospheric pressure);
P_A is the maximum possible acoustic pressure during the oscillation; and
ω is the angular frequency of the ultrasonic wave.
When the ultrasonic wave frequency is >20 kHz, P_A could be predicted by

$$I = P_A^2 / 2\rho c$$

where
I is the ultrasonication intensity;
ρ is the density of liquid;
and c is the velocity of sound in liquid.
The value range of sin ωt is between [−1,1], and normally P_0 is much smaller than P_A. It can be assumed that the total external pressure swings from negative P_A to positive P_A in a very short range of time.

Accompanied by the huge pressure swing during cavitation, bubble temperatures also increase drastically. A simple model was set up under the pre-assumption that adiabatic compression takes place to determine the theoretical maximum temperature within bubbles:

$$T_{max} = T_0 \frac{P_e * (\gamma - 1)}{P_V}$$

where

T_0 is the ambient solution temperature;

P_e is the total external pressure;

γ is the specific heat ratio of the gas/vapor mixture;

and P_V is usually assumed to be the vapor pressure of the liquid.

This explains why cavitation leads to an extreme local high pressure and high temperature. Furthermore, this local extreme temperature will cause water dissociation, and generate reactive hydrogen radicals H• and hydroxyl radicals OH•. These radicals are also one of the important phenomena caused by ultrasonication.

As previously mentioned, the very first step in transient bubble formation is nucleation. This can be well explained with the classic crevice model. In this model, a partially wetted tiny solid particulate containing a gas-filled crevice is regarded as the target (Figure 8.1).

The sketch on the left-hand side of Figure 8.1 shows the gas–liquid interface when the external total pressure is highly positive. The external pressure is much larger than the sum of the gas pressure and vapor pressure. The interface is balanced by the surface tension that draws the interface inward. Thus, the interface becomes concave into the crevice and this suppresses the formation of a gas bubble. The sketch on the right-hand side shows the gas–liquid interface when the external total pressure is highly negative. In this case, the sum of the gas pressure and vapor pressure is much greater than the external total pressure. The interface is pushed outward by surface tension. When the negative pressure is low enough, the gas will be liberated as a free gas bubble. This is the nucleation process. The gas amount in a free gas bubble is constant. When this bubble is exposed to the highest external pressure, it reaches its smallest size. With the decrease in external pressure, the bubble tends to expand, which is the bubble growth process. The gas bubble keeps expanding until the interface cannot withstand the stress force. The bubble will then collapse. During cavitation, thousands of bubbles repeat the same cycle in 1 s. Except for the thermal effect and pressure increase, the sudden collapse of bubbles will also lead to significant hydromechanical shear force. This is also an important influence of ultrasonication.

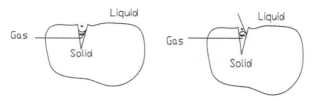

FIGURE 8.1   Gas-filled crevice model.

8.2   APPLICATION OF ULTRASOUND ON SLUDGE

Based on the mechanisms described in the previous section, it is clear that cavitation can cause a drastic change in the target medium. Therefore, ultrasonication was considered applicable for industrial use. Ultrasonication has been used to disintegrate biological cell structures since the 1950s (Davies, 1959; Hughes and Nyborg, 1962; Doulah, 1977; Harrison, 1991). However, waste activated sludge (WAS) ultrasonication did not flourished until the late 1990s. Since then, its mechanisms have been widely analyzed and interpreted. Ultrasonication causes complex effects on sludge samples. During cavitation, thousands of bubbles repeat the same cycle in 1 s. Except for the thermal effect and pressure increase, the sudden collapse of bubbles will also lead to a significant hydromechanical shear force. This is also an important influence of ultrasonication. Hydroxyl radicals are generated during cavitation (Riesz et al., 1985), which may chemically disintegrate the sludge. Meanwhile, local high temperature and high pressure caused by the cavitation effect also contribute to sludge disintegration (Kim and Youn, 2011). Among all these factors, hydromechanical shear force is regarded as the most predominant effect (Tiehm et al., 2001). Therefore, ultrasonication is categorized as mechanical pre-treatment. The possible WAS ultrasonication mechanisms are listed as follows:

- Hydromechanical shear force caused by the rapid collapse of cavitation bubbles
- Hydrophobic volatile decomposition caused by local high bubble temperature
- Increase in temperature sludge during cavitation
- Generation of radicals

The force created by one bubble explosion may be slight; however, the force created by explosions of thousands of bubbles in 1 s is significant. This force may disrupt the microorganism flocs and break cell walls, which may accelerate the subsequent anaerobic process.

The sludge temperature will rise during ultrasonication and sludge solubilization can take place at high temperature. However, results obtained from different research are controversial. Wang et al. (2005) observed that the sludge temperature rises to 80°C after 1 h ultrasonication. However, a very low soluble chemical oxygen demand (SCOD) increase was obtained when the sludge sample was heated to 80°C for 1 h without ultrasonication. On the contrary, Kim and Youn (2011) and Liu et al. (2009) claimed the obvious influence of temperature on sludge solubilization. Liu et al. (2009) indicated that sludge temperature is more important than duration, but results from Kim and Youn (2011) also indicated that longer heated time promotes sludge solubilization.

Ultrasonication can also cause water sonolysis whereby one water molecule is split into one hydrogen radical and one hydroxyl radical (Hua and Hoffmann, 1997). In addition to physical sludge disintegration, many toxic and recalcitrant organic pollutants, such as aromatic compounds, chlorinated aliphatic compounds, surfactants, organic dyes, etc., are also broken down into simpler forms. This is due to the generation of the highly oxidative reactive radicals hydroxyl (OH•), hydrogen (H•), hydroperoxyl (HO$_2$•) and hydrogen peroxide (H$_2$O$_2$) during ultrasound pre-treatment, which leads to the oxidative breakdown of these recalcitrant compounds (Khanal et al., 2007). Hydroxyl radicals can react with organic scavengers in the sample and degrade them into simpler compounds. For example, a formic acid molecule can be degraded to carbon dioxide and water with the help of hydroxyl radicals (Navarro et al., 2011). Wang et al. (2005) tried to eliminate the influence of hydroxyl radicals on sludge solubilization by adding sodium bicarbonate, as it is known to be a very good hydroxyl radical scavenger. The authors found that hydromechanical shear forces were primarily responsible for WAS disintegration when NaHCO$_3$ was added at an ultrasonic density of 0.384 W/mL or lower. However, the contribution of oxidizing radicals was predominant at a higher ultrasonic density of 0.72 W/mL.

Another mechanism that occurs when sludge is sonicated is acoustic streaming. Acoustic streaming has been studied since 1831 (Faraday) and occurs at the solid–liquid (sludge) interface when the solid interface experiences harmonic vibrations. The main benefit of streaming in sludge processing is mixing, which facilitates the uniform distribution of ultrasound energy within the sludge mass, the convection of the liquid and the distribution of any heating that occurs. Overall, there are three regions of acoustic streaming. The largest region is the furthest from the vibrating tool and has circulating currents that are defined by the shape of the container and the size of the wavelength of the acoustic wave in the liquid. The region near the tooling has circulating currents and its size and shape are primarily defined by the acoustic tooling. These circulations are typically called "Rayleigh streaming," with much longer wavelengths than that of the acoustic wave in the liquid. The region nearest to the tool, called "Schlichting streaming," is adjacent to the fluid acoustic boundary layer. This is a region where the tangential fluid velocity is near the velocity of the horn face. This layer is relatively thin. For example, at 20 kHz, the acoustic boundary layer for water at 20°C is <4 μm (Graff, 1988). All three regions play a critical role in mixing the fluid.

It has been shown that low frequencies are more efficient for the degradation of WAS because the mechanical effect facilitates particle solubilization (Tiehm et al., 2001). When ultrasonic waves propagate through a liquid medium, it leads to strong oscillation within the liquid resulting in the acoustic cavitation effect. The latter refers to the expansion and contraction of cavities (i.e., bubbles) due to the passage of acoustic waves through the liquid (Riesz et al., 1985). Hydroxyl radicals are generated during cavitation (Riesz et al., 1985), which may chemically disintegrate the sludge. Meanwhile, local high temperatures (5000 K) and pressures (180 MPa) would also contribute to sludge disintegration (Kim and Youn, 2011). Among all these factors, the hydromechanical shear force is regarded as the most important as it not only disrupts the WAS's extracellular polymeric substances (EPS) matrix, but also causes cell lysis (Tiehm et al., 2001; Wang et al., 2005). Some other merits of ultrasound pre-treatment are as follows:

- Compact design and easy retrofit within existing systems.
- Low-cost and efficient operation compared to several other pre-treatments.
- Production of an in situ carbon source for denitrification plants.

- Complete process automation.
- Potential to control filamentous bulking and foaming in the digester.
- Better digester stability.
- Improved volatile solids (VS) destruction and biogas production.
- Better sludge dewaterability.
- Improved biosolids quality (i.e., biosolids with low residual biodegradable organics, low pathogen counts, etc.).

Ultrasound pre-treatment also faces several challenges. One of the major issues is the high capital and operating costs of ultrasound units. The cost may go down as the technology matures. Similarly, long-term performance data of full-scale ultrasound systems are still limited. This discourages design engineers from recommending ultrasound systems for full-scale application (Khanal et al., 2007).

8.3 QUANTIFICATION OF ULTRASONICATION ENERGY INPUT

Different ways are used to quantify ultrasonication in terms of energy input. Power input (W) is a very direct expression, but it only gives the net energy input regardless of the amount of sample treated. Power density (W/L) is a unit, which equals to power input per volume of sludge treated. It correlates the treated sludge and power input.

Power intensity (W/cm^2) is equal to power input per probe area. As acoustic waves are longitudinal waves, the smaller the probe area, the better the performance. Another widely used unit is the specific energy input (SEI; kJ/kg TS). It involves the treatment time, power input, treated volume and solids concentration of a sludge sample. The calculation is shown as follows:

$$E_s = \frac{P * t}{V * TS}$$

where
E_s is the specific energy (J/g TS);
P is the power input (W);
t is the treatment duration (s);
V is the treated volume (L);
and TS is the total solids concentration in the sludge sample (g TS/L).

This ultrasonication unit is particularly useful for sludge, because it provides the amount of energy required for a given amount of TS. However, this quantification fails to indicate power input, power intensity and power density. Generally, specific energy is used to indicate energy input and it increases with treatment duration. Other parameters are listed as process parameters and do not change with time. At the large scale, the use of kWh per m^3 of sludge treated is more common for energy considerations.

8.4 QUANTIFICATION OF SOLUBILIZATION DUE TO ULTRASONICATION

There are many different quantifications of COD solubilization. For example, an increase in SCOD (i.e., SCOD+) or a change in the SCOD fraction (i.e., SCOD/total COD [TCOD]). But these quantifications vary for different sludge samples. In order to make results inter-comparable, Müller (1998) came up with the term "disintegration degree" (DD). It is assumed that maximum possible COD solubilization is achieved when the sludge is mixed with 1 M NaOH solution in a ratio of 1:2 for 10 min at 90°C. The equation is given as follows:

$$DD = \frac{SCOD_T - SCOD_O}{SCOD_{NaOH} - SCOD_O}$$

where
  $SCOD_T$ stands for the SCOD of a treated sample,
  $SCOD_{NaOH}$ stands for the SCOD of an NaOH-treated sample
  and $SCOD_O$ stands for the SCOD of the original sample.

DD presents the sludge disintegration in a relative value rather than an absolute value, which allows for comparison between different sludge samples.

Another term to indicate the change of SCOD is COD solubilization (Bougrier et al., 2006). It quantifies the change of particulate COD to SCOD. The equation is given as follows:

$$SCOD = \frac{SCOD_T - SCOD_O}{pCOD_O}$$

where
  $SCOD_T$ is the SCOD of the treated sample,
  $SCOD_O$ is the SCOD of the original sample
  and $pCOD_O$ is the particulate COD of the original sample.

COD solubilization can also indicate sludge disintegration in a relative value as DD does. However, DD is more commonly used than COD solubilization.

8.5 EFFECT OF POWER DENSITY ON SLUDGE SOLUBILIZATION

Many papers have reported that power density influences the solubilization process in terms of SCOD and particle size distribution. Chu et al. (2001) observed that higher density can produce more SCOD and better floc size destruction for the same treatment time. At high power densities of 0.33 and 0.44 W/mL, significant floc size reduction was observed in the first minutes of ultrasonication. Floc size reduction became stable after 30 and 20 min, respectively, in both cases. On the other hand, floc size decreased slightly with increasing treatment time at low power densities of 0.22 and 0.11 W/mL. Even after 100 min, the mean diameter of the floc size was still much larger than the smallest particle size obtained at power densities of 0.33 and 0.44 W/mL. Liu et al. (2009) also indicated that high power density can accelerate the solubilization of COD and floc size reduction, but they also mentioned that duration was more important than power density when it comes to the reduction of volatile suspended solids (VSS).

Some researchers showed the beneficial effect of high power density by applying the same SEI. Mao et al. (2004) showed that more COD was solubilized and faster particle size reduction was achieved in both primary and secondary sludge at a higher power density under the same SEI. The tested power densities were 2, 3 and 4 W/mL, which are much higher than other research. Grönroos et al. (2005) reported more COD was solubilized for increasing power densities from 0.29 to 1.7 W/mL at equivalent energy inputs. Zhang et al. (2007a) supported this argument by comparing the DD at different power densities. A high DD was achieved at high power density from 0.1 to 1.5 W/mL instead of long times.

Jiang et al. (2009) observed that when the power density was <0.1 W/mL, the SCOD increased with increasing power density for the same treatment time. When the power density was over 0.1 W/mL, the power density seems to have little influence on SCOD. The duration of these experiments ranged from 10 to 60 s, which may be the reason for this different result.

Although high power density is favored by many researchers, controversial results have been reported. Li et al. (2009b) compared the DD of a treated sample with different power densities but the same energy input. The comparison was conducted at three different power densities 4, 0.8 and 0.4 W/mL corresponding to treatment durations of 1, 5 and 10 min, which resulted in a DD of 10%, 25% and 35%, respectively, indicating that treatment duration is more important at high power densities.

8.6 EFFECT OF PROBE AMPLITUDE ON SLUDGE SOLUBILIZATION

In a laboratory-scale ultrasonicator, the ultrasonic electronic generator transforms an alternating current (AC) line power to a 20 kHz signal that drives a piezoelectric converter/transducer. This electrical signal is converted by the transducer to a mechanical vibration due to the characteristics of the internal piezoelectric crystals. The vibration is amplified and transmitted down the length of the horn/probe where the tip longitudinally expands and contracts. The distance the tip travels is known as the amplitude and is selected by the user. Amplitude is, therefore, a measurement of the excursion of the tip of the probe. For instance, the probe model 4208 provided by Qsonica has a 19.1 mm diameter titanium tip that can move by 60 µm (Figure 8.2, middle).

This particular probe is recommended for treating sample volumes up to 500 mL. The probe was connected to an ultrasonication unit (Q700 model from Misonix, USA) that can deliver a maximum power of 700 W and ultrasound waves at a frequency of 20 kHz. The user can select a percentage of this maximum amplitude. As the amplitude is increased, the sonication intensity will also increase within the sample. In liquid, the rapid vibration of the tip causes cavitation, the formation and the violent collapse of microscopic bubbles. The collapse of thousands of cavitation bubbles releases tremendous energy in the cavitation field. The erosion and shock effect of the collapse of the cavitation bubbles is the primary mechanism of fluid processing.

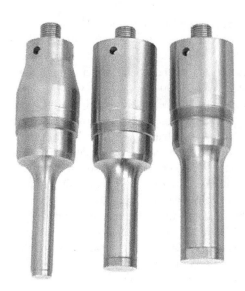

FIGURE 8.2 Laboratory-scale ultrasonication probes used for the pre-treatment of sludge. From left to right: Model #4121, #4208 and #4209 provided by QSonica.

In order to examine the effect of different amplitudes on treating sludge, amplitudes of 10%, 30%, 60% and 90% were chosen while other conditions were kept constant. Specifically, 20 g of a 6% TS sludge was treated using probe 4208 and the energy input was kept constant around 3500 J. The results are shown in Figure 8.3.

Figure 8.3 shows that for a given energy input, the higher the amplitude the greater the COD solubilization. This is mainly because a higher amplitude leads to higher power.

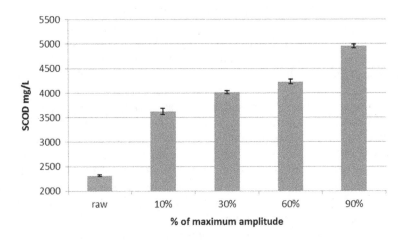

FIGURE 8.3 Effect of probe amplitude (Model 4208 from Qsonica).

8.7 EFFECT OF ULTRASONICATION FREQUENCY

Ultrasonic waves are acoustic waves whose frequency falls in the range between 20 kHz and 10 MHz (Tiehm et al., 1997). When ultrasonic waves propagate through a liquid medium, it leads to a strong oscillation within the liquid. The main effect caused by ultrasonication is the acoustic cavitation effect. Acoustic cavitation refers to the expansion and contraction of cavities also referred to as bubbles, due to the passage of acoustic waves through the liquid (Riesz et al., 1985). The frequency determines how fast the acoustic waves vibrate during propagation. It has been observed that the higher the frequency the more stable the cavitation bubbles. The reason is postulated that the solution is not sufficient to establish standing waves at lower frequency (e.g., 20 kHz), and bubbles will undergo transient cavitation rather than stable cavitation. Cavitation bubble–types determine the effect they will cause. At low frequencies, the mechanical effect, which is effective in disintegrating sludge, is the predominant cavitation effect. However, at high frequencies, radical generation rather than mechanical shear force is promoted, which turns out to be less important in disintegrating sludge (Wang et al., 2005).

Tiehm et al. (2001) proved that sludge DD decreases with increasing ultrasonication frequency (41–3217 kHz) at an ultrasonication intensity of 1.8 W/cm^2 for 4 h. Jiang et al. (2009) compared the change in SCOD at different frequencies (19, 25, 40, 80 kHz) under a power density of 75 W/L for 60 s. An SCOD increase from 19 to 25 kHz was observed; however, with increases from 25 kHz onwards, the SCOD tended to decrease. Similar results were obtained by Jung et al. (2011) by comparing the ultrasonication frequency of 28 and 40 kHz. Under different power densities (40 and 80 W/L) and different treatment durations (10, 15, 20, 25, 30 min), lower ultrasonication frequency always led to higher COD solubilization rates. Jung et al. (2011) also found that dual frequency leads to more COD solubilization than single frequency, but the reasons are not clear. Braguglia et al. (2012a) suggested that high frequency (i.e., 200 kHz) ultrasonication may be an alternative to low frequency ultrasonication in increasing COD solubilization and biogas production; however, comparison data between different frequencies were not provided. Based on up-to-date knowledge, a frequency from 20 to 25 kHz provides the best performance in sludge disintegration.

8.8 EFFECT OF TOTAL SOLIDS CONTENT OF SLUDGE

TS concentration is important for ultrasonication. Small particles act as cavitation nuclei and decrease the cavitation threshold. Tiehm et al. (2001) proved that sludge disintegration can be observed at a power intensity of 0.1 W/cm^2, which is much lower than the water cavitation threshold of 0.4 W/cm^2 (Lormier, 1990).

Neis et al. (2000) reported that DD increases with TS concentration at the same energy input. The TS concentrations tested were 10.5, 15.2 and 34.4 g/kg. None of them had a negative effect on sludge disintegration. Onyeche et al. (2002) compared the biogas production of sludge with 1.8%, 3.6% and 5.4% TS concentrations and reported that a higher TS concentration leads to more biogas production; no negative effect was detected. Controversial results were reported by Akin (2008).

In that research, three different TS concentrations of 2%, 4% and 6% were tested. Ultrasonication with a TS concentration of 2% had a better performance in terms of COD and protein solubilization. These results proved that a high TS concentration may cause adverse effects on sludge disintegration.

Show et al. (2007) recognized that there should be an optimum TS concentration range for ultrasonication based on their results. They tested four different TS concentrations of 1.0%, 1.7%, 2.9% and 3.8%. At the same energy input, 2.9% of TS seemed to have the best ultrasonication efficiency, followed by 3.8%, 1.7% and 1.0% TS.

By combining previous research results and corresponding explanations, the following conclusion is drawn. TS concentration seems to be a double-edged sword for the ultrasonication process. On the one hand, as discussed before, the existence of solids helps to form cavitation nuclei, when TS is too low, the cavitation threshold may increase (Neis et al., 2000). Furthermore, most ultrasonication energy applies to water instead of sludge when the TS concentration is low. This phenomenon also decreases sludge ultrasonication efficiency. On the other hand, high concentrations of TS may hinder the movement of the ultrasonic horn and cause unexpected energy loss (Show et al., 2007; Akin, 2008).

8.9 EFFECT OF ULTRASOUND ON SOLUBILIZATION OF WAS

Ultrasound does not lead to the evaporation or mineralization phenomena, so TS and VS are constant during ultrasonication (Bougrier et al., 2005). Nevertheless, ultrasound reduces VSS and total suspended solids (TSS) by solubilizing both mineral and organic particles. Bougrier et al. (2005) reported that the solubilization of mineral particulates was much lower than that of organic particulates. Less than 3% of mineral particulates were solubilized, whereas 29% of organic particulates were solubilized at an SEI of 15,000 kJ/kg TS. Erden and Filibeli (2010a) observed a maximum VS solubilization of 13.31% and a maximum TS solubilization of 11.45% at an SEI of 11,204 kJ/kg TS.

The solubilization of COD mainly originates from EPS disintegration and bacteria cell lysis. During ultrasonication treatment of WAS, EPS are destroyed, the microorganism floc size is reduced and both simple and complex organics are released. Proteins and carbohydrates, which are the main components of EPS, and intracellular substances such as nucleic acid and enzymes are released to the soluble phase (Dewil, 2006). Cell walls, cell membranes and cell organelles also contain carbohydrates, proteins and nucleic acid. These organics are released and solubilized during ultrasonication.

Although sludge disintegration quantification can be unified, different results were still reported by different research groups. This may be explained by the different sludge characteristics, different operational conditions and measurement errors. The ultrasonication of sludge can be done in continuous or batch mode. Tiehm et al. (1997) used a high power ultrasonicator (3.4 kW) to treat WAS continuously. They reported an SCOD increase from around 300 to around 6000 mg/L in only 96 s. Clark and Nujjoo (2000) used 3×3 kW radial horns to ultrasonicate WAS with a TCOD of around 40,000 mg/L. With a TS concentration of 2.3%, a significant increase in SCOD from 100 to 10,000 mg/L was observed after 20 s. Furthermore, for WAS with a 3.1% TS concentration, more than 30,000 mg/L SCOD was achieved after 60 s. However, no sludge disintegration data were provided by these two papers. Based on the mathematical model created by Wang et al. (2005), power intensity plays a more important role in COD solubilization. Neis et al. (2000) observed a linear relationship between DD and power intensity with a correlation coefficient of 0.99. DD could be doubled by increasing the power intensity from 6 to 8 W/cm^2; however, this is at the cost of higher energy input. Tiehm et al. (2001) showed that the influence of ultrasonic intensity on DD was not significant for the same SEI, especially when the SEI is <20,000 kJ/kg TS.

More papers have been published about the ultrasonication of WAS in a batch mode. An SCOD increase is observed with ultrasonication; however, in different relationships. A linear relationship between the SEI and the sludge DD was observed by Tiehm et al. (2001). Mao et al. (2004) also found a linear relationship between the SEI and the SCOD amount. However, other researchers present a hybrid-order curve (first order at beginning and zero order after a certain time) between the input and sludge DD (Lehne et al., 2001a; Bougrier et al., 2005). The former relation seems unrealistic, because SCOD cannot increase infinitely with increasing energy input. This difference has not been explained by others, but the author speculates that it relates to different power densities and sludge sources.

An increase in SCOD may not be a good quantification of COD solubilization for ultrasonication. Wang et al. (1999) reported an SCOD increase of around 1,000 mg/L with more than 120,000 kJ/kg TS energy input. However, Naddeo et al. (2009) achieved a similar SCOD increase with only 12,718 kJ/kg TS energy input. Erden and Filibeli (2010a) reported an even better SCOD increase of 4000 mg/L with only 9690 kJ/kg TS energy input. Results from different experiments vary from each other significantly. The inter-comparability is better when COD solubilization is quantified in terms of DD. Bougrier et al. (2005) and Erden and Filibeli (2010a) observed consistent results. The DD fluctuates around 50% with an SEI of 6000 kJ/kg TS and beyond. However, according to Tiehm et al. (2001), the DD did not approach 50% until an SEI 40,000 kJ/kg TS was reached. This proves that some sludge may be harder to disintegrate and this may also be an argument for why a linear relationship is sometimes observed between the DD and energy input. It should be noticed that in most cases, a plateau value of DD was around 50%. This can be explained by the fact that alkaline treatment can result in twice as much soluble organics compared to mechanical pre-treatment (Lehne et al., 2001a).

8.10 EFFECT OF ULTRASONICATION ON BIOPOLYMERS AND EXTRA-POLYMERIC SUBSTANCES SOLUBILIZATION

Biopolymers refer to organics with complex polymeric structures and a high-molecular weight (MW). As mentioned in the previous section, during sludge disintegration both simple and complex organics are released into the soluble phase. The solubilization of complex biopolymers is of great importance, because they account for most of the SCOD increase and the hydrolysis limits the rate of the subsequent anaerobic process. Once they are released, they can be more easily hydrolyzed by exoenzymes than when trapped inside microbial flocs.

Carbohydrates and proteins are two predominant biopolymers in an EPS structure, which also contributes a great part of the COD in the sludge (Yu et al., 2008). The solubilization of carbohydrates and proteins provides information about the disintegration of the sludge structure. Another important biopolymer in WAS is nucleic acid. It can be found both outside and inside microbial flocs in WAS, but most of the nucleic acid exists as intracellular substances. They can be used as an indicator of cell lysis during pre-treatment (Yan and Herbert, 2003).

Zhang et al. (2007a) observed a rapid increase in soluble protein to 1500 mg/L in the first 5 min of ultrasonication (0.5 W/mL); thereafter, this increase slowed and steadied. Akin (2008) reported a similar soluble protein change. In that research, the soluble protein concentration rose to 1500 mg/L after 1 min ultrasonication (2.07–3.05 W/mL). These results reinforce the argument that microbial floc disintegration happens in the first minutes of ultrasonication, and large amounts of proteins are released in this process.

However, it is still difficult to pinpoint exactly the transition of floc reduction and cell lysis. Lehne et al. (2001a) proposed that an energy of 3000 kJ/kg TS is enough for floc size reduction only based on the median particle sizes reduction, but no chemical or biological arguments were provided. Zhang et al. (2007a) estimated a floc disruption time of 20 min based on the change in the soluble protein, COD and specific oxygen utilization rate (SOUR); however, in this case the energy input was much higher than previously estimated by Lehne et al. (2001a). Liu et al. (2009) tried to identify the transition point by extracting EPS and comparing its COD to the released SCOD. This seems to be a reasonable way to quantify the degree of floc disruption if physical data about particle size distribution can be provided. It should be noted that different EPS extraction methods may lead to different results (Sheng et al., 2010).

In another study, the soluble protein concentration slightly increased until the specific energy reached 4419 kJ/kg TS. The soluble protein concentration decreased afterward (Erden and Filibeli, 2010a). They explained that protein hydrolysis caused by ultrasonication may contribute to this decrease. However, according to their measurement methods, all proteins and amino acids will be determined as proteins. These proteins were possibly further converted to ammonia.

8.11 EFFECT OF ULTRASONICATION ON DIVALENT IONS SOLUBILIZATION

Divalent ions such as calcium and magnesium ions play essential roles in floc integrity (Bruus et al., 1992). They act as linkage materials to bridge particulates in biological flocs. Chu et al. (2001) observed that ultrasonication led to a release of divalent ions; however, the trend is not necessarily consistent with floc size reduction. Wang et al. (2006b) indicated that divalent ions will decrease after the first phase of increase as they may bind with ultrasonicated small particulates that contain a negative charge.

8.12   EFFECT OF ULTRASONICATION ON ANAEROBIC DIGESTION

An increase in methane is the main target of the WAS pre-treatment. It has been proved that the ultrasonication pre-treatment can lead to a considerable increase in VS destruction and methane production (Tiehm et al., 1997; Wang et al., 1999; Hogan et al., 2004; Bougrier et al., 2006; Erden and Filibeli, 2010a). The ultrasonication pre-treatment has been applied to laboratory-scale reactors, pilot-scale reactors and even full-scale continuous reactors. Some articles report a biogas increase instead of a methane increase, and although methane composition in biogas may vary from case to case, the biogas increase will still be used as an indicator and will not be purposely differentiated. Studies have shown that ultrasonication at specific energies from 1,000 to 10,000 kJ/kg TS resulted in increases in biogas production by up to 40% (Khanal et al., 2007) (see Table 8.1). The broad range of energy required may be the result of the differences in sludge characteristics and operational conditions. Further to these differences in performance, ultrasonication is also often associated with high capital and operating costs (Khanal et al., 2007).

An increase in methane production is accompanied by an increase in VS destruction or COD destruction. An increase in methane can be compared in terms of either total methane production or biodegradability. Biodegradability can be expressed based on either COD or VS. When methane production is expressed in terms of the mass of added COD, a theoretical maximum methane production can be calculated, which equals 395 mL CH_4/g COD at 35°C and 1 atm. This value enables a numerical determination of the biodegradability of one sample.

Dhar et al. (2012) sonicated WAS at 1,000, 5,000 and 10,000 kJ/kg TSS and increased the SCOD/TCOD ratio from 6% to 10%, 18% and 33%, respectively, which resulted in a 15%, 20% and 24% increase in methane production and a 23%, 28% and 30% increase in VSS destruction. The SCOD/TCOD increase from 6% to 33% at 10,000 kJ/kg TS was similar to the increase from 4% to 32% obtained by Bougrier et al. (2005). Wang et al. (1999) presented results by ultrasonically treating thickened WAS (TWAS) with a TS concentration between 3.3% and 4.0% at a power density of 2 W/mL. VS destruction increased from 11% to 46% and methane production (in terms of per milliliter gas per gram VS added) increased from 12% to 69% with the ultrasonication time varying from 10 to 40 min, respectively. They pointed out that this increase is due to the increase in soluble carbohydrates and proteins after pre-treatment. When these biopolymers are solubilized, they are easily converted to VFAs, accelerating the anaerobic digestion and increasing methane production.

Bougrier et al. (2006) reported a 51% increase in methane production from 221 to 334 mL CH_4/g COD_{added} after an ultrasonication dose of 9350 kJ/kg TS was applied to WAS. In this case, treated WAS has a biodegradability of 85% compared to the calculated maximum theoretical value. Xie et al. (2007) reported a full-scale ultrasonication pre-treatment application for a mixture of TWAS and primary sludge. Ultrasonication was conducted with a sonotrode with a maximum power of 6 kW, which under normal conditions operates at <50% of its power capacity. The empty space in the ultrasonication reactor was 3.5 L and a 1.5 s hydraulic retention time (HRT) was used. An average biogas production of 45% with fluctuations between

TABLE 8.1
Ultrasonication Pre-Treatments of WAS

Sludge	Ultrasonication Conditions	Pre-Treatment		Anaerobic Digestion			Scale (HRT, days)	Energy and Costs	References
		COD solubilization	DD (%)	Methane (%)	VS (%)	Dewaterability (s)			
WAS (TS: 1%)	20 kHz, 1 min, 1 W/mL	From 55 to 1700 mg/L	15	+5.6	+34.4	69.4–52.8 (CST)	LS, B	–	Şahinkaya and Sevimli (2013a)
WAS (TS: 1%)	20 kHz, 10 min, 0.5 W/mL 1 W/mL 1.5 W/mL	65 to 1330 mg/L 65 to 1695 mg/L 65 to 1890 mg/L	17.1 22 22.7	–	–	–	LS, B	–	Şahinkaya and Sevimli (2013b)
P (1/3) + TWAS (2/3) (TSS: 11.8–21.2 g/L)	20 kHz, 1.5 s, <3,000 W	N.R.	–	+45	+22 (TS)	N.R.	PS, C	1.44 kWh/ m³ WAS	Xie et al. (2007)
WAS	1,000 kJ/kg TS	SCOD/TCOD from 6% to 10%	–	+15	+23 VSS	81–76 (TTF)	LS, B	–	Dhar et al. (2012)
WAS	5,000 kJ/kg TS	SCOD/TCOD from 6% to 18%	–	+20	+28 VSS	81–75 (TTF)	LS, B	–	Dhar et al. (2012)
WAS	10,000 kJ/kg TS	SCOD/TCOD from 6% to 33%	–	+24	+30 VSS	81–52 (TTF)	LS, B	–	Dhar et al. (2012)
TWAS (TS: 3.3%–4%)	9 kHz, 2 W/mL, Time: 10 min 20 min 30 min 40 min	SCOD 1200 mg/L 2300 mg/L 4200 mg/L 5200 mg/L	[a] 30 33 37 40	10 min: +14 20 min: +45 30 min: +60 40 min: +65	+10 +18 +37 +44	–	LS, B	–	Wang et al. (1999)

(Continued)

TABLE 8.1 (CONTINUED)
Ultrasonication Pre-Treatments of WAS

Sludge	Ultrasonication Conditions	Pre-Treatment		Anaerobic Digestion			Scale (HRT, days)	Energy and Costs	References
		COD solubilization	DD (%)	Methane (%)	VS (%)	Dewaterability (s)			
TWAS (TS: 20 g/L)	20 kHz, 0.45 W/mL, 6,250 kJ/kg TS 9,350 kJ/kg TS	+15% (+12% TS) +16% (+17% TS)	—	+47 +51	—	151–733 (CST) 151–680 (CST)	LS, B	—	Bougrier et al. (2006)
TWAS (25% treated)	31 kHz	—	—	+30.0	+28.6	—	FS	—	Neis et al. (2008)
TWAS (TS: 18.5 g/L)	20 kHz 660–1,355 kJ/kg TS 6,951–14,547 kJ/ kg TS	SCOD/TCOD: 6% to 10.5%–16.1% 6% to 33.1%–41.6%	14 55	+12–24.9 +52–60	—	—	LS, B	—	Bougrier et al. (2005)
WAS	20 kHz 18,000	98–1598 mg/L	18	—	—	—	LS, B	—	Naddeo et al. (2009)
WAS (TS: 2.59%)	41 kHz, Time: 7.5 min 30 min 60 min 150 min	—	0 4.7 13.1 23.7	+0 +15.7 +30.7 +41.6	+5.6 +27 +46 +56.7	—	LS, F-B	—	Tiehm et al. (2001)
WAS (TS: 0.5%–4%)	31 kHz, ≤3,600 W Time: 90 s	—	42.4	+11	+37	—	FS, C	—	Nickel and Neis (2007)
WAS (TSS: 17.5 g/L)	41 kHz, 150 min	20 to 1000 mg/L	—	+25.6	—	—	LS, B	—	Zhang et al. (2010)

(Continued)

TABLE 8.1 (CONTINUED)
Ultrasonication Pre-Treatments of WAS

Sludge	Pre-Treatment Ultrasonication Conditions	Pre-Treatment COD solubilization	Pre-Treatment DD (%)	Anaerobic Digestion Methane (%)	Anaerobic Digestion VS (%)	Anaerobic Digestion Dewaterability (s)	Scale (HRT, days)	Energy and Costs	References
WAS (TS: 0.5%)	24 kHz, 16 s	—	—	+42 +37	+25	—	LB, B PS (20)	30 kWh/m³ 30 kWh/m³	Pérez-Elvira et al. (2009b)
WAS (TS: NR)	24 kHz, 0.51 W/mL 10 min 15 min 20 min 15 min 15 min	50 to 1500 mg/L 50 to 2500 mg/L 50 to 1700 mg/L	—	+6.4 +17.9 +1.3 +30.6–60.6 +15.6–42.2	+24 +39	—	LS, M B B B SC (15) SC (7.5)	—	Apul and Sanin (2010)
WAS (TS: 17.81 g/L)	20 kHz, 1.2 W/mL 3,600 kJ/kg TS 31,500 kJ/kg TS 108,000 kJ/kg TS	<10%	7.8 8 47	+6.7 +36.8 +83.5	+22.3 +50.2 +83.4	—	LS, B, M	—	Salsabil et al. (2009)
WAS (TS: 2.14%)	20 kHz, 0.086 W/mL 9,690 kJ/kg TS Time: 40 min	1200–5280 mg/L	57.9	+44	—	—	LS, B, M	—	Erden and Filibel (2010a)
WAS (TS: 21 g/L)	200 kHz, 0.3–0.33 W/mL, 25,000 kJ/kg TS	—	6.5	+40 (F/I: 0.5) +3.7 (F/I: 1)	+12.9 +34.8	—	LS, B, M	—	Braguglia et al. (2012a)

(Continued)

TABLE 8.1 (CONTINUED)
Ultrasonication Pre-Treatments of WAS

Sludge	Ultrasonication Conditions	Pre-Treatment		Anaerobic Digestion			Scale (HRT, days)	Energy and Costs	References
		COD solubilization	DD (%)	Methane (%)	VS (%)	Dewaterability (s)			
Mixed sludge (13.2% TS)	150 W, 45 min	TOC increase: 82.5% TN increase: 50%	—	+95%	—	—	LS, B, M OLR: 0.9	—	journalMartín et al. (2015)
WAS TS: 35.5 g/L	3,380 kJ/kg TS	DD_{COD}: 21%	—	+42%	+13%	—	LS, B, T-M	—	Riau et al. (2015)
WAS TS: 31.4 g/L	20 kHz, 750 W, 5,742 kJ.kg TS	SCOD/TCOD from 0.02 to 0.1	—		+16.9%	—	LS, B, M	—	Pilli et al. (2016)
WAS TS: 23 g/L	24 kHz, 300 W, ~5,000 kJ/kg TS	DD_{COD}: 9%	—	+35%		—	LS, SC, M (20, 80)	—	Braguglia et al. (2011)
TWAS (43.6 g TS/kg)	100 W, 8 min, 96 kJ/kg TS	SCOD: +1741%	—	+27%		—	LS, SC, M (20, 67)	—	Houtmeyers et al. (2014)

Note: Normalized COD: SCOD-SCODo/TCOD or CODp. P, primary sludge; WAS, waste activated sludge; TWAS, thickened waste activated sludge; LS, laboratory scale; PS, pilot scale; FS, full scale; B, batch; FB, fed-batch; SC, semi-continuous; C, continuous; F/I, feed to inoculum ratio; M, mesophilic; T, thermophilic; OLR, organic loading rate in kg VS/m³.day.

a Wang: $1 - VS1.(TS_0 - VS_0)/VS_0.(TS_1 - VS_1).$

15% and 58% was achieved, and the methane composition did not vary a lot (around 65%). However, no corresponding data on VS destruction or COD destruction after anaerobic digestion were provided in these two cases. This makes it difficult to determine which part contributed to the increase in biodegradability.

It is reported by Tiehm et al. (2001) that methane yield decreases at 7.5 min pretreatment (276 mL/g $VS_{degraded}$) and increases slowly at 30 min (291 mL/g $VS_{degraded}$), 60 min (291 mL/g $VS_{degraded}$) and 150 min (300 mL/g $VS_{degraded}$), but still slightly less than the control sample (303 mL/g $VS_{degraded}$). Nevertheless, the ultrasonication pretreatment still shows a good improvement in VS destruction, total biogas production and methane composition except for the pre-treatment time of 7.5 min. Hence, when the biodegradability in this case is converted to methane volume generated per gram VS added, an obvious increase in biodegradability could still be expected.

Controversial results have also been reported by other researchers. Lafitte-Trouqué and Forster (2002) used an ultrasonicator to treat TWAS with a TS concentration of 20–25 g/L. The treatment time was 90 s, the treated volume was 100 mL and the power input was 47 W. The treated sludge was then fed into three continuous anaerobic digesters, which have different HRTs of 8, 10 and 12 days separately. Taking the standard deviation into consideration, an increase in either VS reduction or methane yield (per gram VS removed basis) was barely observed. Additionally, they did not examine the SCOD concentration or DD after ultrasonication, which makes the case unclear. When converting the given operational parameters to an SEI, an input range of between 1692 and 2115 kJ/kg TS could be calculated out. This energy input seems to be much less than previous research, and this may be the cause of the negligible biodegradability increase. However, a similar SEI has been reported by other researchers. Bougrier et al. (2005) observed a slight increase in biogas production at an SEI of 660 kJ/kg TS, and with increasing energy input more biogas will be generated till 14,547 kJ/kg TS, which is the maximum specific energy they tested. From the data, it appears that biogas production increases with specific energies from 0 to 7000 kJ/kg TS, but the improvement becomes marginal at higher energy. Martín et al. (2015) reported that ultrasonication (150 W, 45 min) effectively improved the batch anaerobic digestion of sewage sludge. The methane yield of sewage sludge after the ultrasonication pre-treatment increased from 88 to 172 mL STP/g VS, increasing by around 95%. The biodegradability of pre-treated sewage sludge reached 81% in VS, which gave a maximum organic loading rate (OLR) of 4.1 kg VS/m^3 day and a methane production rate of 1270 L STP/m^3 day.

It is mentioned in the review by Khanal et al. (2007) that there is no well-defined protocol to indicate the efficiency of ultrasonication. An increase in SCOD, methane production and VS destruction all seem to be beneficial for the anaerobic digestion process. However, based on previous research, there may be a correlation between any two out of three. Tiehm et al. (2001) proved that VS degradation increases linearly with the sludge DD and a correlation coefficient of 0.94 was obtained. Bougrier et al. (2005) figured out that an increase in biogas production after ultrasonication mainly originates from the soluble phase by anaerobically digesting soluble matter and particulate matter separately. They also found a linear relationship between produced biogas and organic soluble matter with a correlation coefficient of 0.93. Despite these results, these correlations are still seldom reported.

Ultrasonication is a very fast process. Xie et al. (2007) reported that a 1.5 s ultrasonication time resulted in a biogas increase of around 35% for a mixture of primary and TWAS in a full-scale plant. An average energy gain of 2.5 was reported, meaning that there was excess energy production as a result of the increment in biogas production. Meanwhile, ultrasonication does not cause the mineralization of the sludge sample. No organic carbon is wasted in the form of carbon dioxide. However, the high capital and operating costs of ultrasonication remain the bottleneck for the popularization of this technology (Khanal et al., 2007).

Nickel and Neis (2007) used ULS at full scale and showed that the continuously operated plug-flow reactor performed significantly better cell disintegration as compared to batch-type sonication tests with the same type of sonotrode. This confirms that the results at large and small scale can be very different. Regarding the energy balance, several authors showed that the energy balance was negative at the laboratory scale (Pérez-Elvira et al., 2009b; Salsabil et al., 2009). However, a net generation of 3–10 kW per kW of energy used was possible at full scale. The three main suppliers of ultrasound technology are Sonico Ltd., UK (Radial Horn US System), Ultra WAVES GmbH (Flat Piezo-Ceramic Transducer) and IWE Tec GmbH (Cascade Sonotrode). Laboratory ultrasonic systems are inefficient because the full-scale sonotrode is much more powerful; therefore, direct use of bench-scale data for full-scale design are misleading. Also, most of the full-scale installations use part-stream sonication (Barber, 2003), which consists of treating only a fraction of the sludge stream, with the objective of reducing costs and enhancing the final sludge dewaterability. Moreover, the higher the sludge concentration, the higher the benefit per unit of sludge volume (Pérez-Elvira et al., 2009b). Therefore, thickening the sludge before ultrasonication is recommended for improving the ultrasonication performance (Onyeche et al., 2002; Pérez-Elvira et al., 2009b). Most studies have not observed any differences in the biogas composition. However, in a few studies the methane percentage increases in the pre-treated reactor (70%) compared to the control (55%) (Apul and Sanin, 2010).

Net energy gain is the ultimate concern of the ultrasonication pre-treatment. Since the ultrasonic process is a cost-intensive process, whether the increased methane can compensate for the energy input is of great importance. Different researchers reported different results. This may be explained by the different sludge sources and the different pre-treatment conditions. Pérez-Elvir et al. (2009b) also pointed out that a positive energy balance is easy to achieve in full-scale ultrasonication because supplied energy in real equipment is much smaller compared to the laboratory scale.

Xie et al. (2007) reported the full-scale application of ultrasonication in a wastewater reclamation plant in Singapore. In that plant, ultrasonication is used to pre-treat a mixture of TWAS and primary sludge. Each cubic meter of biogas was estimated to produce 2.2 kWh electricity. Based on these conditions, a positive net energy gain was obtained for all the operation days. The average ratio of net energy gain to ultrasonication electricity input was 2.5 for the first half observation year and 3.1 for the following observation year. Moreover, they also pointed out that when ultrasound is used to treat only TWAS instead of a mixture of sludge, a one-third increase in net energy gain may be expected.

8.13 PROCESS CONFIGURATIONS AND OPERATING CONDITIONS

The huge hydromechanical shear force generated by cavitation bubbles during ULS is believed to be the predominant effect for sludge disintegration (Tiehm et al., 2001; Wang et al., 2005). In contrast with thermal methods, ULS is, comparatively, a very rapid method that causes solubilization of both the extracellular and intracellular substances leading to an increase in soluble microbial products (Pilli et al., 2011). The microorganisms in WAS degrade the organic matter by producing hydrolytic enzymes that are released into the media. Therefore, a physical treatment such as ULS should be useful to disrupt the flocs and release biopolymers such that the subsequent anaerobic step can be improved (Figure 8.4), although ultrasound has also been applied on the return activated sludge (AS) line at the large scale. Since ultrasonication is an energy-intensive process, its major disadvantage is its high-energy consumption (Khanal et al., 2007). For this reason, it may be more reasonable to apply ultrasound on WAS and not on the primary sludge. Moreover, the ultrasonication of a fraction of WAS should also be considered. The objectives are to improve the solubilization of COD, proteins and carbohydrates and solids destruction and biogas production, while maintaining or even improving the dewaterability of sludge.

A mixture of primary sludge and TWAS (ratio around 1:1 based on dry solids) was collected from a municipal wastewater reclamation plant. The properties of the sludge are listed in Table 8.2.

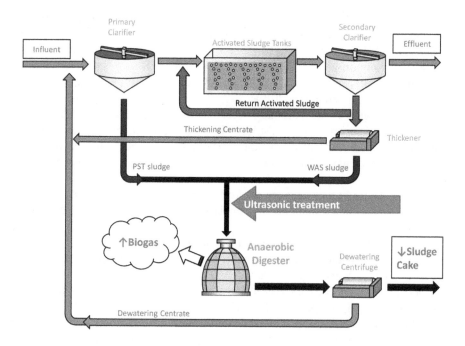

FIGURE 8.4   Typical flow sheet showing where the ultrasonication step can be applied to treat sludge.

TABLE 8.2
Properties of the Sewage Sludge

Parameters (Acronym, Unit)	WAS	Anaerobic Inoculum
pH	5.9–6	7.3
Soluble chemical oxygen demand (SCOD, mg/L)	670–1440	454 ± 8
Total chemical oxygen demand (TCOD, g/L)	18–25	13.75 ± 0.53
Total solids (TS, g/L)	13.6–17.2	9.5 ± 0.3
Volatile solids (VS, g/L)	10.7–13.4	7.1 ± 0.3
Total suspended solids (TSS, g/L)	12.4–15.9	9.3 ± 0.2
Volatile suspended solids (VSS, g/L)	10.3–13.0	7 ± 0.3
Ammonia (mg N/L)	122.97 ± 2.72	NM
Phosphate (mg PO_4^{3-} /L)	24.11 ± 4.71	NM

Note: NM, not measured.

The data presented in the next sections show the effects of ultrasonication. These ultrasonication tests were carried out in batch mode using an ultrasonicator (Misonix, Q700) with a frequency of 20 kHz and a maximum power input of 700 W. A solid tip probe (#4208) with a diameter of 19.1 mm and a maximum amplitude of 60 μm was immersed 1–2 cm below the surface of the sludge; 80% of the maximum amplitude was used and the corresponding power input was around 140 W. During ultrasonication, the temperature was monitored and refrigerated below 30°C with an ice-water mixture if necessary to avoid any thermal effects.

8.14   PERFORMANCE OF ULTRASONICATION PRE-TREATMENT

In this section, the effects of ultrasonication were investigated to illustrate the aforementioned theoretical aspects. Various parameters such as soluble carbohydrates, proteins, SCOD, soluble phosphorus and ammonia are shown in Figure 8.5. Figure 8.5a shows that ULS has a significant impact on soluble biopolymers with an increase in the SCOD, proteins and carbohydrates concentrations to 5.5, 1.6 and 500 mg/L, respectively. The increase in soluble carbohydrates was relatively less obvious because the sludge contained TWAS, which is rich in proteins from bacterial cells and EPS. Figure 8.5b shows that ultrasonication had a significant effect on the soluble phosphorus concentration, which means that ULS was able to break open the cells and release phospholipids from the cell membrane and phosphorus from the DNA into the bulk liquid. The concentration of ammonium in the supernatant was also analyzed, and it was found that it slightly increased from 120 to 170 mg/L during the first 5 min of treatment, but afterward it remained constant. It is possible that some proteins in the sludge were broken down or that ammonium from the cytoplasm was released in the supernatant due to the action of ultrasound.

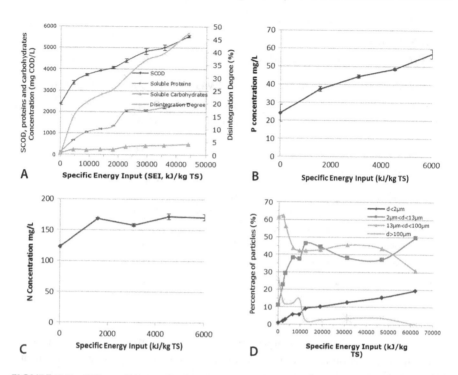

FIGURE 8.5 Effect of ultrasonication specific energy input (SEI) on (a) soluble proteins, carbohydrates and COD, (b) soluble phosphorus concentration, (c) soluble ammonia concentration, (d) evolution of various groups of particles based on the size.

Particle size distribution is an important physical characteristic of sludge. It provides information about the amounts and sizes of biological flocs. The mean diameter, cut diameter and modal diameter are commonly used parameters for indicating particle size distribution. The mean diameter is the average diameter of all the measured particles. The cut diameter is the diameter below which a certain percentage of measured particles is distributed. For example, d_{50} is a diameter below which 50% of measured particles are distributed. It is also known as the median diameter. The modal diameter is the diameter that comes up the most in the measured samples.

Gorczyca (2000) categorized floc sizes and reported the medium diameters of commonly seen flocs. The primary particle was reported to have a diameter of around 2 μm, the diameter of microflocs was estimated to be 13 μm and highly porous flocs have a large diameter around 100 μm. Tiehm et al. (1997) observed a particle size reduction from 165 to 135 and 85 μm after 29.5 and 96 s ultrasonication, respectively. A size reduction was observed, but floc types seemed to remain unchanged. Show et al. (2007) reported a reduction of the mean particle size from 50 to <10 μm at an SEI of 20 kWh/kg TS, thus showing that increasing the energy input does not contribute to an obvious particle size reduction. Similar results were obtained by Chu et al. (2001). They proved that the particle size reduced significantly from 98 to <10 μm during the first 10 min of ultrasonication at a power density of 0.44 W/mL, whereas this reduction became negligible thereafter. Additionally, they also found out that the power density is an important process parameter for floc disruption. Obvious floc reduction can be observed for high power density in a short time, but low power density in a long time did not lead to noticeable floc size reduction. These two studies reported both floc size reduction and a change in floc type. It can be concluded that both continuous and batch ultrasonication lead to size reduction, but batch ultrasonication performs better in changing porous large flocs to micro flocs.

Figure 8.5d shows the evolution of various groups of particles based on their size. The cavitation bubbles caused by ultrasound are known to disrupt floc structures and reduce floc size. Particles larger than 100 µm or cells flocs and aggregates were readily disrupted by ultrasound within the first minutes with the number of large flocs reduced from 26% to 12% and then below 5% as the SEI reached 10,000 kJ/kg TS. At the same time, the number of colloidal particles or small flocs (13–100 µm) also dropped significantly due to physical disruption, while the amount of single cells, small colonies and possibly cell debris (2–13 µm) started to increase markedly from 10% to 50%. This was consistent with Lehne et al. (2001b) who found that an obvious floc size reduction took place below an SEI of 3000 kJ/kg TS. Interestingly, beyond 10,000 kJ/kg TS the impact was negligible. The slow and steady increase in intracellular materials and EPS (<2 µm) showed that ULS could indeed disrupt cell walls and solubilize EPS even at low SEIs, which was consistent with the evolution of soluble biopolymers, ammonia and phosphorus. This contradicts Lehne et al. (2001b) who suggested that cell lysis did not take place until an SEI of 3000 kJ/kg TS was applied. Overall, it can be concluded that ultrasound was more efficient toward large flocs.

ULS pre-treatment resulted in a significant particle size reduction in feed sludge. The mean and median diameter decreased significantly from 45 to 15 µm after 9 kJ/g TS ULS pre-treatment, as shown in Figure 8.6. A further increase in ULS SEI was

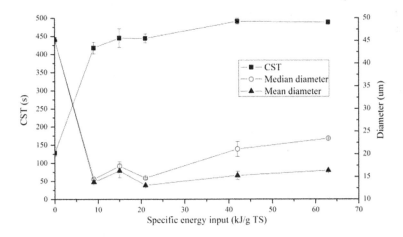

FIGURE 8.6 Change in capillary suction time (CST), median diameter and mean diameter in feed sludge with specific energy input after ultrasonication pre-treatment.

not efficient in reducing the particle size in feed sludge. Additionally, both the mean and median diameter increased slightly when the SEI was over 21 kJ/g TS. Similar observations have been reported previously as a re-flocculation effect due to cell lysate solubilization (Bougrier et al., 2005; Erden and Filibeli, 2010a). Sludge dewaterability is often used as an indicator of the integrity of the sludge structure because a small particle size and a high concentration of soluble biopolymers resulted in poor sludge dewaterability (Wang et al., 2006a). As shown in Figure 8.6, the CST in feed sludge increased significantly from 125 to 430 s after 9 kJ/g TS ULS pre-treatment, indicating the deterioration in feed sludge dewaterability. Afterward, the increase in CST with increasing ULS SEI became negligible, and the CST reached around 490 s after 63 kJ/g TS ULS pre-treatment. These observations suggested that the effects of the ULS pre-treatment on disintegrating the feed sludge structure became inefficient after 9 kJ/g TS.

The MW distribution of organics in the supernatant of the control and ultrasonicated feed sludge is as shown in Figure 8.7a. The MWs of standard polymers are shown as dashed lines. High MW compounds eluted earlier and had a shorter retention time because they did not go as deep into the gel pores as low MW compounds (Trzcinski et al., 2011; Aquino et al., 2006a). According to the change in the UV signal response, the ULS pre-treatment was able to solubilize compounds with an MW over 500 kDa as well as compounds with an MW lower than 106 Da. The supernatant of the control and ultrasonicated feed sludge was further fractioned with the UF fractionation process to obtain the COD fraction in each MW range. As shown in Figure 8.7b, the COD fraction of the organic compounds with an MW over 300 kDa had the most obvious increase after the ULS treatment of feed sludge. It increased from 15.8% to 21.7%, 40.8% and 43.9% after 3, 9 and 21 kJ/g TS ULS pre-treatment had been applied, respectively. This is because the sludge matrix, which consists of high MW EPS and intracellular biopolymers, was solubilized due to ultrasonic disruption (Eskicioglu et al., 2006c).

The effect of ULS alone on anaerobic biodegradability was tested. As shown in Figure 8.8a, the biodegradability of all the ultrasonicated sludge was higher than the control. The ultimate biodegradability increased with increasing SEI. However, ultrasonication at high SEI may not be economical. It was found that 9 kJ/g TS ultrasonication improved the sludge biodegradability by 14.8%, whereas at an SEI seven times higher (i.e., 63 kJ/g TS), the biodegradability increased by 31.8%, which was only 2.2 times higher compared to the improvement induced by 9 kJ/g TS ultrasonication.

In order to evaluate the possibility to use SCOD data to predict biodegradability, the ultimate sludge biodegradability and SCOD concentrations prior to anaerobic digestion tests were plotted as shown in Figure 8.8b. A linear regression was found to be the most suitable model to describe the relationship. The coefficient of correlation (R^2) was 94.8%, indicating a strong correlation, which is in-line with Bougrier et al. (2005) who observed that biogas increase in ultrasonicated sludge originated mainly from the soluble fraction.

A first-order model was used to determine the rate constant of anaerobic digestion with the data obtained (Trzcinski and Stuckey, 2012).

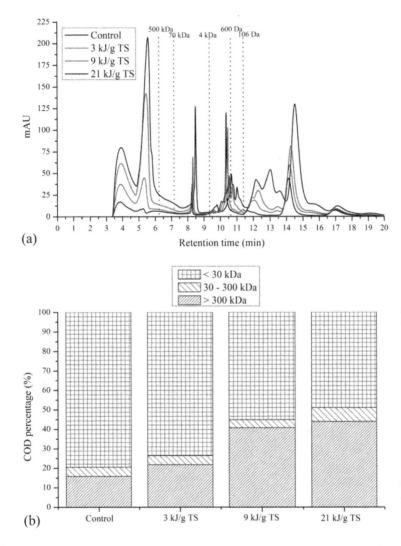

FIGURE 8.7 (a) Molecular weight distribution chromatograms of organics in the supernatant of control and ultrasonicated feed sludge; (b) COD fraction in different MW ranges in control and ultrasonicated feed sludge.

$$Y = Y_{max}\left(1 - \exp\left(k_d t\right)\right)$$

where

Y is the biodegradability at time t (mL CH_4/g COD_{added});

k_d is the rate constant of the anaerobic digestion (day^{-1});

and Y_{max} is the ultimate biodegradability (mL CH_4/g COD_{added}).

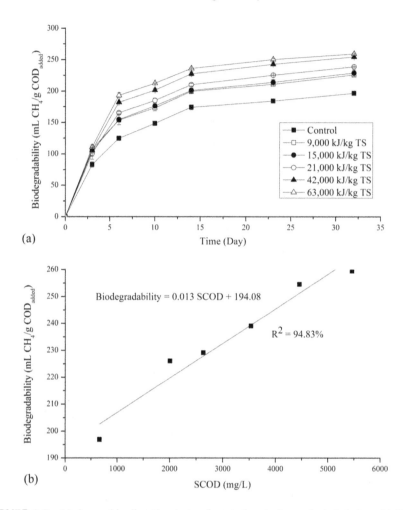

(a)

(b)

FIGURE 8.8   (a) Anaerobic digestion tests of control and ultrasonicated sludge; (b) linear fitting of sludge biodegradability and SCOD concentration after ULS pre-treatment.

The values of the fitted variables and the measured data are compared in Table 8.3. The predicted ultimate biodegradability was consistent with the measured biodegradability and all the values of the coefficient of determination (R^2) are higher than 98%. These indicated that the obtained data fitted the model well and the model can be used for predicting process performance. As shown in Table 8.3, the rate constant "k_d" increased from 0.171 to 0.201 after 9 kJ/g TS ULS pre-treatment, indicating that the methane production rate of feed sludge was accelerated after ULS treatment. However, a further increase in SEI over 9 kJ/g TS did not further increase the methane production rate. This is possibly related to the increase in the COD fraction due to macromolecules with an MW over 300 kDa with increasing SEI. It has been shown in the literature that these macromolecules are relatively slowly biodegradable compared to the smaller ones (Eskicioglu et al., 2006c). Therefore, the solubilization of these high MW

TABLE 8.3

Summary of the First-Order Fitting and Measured Results for Control and Ultrasonicated Sludge

	First-Order fitting			Measured Results	
Sludge	Y_{max} (mL CH4/g COD_{added})	k_d (day)	R^2 (%)	Biodegradability (mL CH4/g COD_{added})	Increase (%)
Control	191.5 (4.2)[a]	0.171 (0.012)	99.33	196.9 (1.2)	N/A
9 kJ/g TS	215.6 (6.2)	0.201 (0.020)	98.64	226.1 (1.7)	14.8
15 kJ/g TS	219.5 (5.6)	0.194 (0.017)	98.97	229.1 (4.8)	16.4
21 kJ/g TS	231.2 (5.3)	0.187 (0.014)	99.18	239.0 (0.4)	21.4
42 kJ/g TS	248.0 (5.7)	0.192 (0.015)	99.19	254.5 (2.1)	29.3
63 kJ/g TS	254.2 (5.5)	0.205 (0.016)	99.22	259.5 (1.0)	31.8

[a] Numbers in parentheses stand for standard deviation.

compounds after ULS treatment of feed sludge could eventually contribute to a higher methane recovery, but did not necessarily further enhance the anaerobic digestion rate.

8.15 EFFECT OF ULTRASOUND PRE-TREATMENT ON SLUDGE DEWATERABILITY

Water in sludge is normally categorized into four types: free water (i.e., water exists freely in sludge without attaching to solids), interstitial water (i.e., water exists inside solid crevices or is trapped inside flocs and organisms), surface water (i.e., water attaches to a solid surface via adsorption and adhesion) and chemically bound water.

Dewaterability is another important parameter for physically indicating sludge structure change. It is related to both particle sizes (Feng et al., 2009) and sludge DD (Li et al., 2009b; Erden and Filibeli, 2010a). A small energy input induces no change (Erden et al., 2010) or just a slight increase in sludge dewaterability (Chu et al., 2001; Li et al., 2009b). An SEI of 800 kJ/kg TS was reported to be the optimal energy for optimal dewaterability (Feng et al., 2009). The increase in dewaterability following ultrasound is postulated by the re-flocculation of small fragments to tighten particles (Li et al., 2009b).

High-energy input significantly deteriorates dewaterability (Bougrier et al., 2006; Feng et al., 2009; Li et al., 2009b; Erden and Filibeli, 2010a; Chu et al., 2002). Bougrier et al. (2006) reported an increase in CST from 151 to 733 s at a specific energy of 6250 kJ/kg TS, which implies a significant decrease in dewaterability.

This deterioration in dewaterability is attributed to the fact that ultrasonication produces more small particles that provide more surface area for water to attach to (Chu et al., 2002).

8.16   EFFECT OF HIGH ULTRASOUND DOSAGE

Another interesting issue is whether excessive energy inputs cause adverse effects on anaerobic digestion. Based on most of the past research, methane production increases all the way with increasing energy input (Wang et al., 1999; Tiehm et al., 2001; Pérez-Elvira et al., 2009b), and no decrease in methane production was observed at a higher energy input. However, Erden and Filibeli (2010a) reported a decrease in sludge DD after an SEI of 9690 kJ/kg TS, but no corresponding biochemical methane potential (BMP) results were provided. Apul and Sanin (2010) observed a decrease in SCOD after 15 min ultrasonication. Methane production also decreased following the pre-treatment, but no explanation was given. Although the super-critical condition created by cavitation bubbles only lasts for <1 s, the instantaneous extreme condition may also be harmful. Cavitation effect creates local high temperatures in a sludge sample, and under high temperature some unexpected reactions may take place. For example, Stuckey and McCarty (1984) detected a decrease in biodegradability for the thermal pre-treatment of glucose, deoxyribose and ribose; some biodegradable components may convert to refractory compounds. Hence, if the ultrasonication is wrongly operated or is overused, an increase in toxicity might also be expected.

8.17   EFFECT OF ULTRASOUND ON BACTERIAL ACTIVITY

WAS is aerobic sludge that contains plenty of microorganisms. Its bacteria activity is generally quantified by SOUR. The higher the SOUR, the higher the aerobic bacteria activity of the sample. In addition to providing information about the activity of aerobic bacteria, SOUR also suggests the influence of ultrasonication on biological flocs, single bacteria inactivation and even cell lysis.

The influence of ultrasonication on bacteria activity can be described in two stages. In the first stage, the EPS structure loosens and makes the oxygen transfer much easier. This leads to a slight increase in SOUR. In the second stage, after the floc structure is destroyed, single bacteria cells are exposed to ultrasound and will lyze open. Li et al. (2009b) correlated sludge DD with SOUR. A maximum 22% increase in SOUR at a DD of 10% was reported, with SOUR decreasing with increasing DD thereafter. However, this result may vary from sample to sample. For example, a decrease in SOUR was observed by Tiehm et al. (2001) only at a DD of 4.7%.

8.18   FULL-SCALE ULTRASONICATION TREATMENT OF SLUDGE

Ultrasound is used in a wide range of industrial applications, in many different frequencies and forms; therefore, there is an extensive list of ultrasound manufacturers in North America and Europe. However, the number of manufacturers with equipment designed for wastewater treatment applications and with experience in this field is limited. The two main suppliers are Ultrawaves, represented in North America by Dorr Oliver Eimco's Sonolyzer™ equipment, and Sonico with the Sonix™ system. Ultrawaves is a German company that was established as a spin-off from ultrasound research conducted at the Technical University of Hamburg-Harburg, which purchases its ultrasound equipment from Sonotronic, a local manufacturer. Sonico is a UK company that was established as a joint venture company between Purac, an equipment supply company, and Atkins, an engineering consulting company that conducted applied research in ultrasound wastewater applications. Sonico purchases its ultrasound equipment from Branson, a company based in the United States. These two companies have the largest number of installations in wastewater applications, with over 30 installations in Europe, Asia and Australia. The largest installation is the Sonico installation for WAS pre-treatment for enhanced digestion at the 400 MLD Mangere wastewater treatment plant (WWTP) in New Zealand. Two smaller companies also have a few smaller wastewater installations, Ultrasonus from Sweden and VTA from Austria.

Ultrasound may be generated by two different methods: magnetostrictive and piezoelectric. The former uses electric energy passed through a magnetic coil attached to a vibrating piece to produce mechanical energy, or vibration. The latter uses electrical energy, converted to high frequency electric energy, which is applied to piezoelectric crystals that vibrate at the same frequency. The crystals are attached to a vibrating piece (known as the sonotrode, probe or horn), causing the vibration to be transferred to the liquid. Magnetostrictive systems typically have a longer life but a lower energy efficiency as the electrical energy applied is converted to magnetic energy prior to being converted to mechanical energy. For wastewater applications, it appears that economics favors the use of piezoelectric systems due to the high-energy intensity required to lyze the cellular material in the sludge. There are differences in the ultrasound systems available for wastewater treatment. Each manufacturer uses a unique shape for the vibrating ultrasound piece; for example, Ultrawaves uses short rod-shaped sonotrodes that project into the flow path, while Sonico uses ring-shaped horns that sit within a pipe spool. Ultrawaves treats a smaller portion of the flow for a longer retention time, while Sonico typically treats a greater portion of the sludge flow for a considerably shorter retention time. Each probe in the Ultrawaves system is rated for 2 kW power input, but typically operates at 1 kW. Sonico has 3 and 6 kW horns, with the latter as their standard unit, operating at 50%–60% of rated power. The majority of ultrasound applications to date have been for WAS pre-treatment prior to anaerobic digestion. However, both Ultrawaves and Sonico have European installations on RAS streams within the WAS process (Roxburgh et al., 2006).

Ultrasound installations in Europe show that treating a sidestream of the RAS flow can reduce WAS production from the secondary treatment process by 25%–50%, depending on the characteristics of the sludge, the operation of the secondary treatment process and the amount of power input to the system. A number of full-scale trials and installations using Sonix have taken place around the world (Hogan et al., 2004). This section presents some of the results to date and explores the main drivers behind a selection of these projects.

8.18.1 AVONMOUTH, WESSEX WATER, UNITED KINGDOM

This treatment plant is owned by Wessex Water and serves a population equivalent of 1,200,000, treating a mixture of domestic and industrial effluents. The plant has six 2700 m³ mesophilic anaerobic digesters, each fed with a mixture of thickened primary sludge and TWAS through a manifold, at a ratio varying between 100/0 (primary/TWAS) and 80/20. If the portion of TWAS fed to the digesters was above 20%, the digesters became unstable and started to foam due to the low digestibility of TWAS. One digester was dedicated to the trial, with the other five in normal operation. The typical retention time of these digesters was 12 days, with a daily feed of about 180 m³/day. As TWAS is less digestible than primary sludge, the trial concentrated on sonicating only the TWAS portion of the feed to the test digester. The control digester was fed 180 m³/day of a mixture of primary and unsonicated TWAS through the manifold, the ratio being determined by the conditions on-site. The test digester was fed a similar mixture of primary sludge and unsonicated TWAS through the manifold, plus a determined amount of sonicated TWAS directly to the digester through its recirculation loop. It was shown that the TS and VS destruction of the control digester (fed primary sludge) was typical of such sludges, with 40% and 50% destruction, respectively. On the other hand, the TS and VS destructions obtained in the test digester for the sonicated TWAS were, respectively, 60% and 70%. Expected values for non-sonicated TWAS were of the order of 20% and 30%. The gas production data were also in concordance with this. Therefore, sonicating TWAS prior to anaerobic digestion at Avonmouth enabled this sludge to be destroyed up to three times more than if unsonicated.

8.18.2 SEVERN TRENT WATER, UNITED KINGDOM

A five-horn plant was installed on a trial basis at a Severn Trent Water WWTP in summer 2003. The plant serves a population equivalent of approximately 150,000 of which about one-third is industrial load. The plant has two mesophilic anaerobic digesters of 1975 m³ capacity. The digesters had an average retention time of 14 days. The Sonix unit treated the secondary sludge feed to the digesters and delivered sonicated TWAS into a digester feed tank where the treated TWAS was mixed with primary sludge and imported sludge prior to feeding into the digesters. The trial ran from the end of June until the end of October 2003. The work was undertaken to assess the ability to improve digester performance and increase the available biogas for use in a proposed combined heat and power system at the WWTP. The increase in VS destruction across the digesters was demonstrated once Sonix was installed. The average VS destruction increased from approximately 38% to an average of approximately 54%. This was achieved by treating only 20%–25% of the digester feed load by mass. Once the trial was completed and the Sonix was switched off, the VS destruction was seen to decrease. In addition, the higher solids destruction across the digestion plant resulted in more gas production. An increase of approximately 40% was seen once Sonix was fully optimized.

8.18.3 Anglian Water, United Kingdom

Sludge minimization tests were conducted by Anglian Water in the United Kingdom using a Sonico ultrasound system in parallel AS lanes. The control lane was operated at a solids retention time (SRT) of 12–15 days and a mixed liquor concentration of 3500 mg/L. The test showed that a titanium ultrasound horn system treating approximately one-sixth of the RAS flow provided up to a 30% reduction in WAS production. The ultrasound system was also found to have an additional advantage in reducing the impact of foaming and maintaining good settleability (95 mL/g) during a period when filamentous organisms reduced the settleability of the control lanes (140 mL/g). No increase in aeration demand was noted during the test (Roxburgh et al., 2006).

8.18.4 Kavlinge, Sweden

The Sonix installation at Kavlinge WWTP in Sweden was completed in conjunction with PURAC AB, the licensee for Sonix in Scandinavia. It is a four-horn installation and the plant was commissioned in December 2002. The drivers for the installation of the Sonix unit were linked to the nature of the TWAS from the AS plant, which received high-strength industrial waste from an abattoir. The waste was pre-settled at source resulting in a highly organic low particulate waste stream being discharged directly into the AS plant. This configuration resulted in a high TWAS yield from the AS plant. Historically, this TWAS was mixed and dewatered with digested primary sludge from the domestic sewage load to the works, resulting in a sludge ratio of 25:75 digested primary:secondary sludge in the centrifuge feed make-up. Primary sludge was digested in two mesophilic anaerobic digesters. As a result of this high TWAS ratio, the centrifuges were producing a volatile sludge that was unable to be dewatered to >15% dry solids leaving a fluid-like sludge product that was malodorous and unacceptable for an agricultural disposal route. In addition, the mesophilic digesters were unable to cope with the increase in TWAS load so the digesters were converted to 9.5 days HRT thermophilic digesters to handle the increased sludge volumes. The Sonix plant was commissioned to break open the biological sludge, ensuring stability and digestion of the unusual sludge feed make-up. Improvements in sludge quality and increased biogas production have been significant due to the digestion of the TWAS. The combined digested sludge is now able to be dewatered to a cake with 25% TS. There has been a significant drop in tanker movements in the village and excess biogas is now used to heat water at a local sports hall. Sonix operation was seen not only to increase biogas production from the plant but also to allow the digester to receive and treat an increased sludge load. The recorded biogas correlated to the expected gas production based upon solids destruction across the digestion plant.

8.18.5 Orange County Sanitation District, United States

Orange County Sanitation District (OCSD) operates two treatment facilities. Both facilities include AS plants, and between them, the two sites have a total of 26 mesophilic anaerobic digesters. Under normal operation, the feed to each digester includes no more than 30% TWAS. The Sonix demonstration trial at OCSD ran for 5 months, from February to July 2002. In order that a direct comparison could be made between sonicated and unsonicated feed sludge, two mesophilic anaerobic digesters (test and control) were chosen for study. The size of each of these was 4507 m³. They were operated under identical conditions, and the proportion of TWAS fed to both digesters was maintained at approximately 50% on a solids loading basis. This enabled the results to be compared both against baseline conditions (longitudinal study) and between digesters (cross-sectional study) so that conclusions could be drawn for the full installation of ultrasound at all of OCSD's facilities. A single V5 unit was located on the TWAS feed line to the test digester. The daily TWAS flow to the test digester did not exceed 190 m³ (the control digester was fed identical volumes each day) and typically all five ultrasonic horns were operational. Compared to the control digester, significant enhancements in gas production and solids destruction were achieved by sonicating the secondary sludge. Once the sonication threshold was overcome by turning up the power delivered by the ultrasonic horns, the gas production in the test digester increased then stabilized. Midway through the study, new gas meters were installed. The differential gas production continued. This further confirmed that the measured difference in gas production between the two digesters was real and not a result of measurement or calibration error. A gas production increase of approximately 50% was seen for the test digester. After the study was complete, sonication was stopped and the digester operation was returned to pre test conditions; that is, the digesters were fed only 20%–30% TWAS. The gas production from both digesters became essentially identical, further verifying that sonication was responsible for the gas production increase during testing.

Dewatering tests of the residual solids were carried out using a bench-scale belt press. Nine separate runs were performed on different days, and improvements of 1.2%–2.6% were seen in the biosolids cake from the test digester compared to the control. The mean increase was 1.64% with a 95 percentile confidence interval of ±0.32%. The polymer dose for the sludge from both digesters was identical.

8.18.6 BEENYUP WWTP, AUSTRALIA

The Beenyup plant contract in Perth was awarded in early July 2003 after a competitive tendering process against alternative ultrasonic technologies. Commissioning was due to be carried out in December 2003. The Beenyup sewage works has a population equivalent of approximately 700,000 though not all secondary sludge will be pre-treated by Sonix. The plant at Beenyup is a seven-horn unit, and was installed on the TWAS feed to the digesters. The key drivers for the scheme were greater gas production, more stable digester performance and improved dewatering.

The development of the novel Sonix horn has been critical. Its radial shape allows the ultrasonic energy to be focused and the intensity of the cavitation to be increased. The ultrasonic horns are positioned in-line with the sludge flow and are contained within a reaction chamber. The equipment is designed for easy installation or retrofit within existing treatment facilities. Improved solids destruction, substantial increases in gas production and better residual solids dewatering were the primary benefits observed with sludge treatment using Sonix. In addition, stable digester operation was maintained at a high feed ratio of TWAS to primary sludge. The ability to treat 100% TWAS through digestion with Sonix provides an alternative for plants that do not have primary treatment, which are common in Australia. This provides the benefits of lower operating costs, lower sludge production, better stability and the ability to recover power from digester gas.

8.18.7 MELDORF WWTP, GERMANY

A full-scale installation was implemented at Meldorf WWTP (65,000 PE) in February 2005 after a 3-month test period. The plant was experiencing problems with foaming in the anaerobic digester as a result of excessive growth of filamentous bacteria (*Microthrix parvicella*) in the WAS. The purpose of the ultrasound installation was to eliminate the source of the foaming problems, thereby ensuring an undisturbed anaerobic digestion. Sonication was applied to 100% of the TWAS flow before it was sent to the two anaerobic digester tanks present at the plant.

A short time after the installation of the ultrasound equipment at the Meldorf WWTP, the problems with foaming sludge were no longer an issue. This, in turn, provided the conditions necessary for the smooth and effective digestion of the sludge when it reached the fermenter. The VS, expressed as a percentage of dry solids, were reduced from 60% to 45% in the stabilized sludge. With regard to biogas production, a 30% increase after the ultrasound installation was noted as compared to before the installation. These improvements correspond to improved self-sufficiency with regard to energy supply and a reduced sludge mass for disposal. In addition, the feeding of co-substrates was made possible as a result of the improved stability of the anaerobic digestion process. The plant was able to accept process liquids from a local food producer, which served the interests of both parties.

8.18.8 BAMBERG WWTP, GERMANY

A full-scale installation with two ultrasound reactors was completed in August 2004 at the WWTP in Bamberg, Germany (actual load 330,000 PE). The ultrasound application had previously been successfully tested at the plant during a trial period that lasted for 4 months. The purpose of the ultrasound application was to enhance the VS degradation to a minimum of 45% in order to avoid the costly task of constructing another anaerobic digester. In order to achieve this, 25% of the TWAS was sonicated before it was sent to the anaerobic digester. Ideally, organic material that is available to the organisms in the sludge is converted to methane in the anaerobic food chain. The gas production, compared to conventional digestion, increases as a greater amount of organic material is made bio-available through sonication. By feeding the digester with ultrasonically treated sludge, the gas production has been shown to increase by almost 30% at the WWTP in Bamberg. The methane content also increased slightly, making the biogas a more attractive and energy-rich product. The residual VS content in the digested sludge was reduced from 60% to 54%. The desired goal to reach a minimum of 45% VS degradation was met and surpassed. The increase in VS destruction implies that more of the organic matter in the sludge was metabolized in the digestion process. This coincides with the increase in gas production; less organic matter in the sludge means that the carbon source must have been metabolized in the digestion process to yield biogas. Through the increase in gas production, the self-sufficiency of the plant with regard to energy supply was improved. To a larger extent, the plant was able to supply its operative processes with energy obtained from its anaerobic digestion process. This is, of course, advantageous from an economic perspective.

8.18.9   LEINETAL WWTP, GERMANY

The Leinetal WWTP in Heiligenstadt, Germany, tested an Ultrawaves ultrasound system prior to full-scale installation. The AS process was operated as an extended aeration process, with an SRT of 18 days and experienced problems associated with bulking sludge. The ultrasound system was installed to treat 30% of the thickened secondary sludge flow, which was then returned to the aeration basins. The test showed that the ultrasound system reduced WAS production by 30%, improved the settleability of the secondary sludge and increased the dewatered cake dryness by two percentage points. Foaming and bulking problems were also eliminated. The full-scale installation has been operational since June 2003 and the plant has been able to avoid the construction of a new aeration tank. Treating a thickened sludge stream, rather than unthickened RAS, may improve the cost-effectiveness of ultrasound installations for AS applications, as the ultrasound systems are designed to maintain a specified hydraulic retention time within the sonication chamber to achieve cell lysis. Ultrasound applications for WAS pre-treatment prior to anaerobic digestion have shown that with appropriate levels of sonication, WAS will achieve similar digestion rates to primary sludge, typically increasing the VSr of the WAS from 40% to 60%. The overall improvement in digester VSr will depend on the ratio of primary sludge to WAS in the digester feed, as well as other site-specific conditions, such as the baseline performance of the digesters and the sludge characteristics. As ultrasound increases the hydrolysis rate, the impact will be greatest for digesters at short HRTs where cell lysis is rate limiting. Excess polymer dosing for WAS thickening will reduce the effectiveness of ultrasound as particles may re-agglomerate if active polymer is still present (Roxburgh et al., 2006).

Ultrasound systems are typically manufactured in modular units, with different suppliers providing different configurations and sizes. Ultrasound systems are sized to treat a unit flow for a certain retention time. Therefore, the systems will be more cost-effective on thicker sludges. However, over around 6–7 percentage solids concentration, the energy required to overcome the sonication threshold or cavitation attenuation caused by higher viscosities and solids concentrations will rise. Treating 30% of the WAS through the unit at approximately 30 m^3/day would typically provide a 15% improvement in sludge reduction while a flow rate of 15 m^3/day would typically provide a 30% improvement in anaerobic digestion. To accommodate higher flows, additional units would be installed in parallel or in sequence.

Systems can be customized to provide the required number of horns, either in series or in parallel. The Sonico system supply typically includes the feed pumps, system instrumentation, ultrasound hardware, control cabinet and automated supervisory control and data acquisition (SCADA) system with a PC, housed within a shipping container. Costs quoted in 2005 ranged from around $500,000 for a 4000 L/h unit to $900,000 for a 15,000 L/h unit and $2.2 million for a system to treat 37,000 L/h (Roxburgh et al., 2006). Future costs will change to reflect changing commodity and labor prices among other factors. Alternatively, the units can be installed within pipe galleries or existing buildings, as the units are designed to connect directly to 6-inch flanged pipes. Sound enclosures are important for such installations as all ultrasound systems create subharmonics that are within the range of human hearing. The installation of an ultrasound system would entail other costs, including the storage of TWAS; piping; additional pumping to or from the system; electrical, instrumentation and control costs; site preparation; utilities; maceration or pre-screening (if fine screens are not used at the headworks); construction costs; and costs for an alternative building if a shipping container is not the preferred means of housing. The life cycle cost–benefits will be more favorable for thicker sludges. Electricity, natural gas and biosolids handling costs will also affect the economic balance and payback period. Actual economics and payback periods will be site specific. The economic benefits of ultrasound pre-treatment are greater for plants facing higher unit power and biosolids disposal costs. The economics for the installation of an ultrasound system on a RAS recycle for the reduction of WAS production from the AS process would be considerably different. Gas production from the digesters would be reduced due to the reduction in the quantity of WAS produced. However, there would be potential savings in operating costs and future expansion capital costs for thickening, digestion and dewatering facilities and biosolids management. As shown in the Anglian Water trial, the ultrasound system allowed the AS system to operate effectively at lower mixed liquor suspended solid (MLSS) concentrations and minimized the impact of filamentous organisms on sludge settleability. This could provide an advantage at WWTPs where the secondary clarifiers may be solids overloaded. As noted at the Leinetal WWTP test, the facility was able to avoid the construction of another aeration lane by installing an ultrasound system, as well as reducing WAS production by 30%.

9 Combination of Ultrasound and Thermal Pre-Treatments

9.1 INTRODUCTION

While there are many papers on the ultrasonic and thermal pre-treatment of waste activated sludge (WAS) applied individually, there are very few papers on their combined application. A few authors have reported significantly improved sludge disintegration due to the synergistic effects of the two mechanisms (Dhar et al., 2012; Şahinkaya and Sevimli, 2013a). The sludge flocs were first disintegrated by hydromechanical shear forces created by ultrasonication and then, the increased temperature caused cell lysis by disrupting the chemical bonds in the cell membrane. Long pre-treatment times from several hours to a few days have been reported if thermal pre-treatment was applied alone at temperatures below 100°C. Therefore, ultrasonication has the potential to significantly shorten treatment times when using low-temperature thermal pre-treatment.

Şahinkaya and Sevimli (2013a) reported a low disintegration degree (DD) of 7.2% and a soluble chemical oxygen demand (SCOD) increase from 55 to 850 mg/L using thermal treatment alone (80°C, 1 h). The methane increase was only 4.2%. Ultrasonication for 1 min (1 W/mL) yielded a DD of 15% and the SCOD increased from 55 to 1700 mg/L. The increase in methane was 5.6%. However, when both treatments were combined, the DD was 25% and the SCOD increased from 55 to 2700 mg/L. The methane improvement was 13.6%, which was more than the sum of individual percentages. The volatile solids (VS) destruction reached 37.8% and the capillary suction time (CST) was reduced from 69.4 to 47.5 s. A further rise in temperature beyond 80°C was reported to slightly improve the synergistic effect. It was also shown that the synergistic effect was promoted with an increase in both ultrasonication density and sonication times. However, based on an economic assessment, the combination was found to be not feasible due to the high-energy requirement. Combining the pre-treatments did not improve dewaterability beyond that obtainable by individual pre-treatment.

The improved performance could have been due to the disruption of flocs containing active extracellular enzymes or the breakdown of cells containing intracellular hydrolytic enzymes. These enzymes then became more active at thermophilic and hyper-thermophilic temperatures. Nabarlatz et al. (2012) showed that it was possible to extract enzymes from WAS using ultrasonication. It has been shown that free enzymatic activity present in the liquid phase is almost negligible as the enzymes would be immobilized on flocs (connected to the polymeric extracellular substances) or attached to the cellular walls by ionic and hydrophobic interactions (Cadoret et al., 2002; Gessesse et al., 2003). Therefore, a physical treatment that can disrupt the flocs thereby mobilizing the intracellular enzymes is useful. Nabarlatz et al. (2012) also reported that the activity of extracellular protease in an activated sludge tank was much lower than intracellular protease; therefore, it is sensible to use ultrasound (ULS) prior to thermal treatment. Because the combined ULS/thermal treatment involves these enzymatic reactions, the type of sludge will affect the synergy between the ULS and thermal treatments. Secondary sludge, which contains more proteins and is more likely to contain hydrolytic and proteolytic enzymes, is more preferable for such pre-treatment than primary sludge, which is rich in carbohydrates.

Dhar et al. (2012) and Bougrier et al. (2006) reported that ULS was more efficient than thermal pre-treatment in terms of biogas production, although better COD solubilization was obtained with thermal pre-treatment. Such greater solubilization with thermal pre-treatments would have resulted from the solubilization of colloidal COD, which is not affected by ULS. It is possible that the lower gas production associated with thermal pre-treatment is due to the formation of agglomerates and hence an increase in particle size, while the opposite occurs with ULS. Combining the pre-treatments did not significantly improve dewaterability when compared with thermal pre-treatment alone.

The sludge temperature rises during ultrasonication. Sludge solubilization occurs at high temperatures. However, results obtained from different research are controversial. Wang et al. (2005) observed that the sludge temperature rises to 80°C after 1 h ultrasonication. However, a very low SCOD increase is obtained when the sludge sample is heated at 80°C for 1 h without ultrasonication. On the contrary, Kim and Youn (2011) and Liu et al. (2009) stated the obvious influence of temperature on sludge solubilization. Liu et al. (2009) indicated that the sludge temperature is more important than the heated time, but results from Kim and Youn (2011) also indicated that a longer heated time promotes sludge solubilization.

Biological treatment encompasses a broad range of processes that can include both aerobic and anaerobic processes. Biological pre-treatment aims at intensification by enhancing the hydrolysis process in an additional stage prior to the main digestion process. The most common type is temperature-phased anaerobic digestion (TPAD), which uses a higher stage in either thermophilic (around 55°C) or hyper-thermophilic (between 60°C and 70°C) conditions, anaerobic or aerobic. Thermophilic processes, particularly the thermophilic hydrolytic activity of bacterial populations, were investigated 80 years ago, mainly at a temperature of 55°C (Carrere et al., 2012).

A number of configurations have been tested, including short pre-treatment prior to mesophilic digestion (Mottet et al., 2009), dual digesters: thermophilic and mesophilic (Erden and Filibeli, 2010b), single-stage digesters (Lin et al., 2012; Gianico et al., 2013) and recently, temperature co-phase processes (Jang and Ahn, 2013; Lu et al., 2012). Thermophilic conditions generally result in an increase in the organic solids destruction rate, attributed to increased hydrolytic activity. Ge et al. (2011) evaluated thermophilic against mesophilic pre-treatment (hydraulic retention time [HRT] of 2 days) prior to mesophilic anaerobic digestion (HRT of 13–14 days) for primary sludge. An increase of 25% on methane production and solids destruction was observed. Model-based analysis indicated that the improved performance was due to an increased hydrolysis coefficient rather than an increase in inherent biodegradability (Malliaros and Guitonas, 1997). Elevating the temperature above 55°C did not provide additional benefits. Bioaugmentation by specific thermophilic hydrolytic anaerobic bacteria has been attempted, but not successfully (Tyagi and Lo, 2012b).

Several papers reported that the combination of sonication and thermal treatment resulted in a significant improvement in sludge disintegration due to the synergistic effects of the two mechanisms; the sludge flocks are first disintegrated by hydromechanical shear forces created by ultrasonic irradiation and then, the increasing temperature breaks down the undisrupted microorganism cells into soluble organic matters by destroying the chemical bonds of the cell walls and membrane.

Dhar et al. (2012) sonicated WAS at 1,000, 5,000 and 10,000 kJ/kg total suspended solids (TSS) and increased the SCOD/total COD (TCOD) ratio from 6% to 10%, 18% and 33%, respectively, which resulted in a 15%, 20% and 24% increase in methane production and a 23%, 28% and 30% increase in volatile suspended solids (VSS) destruction. Their thermal treatment alone at 50°C, 70°C and 90°C for 30 min resulted in an SCOD/TCOD increase from 6% to 18%, 20% and 36%, respectively, while methane increased by 14%, 19% and 13%, respectively. The combined treatment (10,000 kJ/kg TSS and 90°C for 30 min) increased the SCOD/TCOD from 6% to 37%, while methane production and VSS destruction increased by 30% and 36%, respectively. The dewaterability was also improved as the time to filter (TTF) was reduced from 81 to 58 s. Longer thermal treatment time may be more beneficial for the combination. However, the authors showed that ULS was economically feasible with a specific energy input (SEI) of 1000 kJ/kg TSS with a net saving of $54/ton total solids (TS). All combined treatments (1000 kJ/kg TSS and 50°C, 70°C and 90°C) were feasible giving a net saving of $44–$66/ton TS. Because ultrasonication is energy-intensive, it is important to find out the solids content that maximizes the net energy gain. Ultrasonication at low TS content will not be economical because of the large volume and the energy that will be wasted on heating water instead of decomposing the recalcitrant compounds and bacterial cells. Therefore, a centrifugation step is likely to be necessary before applying ultrasonication and thermal treatment.

The sequence of treatment may also affect the performance, but there is still a lack of comparative data regarding this aspect. The sequential application of ultrasonic and thermal pre-treatments is suggested based on the available information on the individual pre-treatments. The benefit of potential enzyme mobilization has been discussed previously. In addition, if thermal pre-treatment was conducted prior to ultrasonication, significant solids solubilization induced by thermal treatment would leave fewer solids as nuclei for the formation of cavitation bubbles. Therefore, the sequence ultrasonication–thermal may be better, but further tests need to confirm this. The other reason for this combination sequence is that a rising sludge temperature has a negative influence on the ultrasonic cavitation effect (Kim and Youn, 2011).

9.2 EFFECT OF TEMPERATURE DURING THE ULS-THERMAL PRE-TREATMENT OF SLUDGE

At the laboratory scale, it is commonly observed that the temperature increases up to 70°C during ULS if a small volume of sample (<50 mL) was used and if the sample was not cooled down. Using a pulse mode could reduce the heat generated, but it was not efficient to solubilize more COD (data not shown). It was then decided to investigate the effect of ULS and heat separately and by combining ULS and thermal treatment in sequence. The sequence of thermal–ULS was tested and it was found that the thermal energy could lyze cells that then released soluble materials, such as colloids and proteins, leading to an increase in the SCOD concentration (data not shown). However, during the subsequent ultrasonication, the SCOD, soluble proteins and carbohydrates concentration increased by only 700, 60 and 145 mg/L, respectively, which was deemed insignificant. It was postulated that the propagation of ultrasound waves was hindered and could not reach intact cells. Therefore, the focus is on the ULS-thermal sequence in this chapter.

First, a specific percentage of sludge (0%, 25%, 50%, 75% and 100%) was ultrasonicated and the SCOD concentration obtained after mixing with the non-treated fraction was measured. The sludge was then incubated at a specific temperature without mixing and at neutral pH. From Figure 9.1 (top), it can be seen that after 24 h incubation at 30°C the SCOD increased to ~3 g/L for the 25% ULS-treated sludge and decreased for the higher ratios presumably because the SCOD was consumed by mesophilic microorganisms at 30°C and converted to CO_2.

Interestingly, the situation was very different at 55°C, as shown in Figure 9.1 (bottom). When the sludge was not ultrasonicated and heated to 55°C (100% raw), the SCOD increased to 5.35 g/L, whereas with 25% ULS-treated sludge the SCOD increased to 7.1 g/L. The improvement in the solubilization of the sludge due to the combination of thermal treatment and ultrasonication is in-line with previous studies (Şahinkaya and Sevimli, 2013a).

Moreover, the use of ultrasonication on 25% of the sludge further improved the performance of thermal treatment. This was due to the disruption of flocs and the breakdown of cells containing intracellular hydrolytic enzymes. This is in-line with researchers who showed that enzymes could be extracted from WAS using ULS (Nabarlatz et al., 2012). The foregoing data confirmed that a physical treatment was useful to disrupt the flocs and release the enzymes. Yu et al. (2008) showed that ultrasonic pre-treatment enhanced enzymatic activities and promoted the shifts of extracellular proteins, polysaccharides and enzymes from inner layers of sludge flocs (i.e., pellet and tightly bound EPS) to outer layers (i.e., slime) and this increased the contact and interaction among extracellular proteins, polysaccharides and enzymes that were originally embedded in the sludge flocs, resulting in improved efficiency in the subsequent aerobic degradation. Heat and ultrasonication can also be used to rupture cells and release the intracellular proteases that can hydrolyze proteins in the sludge. Nabarlatz et al. (2012) also found that the activity of extracellular protease in an activated sludge tank was much lower than that of intracellular protease; therefore, it is sensible to use ULS prior to thermal treatment. At 100% ULS-treated sludge, the final SCOD concentration reached almost 8 g/L. However, at higher ratios of ULS-treated sludge, the improvement in ultrasonication became marginal.

Based on the previous experiment, a 25% ratio was used to determine the optimum temperature for the enzymatic treatment. Figure 9.2 shows that the higher the temperature, the more COD was solubilized (up to ~11 g/L). A higher SCOD (up to 14 g/L) could be obtained depending on the initial solid concentration of the sludge (data not shown). Temperatures >65°C resulted in a marginal increase. It was also found that mixing during the thermophilic treatment resulted in a 20% increase in the SCOD concentration (data not shown).

To investigate further the effect of heat, a sample was autoclaved (121°C for 20 min) and the final SCOD was only 6700 mg/L. As the temperature slowly increased in the autoclave, the enzymes were still active but could have been deactivated at high temperatures (>85°C), which limited the extent of solubilization compared to a milder thermal process. This showed further that heat was not the only phenomenon taking place. The solubilization of WAS by heat treatment can be induced by sludge lysis and further cryptic growth (lysis-cryptic growth) (Wei et al., 2003). In the lysis-cryptic growth, sludge reduction is achieved because some portions of lysates are consumed for catabolism and finally emitted as CO_2. This was

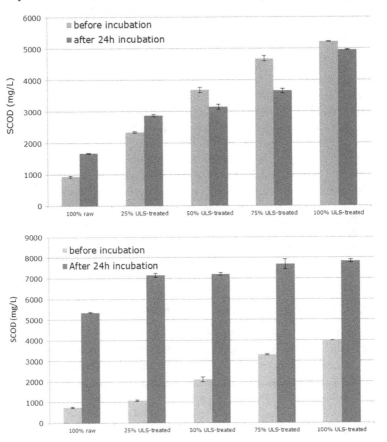

FIGURE 9.1 Effect of ultrasonication of 0%, 25%, 50%, 75% and 100% of sludge (30 s, ~5000 kJ/kg TS) followed thermal treatment at 30°C (top) and 55°C (bottom). (Reproduced from Trzcinski et al., 2015, with permission from Taylor & Francis.)

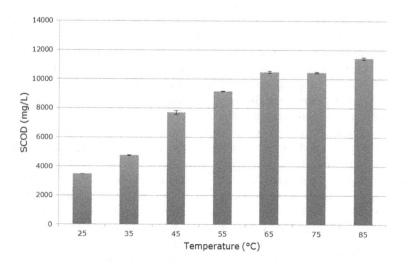

FIGURE 9.2 Effect of the incubation temperature on the SCOD during enzymatic treatment following ultrasonication of 25% of sludge (30 s, ~5000 kJ/kg TS). (Reproduced from Trzcinski et al., 2015, with permission from Taylor & Francis.)

confirmed as a CO_2 production of 4.4 and 6 mL recorded after 1 h incubation at 55°C and 65°C, respectively. After 24 h, the cumulative CO_2 production reached 9.9 and 10.2 mL, respectively, indicating the consumption of SCOD for the growth of both thermophilic and hyper-thermophilic bacteria.

Yan et al. (2008) used a simple heat treatment process (700 mL was incubated at 60°C, 120 rpm for 24 h in a 1 L Erlenmeyer flask) and showed that there was a rapid increase in the population of thermophilic bacteria at the early stage of heat treatment and the emergence of protease-secreting bacteria. Hasegawa et al. (2000) showed that the hyper-thermophilic aerobic microbes were identified as belonging to *Bacillus*. Therefore, the potential for increased performance is inherent in the sludge itself (Sakai et al., 2000) and although heat treatment is beneficial for solubilization, long thermal treatment is not interesting from a process point of view but also because some of the lyzate is consumed by thermophilic bacteria and lost as CO_2 and cannot be used to produce methane.

9.3　ULTRASOUND-THERMAL PRE-TREATMENT AT 55°C

In this experiment, a specific percentage of sludge (0%, 5%, 10%, 20%, 50% and 100%) was ultrasonicated, then mixed with the remaining non-ultrasonicated fraction and incubated in a waterbath to study the kinetics of solubilization at 55°C and 65°C. As 75°C and 85°C were shown to result in a marginal SCOD increase in the previous section, these temperatures were not tested further in detail in this study. Carbohydrates and proteins are two predominant biopolymers in an EPS structure, which also contribute a great part of COD to sludge (Yu et al., 2008). Therefore, the solubilization of carbohydrates and proteins provides essential information about the disintegration of the sludge structure. The SCOD, proteins and carbohydrates concentrations obtained at 55°C are shown in Figure 9.3.

It can be seen that the thermophilic treatment alone resulted in a final SCOD of 7.8 g/L, whereas a significant increase to 8, 8.7 and 9.3 g/L was observed when 20%, 50% and 100% of sludge was ultrasonicated prior to thermal treatment, respectively. Below 20% of ULS-treated sludge there was a small effect as indicated by close SCOD values. The results indicated that as the percentage of ULS increases, more cells are broken down and more intracellular materials are released into the bulk, as shown by higher SCOD concentrations. However, the effect of ULS was not linear, meaning that 100% ULS treated did not result in twice the solubilization of 50% ULS-treated sludge. This shows that treating 100% of sludge by ULS is not an interesting option; however, 20% and above had a positive impact on the subsequent thermal treatment. It was also found that ultrasonication increased the COD solubilization rate of the overall pre-treatment. For instance, the thermophilic treatment took 24 h to reach 7.8 g/L SCOD, whereas only a 3 h thermal treatment was required when 100% sludge was ultrasonicated. This demonstrated that the thermal treatment time can be significantly reduced by combining ULS.

It can be seen from Figure 9.3b and c that ultrasonication improved the solubilization rate of proteins and carbohydrates compared to the thermophilic treatment alone. The concentration increased during the first 6 h of thermophilic treatment and decreased afterward due to the consumption of nitrogen and carbohydrates by thermophilic bacteria. Proteins and carbohydrates solubilization might have continued after 6 h, but it could not compensate for the uptake by opportunistic thermophilic microorganisms, resulting in a net decrease after 6 h of treatment. This net decrease was, however, not observed in SCOD concentrations (Figure 9.3a) as COD analysis encompassed various biopolymers including proteins and carbohydrates, and also lipids, phosphates, ammonia, humic and fulvic acids that were solubilized.

9.4 ULTRASOUND-THERMAL PRE-TREATMENT AT 65°C

The SCOD, proteins and carbohydrates concentrations obtained at 65°C are shown in Figure 9.4. As expected, the extent and rates of COD, proteins and carbohydrates solubilization were enhanced at 65°C compared to 55°C. This is due to improved cell lysis and possibly higher enzyme activity. 428Yu et al. (2008) also showed that enzymatic activities (proteases, α-amylase, α-glucosidase, alkaline-phosphatase and acid-phosphatase) were markedly increased after ultrasonication. In terms of the final SCOD concentration, 100% ULS was equivalent to 1 h hyper-thermophilic treatment. Both conditions resulted in ~5 g SCOD/L. When 100% of sludge was ultrasonicated, <1 h of hyper-thermophilic condition was required to reach 8 g SCOD/L. However, 24 h were required to reach that level in an individual hyper-thermophilic pre-treatment. Therefore, ULS significantly shortened the hyper-thermophilic treatment. Şahinkaya and Sevimli (2013a) reported that the SCOD increased from 55 to 3500 mg/L after 10 min ultrasonication (1.5 W/mL) at 80°C for 1 h, which was found to be the optimum temperature. Ultrasonication alone resulted in a concentration of 2250 mg/L. This confirmed the better results using a combination of ultrasound and thermal treatments. However, these concentrations were much lower than in this study due to the lower TS level (4 g/L) in the raw sludge.

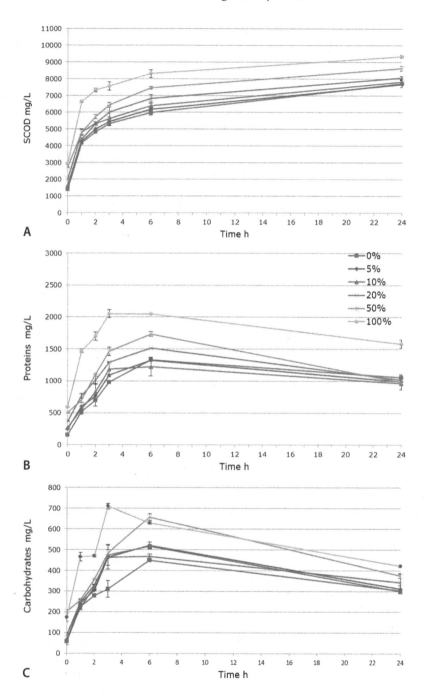

FIGURE 9.3 Evolution with time of the SCOD (a), soluble proteins (b) and carbohydrates (c) during enzymatic treatment at 55°C following ultrasonication of 0%, 5%, 10%, 20%, 50% and 100% of sludge (30 s, ~5000 kJ/kg TS). (Reproduced from Trzcinski et al., 2015, with permission from Taylor & Francis.)

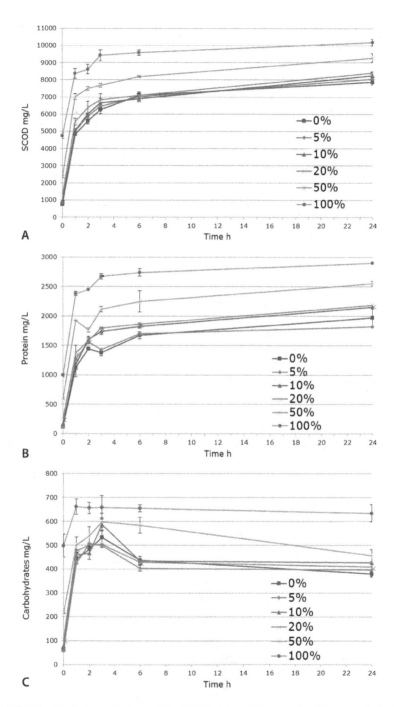

FIGURE 9.4 Evolution with time of the SCOD (a), soluble proteins (b) and carbohydrates (c) during enzymatic treatment at 65°C following ultrasonication of 0%, 5%, 10%, 20%, 50% and 100% of sludge (30 s, ~5000 kJ/kg TS). (Reproduced from Trzcinski et al., 2015, with permission from Taylor & Francis.)

It can be seen that the extent of protein solubilization increased as the percentage of ULS-treated sludge increased. This is in-line with the previous observations at 55°C. However, at 65°C, the effect of ULS was more dominant, as shown by a significantly higher solubilization rate at percentages as low as 10%. Interestingly, there was no net decrease in soluble proteins concentrations at 65°C in contrast to what was observed at 55°C. This indicates that the rate of proteins solubilization was higher than the rate of proteins degradation and consumption by hyper-thermophilic bacteria. Soluble carbohydrates, however, were consumed by hyper-thermophilic bacteria as indicated by a net decrease in concentration after 6 h. The existence of such hyper-thermophilic bacteria was previously documented and was found to belong to *Bacillus* (Hasegawa et al., 2000). The net decrease was insignificant for the 100% ULS-treated sample showing that ULS could also inhibit to some extent the growth of the hyper-thermophilic bacteria and avoid the consumption of soluble carbohydrates. The visual effect of the combined ULS enzymatic treatment can be seen in Figure 9.5.

From Figure 9.5, we can clearly see that after 24 h incubation at 65°C the supernatant becomes turbid and yellowish due to cell debris and intracellular compounds released into the medium. Also, the solid level is lower than the untreated WAS because some cells were disrupted and solubilized by the pre-treatment. The ULS and thermal treatment disrupted flocs so that after centrifuging, the sludge was more compact. The total volume increased due to the release of interstitial water and intracellular water into the medium.

The kinetics of COD solubilization due to enzymatic degradation is shown in Figure 9.6. For this purpose, 20 mL of WAS was treated with an energy input of 3500 J. Then, 0.5 mL of ultrasonicated sludge was mixed with 9.5 mL (5%) of raw sludge, vortexed for 10 s and incubated at 65°C. Identical samples were prepared and one sample was sacrificed for each sampling time (0, 30 min, 1 h, 1.5 h, 2 h, then every 2 h till 24, 48 and 72 h). It can be seen that initially the enzymes are relatively

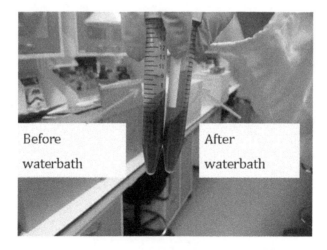

FIGURE 9.5 Visual effect of ultrasonication–thermal pre-treatment of sludge. Five percent of the sludge was ultrasonicated (3500 J) and then mixed with 95% of raw mixture. The mixture was then incubated in a waterbath at 65°C for 24 h.

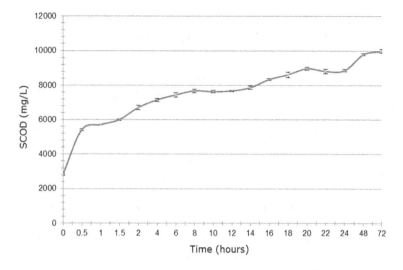

FIGURE 9.6   Kinetics of COD solubilization during ultrasound thermal treatment at 65°C.

active and start degrading the organic matter as soon as it is incubated at 65°C. In fact, the SCOD almost doubled within 30 min. After 6 h, the SCOD still continued to increase but at a lower rate. After 24 h, the SCOD increased from 9,000 to 10,000 in 2 days, showing that the enzymes were still active.

Furthermore, Figure 9.7 shows that mixing during the enzymatic treatment at 65°C resulted in a 20% increase in SCOD compared to the non-mixed sample. This indicates that mixing provides a better enzyme–substrate interaction, resulting in enhanced solubilization.

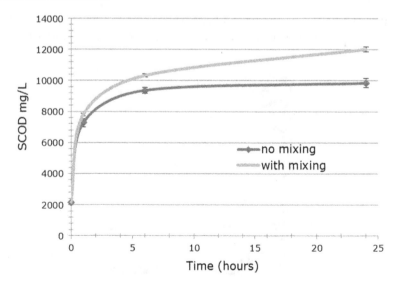

FIGURE 9.7   Effect of mixing during enzymatic treatment at 65°C following ultrasonication of 100% of WAS at 5800 kJ/kg TS.

9.5 QUALITATIVE ANALYSIS OF INDIGENOUS ENZYMES DURING ULTRASOUND AND THERMAL PRE-TREATMENT

In this section, more qualitative evidence of proteolytic enzymes in WAS is reported. For this purpose, petri dishes with agar (10 g/L), potassium phosphate (50 mM) and skimmed milk (20 mL/L) were prepared. The method of Lin et al. (1992) was followed; casein in skimmed milk gives a whitish color to petri dishes. The addition of this protein to the agar medium allows us to observe colonies that are able to degrade the casein. If the colony produces proteases that degrade casein, then there will be a clear patch around that colony. Several dilutions of raw WAS were prepared in order to be able to observe single colonies on the petri dishes, which were incubated at 37°C. Figure 9.8 shows the growth of several colonies; a few colonies were able to degrade casein leaving a clear patch around it. This shows evidence that WAS contains proteolytic bacteria.

It can be seen from Figure 9.9 that proteolytic bacteria were able to grow even when the petri dish was sealed with parafilm in order to reduce the amount of oxygen during the incubation period. This shows that these bacteria are probably facultative and the enzymatic degradation of casein was not affected by these lower oxygen conditions.

Figure 9.10 shows that when WAS was not diluted, then almost all the casein in the petri dish was consumed, showing evidence of the powerful action of proteases in WAS. Furthermore, a colony from Figure 9.9 was isolated on a new petri dish using a sterile loop and then incubated at 55°C in order to reveal if these proteolytic bacteria were able to grow in the thermophilic range. Interestingly, Figure 9.11 shows that these bacteria could not grow at 55°C, but their enzymes (isolated together with the loop) were still active at 55°C and could degrade casein, leaving a clear patch on the petri dish.

In the next step, we tried to isolate single colonies onto new petri dishes (Figure 9.12). This isolation step was successful except for the bottom left picture

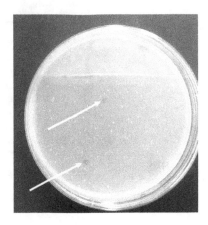

FIGURE 9.8 Growth of microorganisms from WAS onto an agar-skimmed milk petri dish after 24 h and evidence of proteolytic bacteria (10^3 dilution).

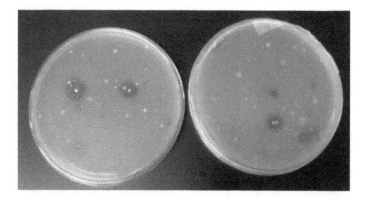

FIGURE 9.9 Growth of proteolytic bacteria from WAS after 48 h. The petri dish on the right-hand side was sealed with parafilm.

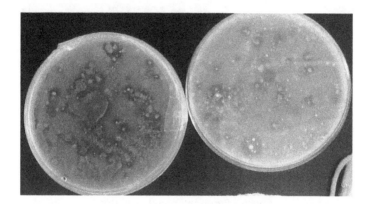

FIGURE 9.10 Growth of proteolytic bacteria from WAS (non-diluted) after 48 h.

FIGURE 9.11 Isolation of a proteolytic colony at 55°C. The colony did not grow, but the enzymes were still active.

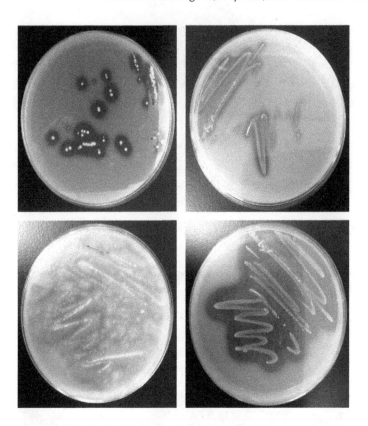

FIGURE 9.12   Isolation of several colonies onto new petri dishes.

of Figure 9.12 where a fungus was growing together with a bacterium. Interestingly, these two microorganisms were still able to degrade casein.

In Figure 9.13a, it can be seen that proteases were more active when WAS was ultrasonicated. Figure 9.13b and c shows that proteases producers were more active, grew faster (bigger colonies) and produced more enzymes (larger plaques) when WAS was ultrasonicated. These qualitative results support Figure 9.4 showing that 100% ULS resulted in a higher SCOD compared to 0% ULS.

9.6   SOLIDS REMOVAL DURING ULTRASOUND AND THERMAL PRE-TREATMENT

Table 9.1 shows the TSS and VSS removal during the individual and combined pre-treatments. It can be seen that ULS alone resulted in TSS removals lower than 10%, while the thermal treatment resulted in TSS removals in the range 20%–23%. When 50% of the sludge was ultrasonicated and treated at 65°C, then a maximum of 27% TSS and VSS removal was obtained. Treating 100% of the sludge by ultrasonication did not increase this removal, confirming that ultrasonication of a high proportion of sludge is not required. Yu et al. (2008) obtained 11.8% TSS reduction after an ultrasonication treatment (10 min, 3 kW/L) of WAS and attributed this to the release of soluble organic carbon sources and extracellular enzymes, and the enhanced contact between them. The sludge reduction for TSS was 30.9% after aerobic degradation (compared with 20.9% in the control), showing that the ultrasonic pre-treatment could significantly enhance aerobic digestion efficiency and the extent of sludge biodegradability.

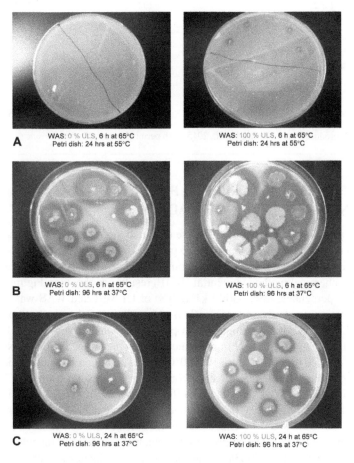

FIGURE 9.13   (a) Samples taken after 6 h of hyper-thermophilic enzymatic pre-treatment at 65°C with and without ultrasonication and pipetted into wells on petri dishes incubated at 55°C. The top half of the petri dish contains replicated wells with sludge samples. The bottom half contains wells with only the clear supernatant from the sludge. (b) Samples taken after 6 h of enzymatic pre-treatment at 65°C with and without ultrasonication and pipetted into wells on petri dishes placed at 37°C. (c) Samples taken after 24 h of hyper-thermophilic enzymatic pre-treatment at 65°C with and without ultrasonication and pipetted into wells on petri dishes placed at 37°C.

TABLE 9.1
TSS and VSS Removal during the Combined ULS/Thermal Pre-Treatment

	TSS Removal (%)	VSS Removal (%)
Raw	0	0
ULS 20% (5000 kJ/kg TS)	4.6	2.36
ULS 50% (5000 kJ/kg TS)	7.28	5.91
ULS 100% (5000 kJ/kg TS)	8.62	6.86
Raw + 55°C for 24 h	20.5	19.15
Raw + 65°C for 24 h	22.22	22.93
ULS 20% + 55°C for 24 h	21.65	23.4
ULS 50% + 55°C for 24 h	21.46	20.57
ULS 100% + 55°C for 24 h	22.8	22.46
ULS 20% + 65°C for 24 h	23.75	23.4
ULS 50% + 65°C for 24 h	27.2	26.95
ULS 100% + 65°C for 24 h	24.33	24.35

Source: Reproduced from Trzcinski et al., 2015, with permission from Taylor & Francis.

9.7 EFFECT OF THE COMBINED PRE-TREATMENTS ON ANAEROBIC BIODEGRADABILITY

Since it was found that higher SCOD were obtained by combining ULS and thermal treatment (up to 14 g/L), a higher methane production was expected to be found using the combined pre-treatments. Several combinations of pre-treatments were tested and the biochemical methane potential (BMP) results are shown in Figure 9.14.

A small percentage (5%, ~5000 kJ/kg TS) of ULS-treated WAS was combined with the thermal treatment at 65°C for 24 h and it was found that methane production increased by 20%. This was higher than previous studies (Şahinkaya and Sevimli, 2013a), where a 13.6% increase in methane was obtained after 1 min ultrasonication (1 W/mL) and 1 h thermal treatment at 80°C. It was also found that methane production was greater with the combination compared to ultrasonication of 5% or even 100% alone. The methane percentage in the biogas was up to 6% higher indicating a higher calorific value due to the combined pre-treatments. However, a lag phase of 8–12 days was observed following the combined treatment, which may be the result of a higher SCOD and its components to which the anaerobic inoculum was not acclimated. Gavala et al. (2003a) found that there are indigenous microorganisms in primary sludge capable of methane production and incubation at 70°C for 1 day or more as a pre-treatment resulted in their inactivation. Furthermore, they found that the thermal pre-treatment of both primary and secondary sludges led to increased hydrogen levels that can inhibit methane production. Our anaerobic inoculum may not have contained enough hydrogenotrophic species, which led to some inhibition and the observed lag phase.

Moreover, this combination (+20%) was more efficient than ULS alone at high SEI (100%, 9000 kJ/kg TS), as shown in Figure 8.8a (+14.8%). This is due to the

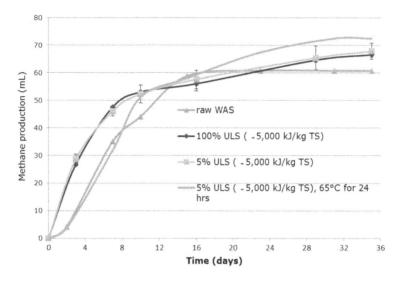

FIGURE 9.14 Cumulative methane production after the combined ultrasonication and thermal pre-treatment of sludge. (Reproduced from Trzcinski et al., 2015, with permission from Taylor & Francis.)

COD solubilization obtained after the pre-treatment. After 100% ULS, typical SCOD concentrations are in the range 4–5 g/L (Figure 8.5a), whereas the combination of ULS and thermal treatment resulted in 10–11 g/L SCOD.

9.8 CONCLUSIONS

In this chapter, we investigated the pre-treatment of sewage sludge using ultrasonic and thermal treatments. The optimum temperature during the thermal treatment was found to be 65°C. It was also found that the combination of ULS (30 s, 5000 kJ/kg TS) and thermal treatments resulted in greater solubilization of COD (760–10,200 mg/L), proteins (115–2,900 mg/L) and carbohydrates (60–660 mg/L) than individual treatments. During ultrasonication treatment alone (30 s, 5000 kJ/kg TS), SCOD, soluble proteins and carbohydrates concentrations increased to 4700, 1000 and 500 mg/L, respectively. The ultrasonication of 50% of the sludge followed by incubation at 65°C could increase the SCOD from 760 to 9300 mg/L. It was also found that ultrasonication increased the COD solubilization rate of the subsequent thermal treatment at 65°C and treatment time could then be reduced to a few hours (1–6 h) instead of 24 h or several days. The combined treatment resulted in a 20% increase in biogas production.

ULS is a fast method, but relatively inefficient to solubilize COD and it is expensive to treat 100% of WAS. Thermal treatment is efficient to solubilize COD, but it is a slow process. Thermal treatment could be a viable option to consider if waste heat is available on-site. It was found that these disadvantages can be alleviated when both methods are combined, while methane production is improved. Further work is needed to find an optimum combination for each type of sludge.

10 Physicochemical Treatment

Application of Microwave and Chemical Treatment of Sludge

10.1 INTRODUCTION TO MICROWAVE TREATMENT OF SLUDGE

Like ultrasonication, microwave (MW) irradiation is considered a popular alternative to conventional heating (CH) technology. In the electromagnetic spectrum, MW irradiation operates in wavelengths of 1 mm–1 m with the corresponding oscillation frequencies 0.3–300 GHz. In industry, a shorter frequency either close to 900 MHz or at 2450 MHz is often adopted. Damage to sludge cells with MW irradiation may occur in two ways: (i) thermal effect that is generated through the rotation of dipoles under oscillating electromagnetic fields, which heats the intracellular liquor to boiling point and brings about the break up of bacterial cells; and (ii) athermal effect that is induced by the changing dipole orientation of polar molecules, giving rise to the possible breakage of hydrogen bonds and the unfolding and denaturing of complex biological molecules, which kills microorganisms at lower temperatures.

As an alternative to CH, it has been suggested that MW irradiation can be used to disrupt biopolymers by dipole rotation or orientation effects. The MW orientation effect is mainly caused by polarized parts of macromolecules lining up with the poles of the electromagnetic field resulting in the possible breakage of hydrogen bonds (Eskicioglu et al., 2006a). Polysaccharides, proteins and smaller amounts of DNA and RNA can be released in the soluble phase due to the orientation (athermal) and subsequent heating (thermal) effects to break the polymeric network. MW-irradiated microbial cells showed greater damage than conventionally heated cells at similar temperatures (Eskicioglu et al., 2007c).

Beszédes et al. (2011) concluded that the MW pre-treatment of food industrial sewage sludge resulted in 3.1-fold higher solubility and 1.7-fold increased biogas production. Ahn et al. (2009) reported that the solubilization degree (soluble chemical oxygen demand/total chemical oxygen demand [SCOD/TCOD]) of sludge increased from 2% (control) to 22% and the biochemical acidogenic potentials increased from 3.58 to 4.77 g COD/L after MW pre-treatment (2450 MHz, 700 W for 15 min). MW can destroy fecal coliforms and solubilize the particulate COD of waste sludge (Hong et al., 2004). SCOD/TCOD ratios in waste activated sludge (WAS) increased from 8% to 18% after MW to 72.5°C (Hong et al., 2006) and from 6% to 18% at 96°C (Eskicioglu et al., 2007b) (see Table 10.1). SCOD/TCOD ratios of 19% and 21% were also reported for WAS irradiated to 91°C and 100°C, respectively (Park et al., 2004). WAS microwaved to 96°C resulted in 15% and 20% increased biogas production during batch digestion at 1.4% and 5.4% total solids (TS), respectively (Eskicioglu et al., 2007b). In a continuous flow digester, WAS treated to 96°C by MW and CH achieved 29% and 32% higher TS and 23% and 26% volatile solids (VS) removal efficiencies compared to the control (Eskicioglu et al., 2007a).

In another study, WAS was treated by MW at 91°C and 64% and 37% higher COD removal and methane production were reported, respectively (Park et al., 2004). Eskicioglu et al. (2009) treated thickened WAS (TWAS) with MW at various temperatures and observed that the rate and extent of biogas production improved after the pre-treatment. In fact, in the pre-treatment range of 50°C–96°C, WAS treated by MW and CH resulted in similar SCOD and there was no discernible MW athermal effect on the COD solubilization of WAS (Eskicioglu et al., 2007c). However, MW-treated WAS produced 16% more biogas, which was more than the CH, indicating the beneficial impact of MW on mesophilic anaerobic digestion (AD). The authors also reported an improvement in the dewaterability of sludge. In contrast to this study, Mehdizadeh et al. (2013) compared the influence of MW (2.45 GHz, 1200 W) and CH pre-treatments to dewatered sludge solubilization and AD. It was reported that the heating method (CH vs MW) had no statistically significant effect ($P > 0.05$) on biosolids solubilization and methane production. There was no pattern of MW-heated digesters showing enhanced performance over conventionally heated digesters due to the athermal effect.

The different results can be explained by the different temporal heating profiles applied by various researchers. This parameter can be more important than the final pre-treatment temperatures reached. Appels et al. (2013) reported an energy consumption of 336 kJ/kg TS during a pilot-scale experiment. Although an increase of 50% in biogas was obtained, the additional amount of biogas could not make the process economically viable. The application of ultrasound with MW irradiation for sludge pre-treatment also has a positive effect on biodegradability and biogas production. Yeneneh et al. (2013) applied MW at 2450 MHz for 3 min, followed by ultrasonication at a density of 0.4 W/mL for 6 min, to different types of sludge. They found that the pre-treatment increased the cumulative methane production (66.5 mL/g TCOD) to a greater degree in excess activated sludge, as compared to the methane yield from mixed sludge.

10.2 MICROWAVE/H_2O_2 PRE-TREATMENT

Eskicioglu et al. (2008) observed a synergistic effect when MW and hydrogen peroxide (H_2O_2) were combined. The combined effects disintegrate the complex floc structure of WAS and solubilize the intra- and extracellular polymeric compounds (proteins, sugars and humic acids). H_2O_2 as a strong oxidizer converts some of the sensitive organic compounds to CO_2 and water. Heating increases the decomposition of H_2O_2 into OH• radicals, thereby enhancing the oxidation process. In a temperature

TABLE 10.1
Microwave Pre-Treatments of Sludge

	Solubilization			Anaerobic Digestion				Scale (HRT, days)	References
Sludge	Microwave Conditions	COD	DD (%)	CH₄ Production	VS Destruction	COD rem. %	Dewaterability		
WAS (4.6%)	2,450 MHz 50°C 75°C 96°C 120°C 150°C 175°C	SCOD/TCOD: 9%–12% 9%–21% 9%–24% 9%–24% 9%–28% 9%–35%	54	+31%	–	–	–	LS, B	Eskicioglu et al. (2009)
WAS (4.6%)	2,450 MHz 96°C	SCOD/TCOD: Times 3.6	–	+20%	+29%(TS) +23%(VS)	–	–	–	Eskicioglu et al. (2007a)
WAS (4.6–5.5%)	2,450 MHz 50°C 75°C 96°C	SCOD/TCOD: 5%–14% 5%–19% 5%–22%	–	+5% +8% +16%	–	–	–	LS, B	Eskicioglu et al. (2007c)
WAS (4.7–5.9%)	2,450 MHz 96°C	SCOD/TCOD: 6%–15%	–	+211%	–	–	–	LS, B	Eskicioglu et al. (2006a)
WAS (6.4%)	2,450 MHz 60°C 80°C 100°C 120°C	SCOD/TCOD: 3%–12% 3%–15% 3%–16% 3%–15%	–	–6% +1.6% +29% +0.6%	–	–	CST: 180.7–171.9 180.7–166.4 180.7–193.3 180.7–158.2	–	Eskicioglu et al. (2008)
WAS	2,450 MHz, 72.5°C	SCOD/TCOD: 8%–18%	–	–	–	–	–	LS, B, M	Hong et al. (2006)

(Continued)

TABLE 10.1 (CONTINUED)
Microwave Pre-Treatments of Sludge

Sludge	Microwave Conditions	Solubilization		Anaerobic Digestion				Scale (HRT, days)	References
		COD	DD (%)	CH_4 Production	VS Destruction	COD rem. %	Dewaterability		
WAS (30 g/L)	2,450 MHz, 91°C	SCOD/TCOD: 2%–19%	–	+28.6 +29.8%	25.5 (23.2) 25.9 (23)	19.8 (13.8) 23.6 (14.4)	–	LS, C, M (10) (15)	Park et al. (2004)
WAS (TS: 4%)	800 W, 3.5 min, 336 kJ/kg TS, 80°C	1353–4247 mg/L (+214%)	–	+50	70% (51)	–	–	PS, SC, M (20)	Appels et al. (2013)
TWAS	1,250 W, 2,450 MHz, 100% intensity	SCOD/TCOD: from 0.06 to 0.2	–	+106%	53.1%	–	–	LS, SC, T-M	Coelho et al. (2011)
TWAS (4.4%)	2.45 GHz, 800 W, 1 min, 96 kJ/kg TS	SCOD: +117%	–	+20%	–	–	–	LS, SC, M (20,67)	Houtmeyers et al. (2014)
WAS (1.4%)	14,000 kJ/kg TS	SCOD/TCOD: 2%–21%	–	+570%	–	–	–	LS, B, M (35)	Ebenezer et al. (2015)
Dairy activated sludge (11.7 g TS/L)	2,450 MHz, 900 W, 12 min, 1,814 kJ/L	SCOD: +19%	–	+57%	+64%	–	–	LS, SC, M (15)	Rani et al. (2013)

Note: WAS, waste activated sludge; LS, laboratory scale; PS, pilot scale; B, batch; SC, semi-continuous; C, continuous; M, mesophilic anaerobic digestion; T, thermophilic.

environment of 60°C–120°C, the addition of H_2O_2 (2.6 and 5.4 g H_2O_2/g TS) was found to enhance the release of ammonia (Wong et al., 2006).

Eskicioglu et al. (2008) treated TWAS (6.4% TS) with H_2O_2 (1 g/g TS) and MW at temperatures of 60°C, 80°C, 100°C and 120°C. The SCOD/TCOD ratio increased from 3% in the control to 15%, 18%, 21% and 24% at these temperatures, respectively, which was more than MW (12%, 15%, 17% and 15%) and H_2O_2 (17%) treatment applied alone. The COD solubilization was consistent with other MW/H_2O_2 studies on secondary sludge (Wong et al., 2006; Yin et al., 2007). Despite COD solubilization, the anaerobic biodegradability varied from 308.4 mL CH_4/g VS_{added} in the control to 290.8, 302, 308.1 and 316.8 mL CH_4/g VS_{added} in the combined pre-treated samples. This was due to the decrease in the methane potential of WAS via advanced oxidation. The authors suggested using a lower dosage in order to minimize the loss of methane potential. Dewaterability using the combined pre-treatment improved (CST of 180.7 to 154–171 s) compared to raw TWAS, but was not significantly better than the individual pre-treatment.

10.3 MICROWAVE/ALKALINE PRE-TREATMENT

MW irradiation is recognized as an efficient technology due to its considerable advantages, e.g., low cost, flexible control and high COD. Recently, it has been suggested that MW pre-treatment should be combined with further methods to improve the disintegration of the sludge and sanitization from pathogens (Anjum et al., 2016). Various studies have revealed the synergistic effect of MW combined with alkali pre-treatment (Chang et al., 2011; Doğan and Sanin, 2009; Qiao et al., 2008).

Chang et al. (2011) found that the pre-treatment of sludge with MW-alkali increased the solubilization rate by 20%, in comparison to that in a single treatment. This increased the aerobic treatment efficiency up to a point at which a 93% reduction in SCOD was achieved in just 16 days of retention. Thus, the pre-treatment can also decrease the retention time in the reactor. When it comes to MW-alkali, various types of bases can be used, including NaOH, KOH and Ca(OH)$_2$. Tyagi and Lo (2012) have established that the use of NaOH with MW demonstrates a high rate of sludge solubilization in comparison to KOH. Sodium hydroxide achieves sludge solubilization up to 52.5%, approximately 4% higher than with KOH. However, both bases demonstrated a 20% higher rate of efficiency in comparison to a single treatment. In addition to alkali, MW can also assist with ultrasound pre-treatment to increase the AD and biogas production, i.e., Yeneneh et al. (2015) achieved a 31% higher removal of TS in comparison to MW alone. Similarly, the biogas production increased by as much as 15%. Overall, the positive aspects of combined MW pre-treatments include the high rate of sludge solubilization and biogas production and lower retention times.

Doğan and Sanin (2009) found that MW treatment could be an alternative to CH treatment in combination with alkaline for a larger impact. The SCOD/TCOD ratio increased from 0.005 to 0.18, 0.27, 0.34 and 0.37 when MW pre-treatments were combined with alkaline pre-treatments at pH of 10, 11, 12 and 12.5, respectively. Although no synergistic COD solubilization was observed, the 34% solubilization at MW-alkali (pH 12) was similar to the sum of the solubilization obtained with each individual treatment (38%). Another advantage of this combined treatment lies in the improvement in dewaterability. Although the dewaterability was deteriorated after individual MW or alkaline pre-treatment, the combined MW-alkali treatment resulted in a better dewaterability. In addition, a 43.5% and 53.2% increase in total biogas and methane production from the semi-continuous anaerobic digester (solids retention time [SRT]: 15 days) was observed when sludge was pre-treated at MW-alkali (pH 12). They suggested that prior alkaline treatment could weaken the cell walls and make them easily lyzed in the subsequent MW treatment.

In contrast, synergistic COD solubilization was observed by Chang et al. (2011). Individual MW and alkali pre-treatment only resulted in 8.5% and 18% COD solubilization; whereas combined treatment induced 46% COD solubilization, which was higher than the sum of the solubilization value in individual treatments (26.5%). Such observation is a result of the synergistic mechanism of MW and alkali treatments. Furthermore, they indicated that the MW-alkali pre-treatment was more effective in COD solubilization than the MW/H_2O_2 pre-treatments.

Yang et al. (2013a) used response surface methodology to optimize the combined MW-alkali treatment in terms of disintegration degree. The optimal dosage of MW and alkali treatments were 38,400 kJ/kg TS and pH 11 to obtain the maximum disintegration degree (DD) of 65%. In addition, they found that volatile fatty acid (VFA) accumulation during AD increased significantly after combining alkali with MW irradiation. The maximal VFAs accumulation was 1,500 mg COD/L with MW treatment of 28,800 kJ/kg TS and fermentation time of 72 h, which was higher than the VFAs yield of alkali treatment without MW radiation (850 mg COD/L) at 96 h. Jang and Ahn (2013) also successfully applied a combination of alkali and MW, which resulted in a 228% increase in methane production during a semi-continuous trial at 5 days SRT (see Table 10.2). This application was relatively new compared to the combination of conventional heat treatment with alkali, which had been reported since 1978 (Stuckey and McCarty, 1978a). However, a performance comparison between a combination of alkali treatment with MW and CH has not been investigated. In addition, compared to CH, MW is harder to apply continuously in full scale. Therefore, further studies are needed to overcome the aforementioned bottlenecks.

TABLE 10.2

Combined Microwave and Alkaline Pre-Treatments of Sludges

Substrate	Microwave Conditions	Alkaline Conditions	Solubilization		Anaerobic Digestion			Scale, Mode (HRT, days)	References
			SCOD/TCOD	DD (%)	CH$_4$	VS Removal	TCOD Removal		
WAS (5.7 g/L)	30 min sequential heating: 100°C (2 min) 148°C (8 min) 171°C (6 min) 168°C (14 min)	pH 10 pH 11 pH 12 pH 12.5	0.005–0.18 0.005–0.27 0.005–0.34 0.005–0.37	–	+18.9% +18.9%		–	LS, B (49)	Doğan and Sanin (2009)
WAS (5.7 g/L)	30 min sequential heating: 100°C (2 min) 148°C (8 min) 171°C (6 min) 168°C (14 min)	pH 12	0.005–0.34	–	+53.2%	+35.4%	30.3%	LS, SC (15)	Doğan and Sanin (2009)
WAS (19 g/L)	135°C (10 min)	0.02 mol/L NaOH	0.03–0.532	–	+102% +155% +157% +228%		–	LS, SC (15) (15) (10) (7) (5)	Jang and Ahn (2013)

Note: LS, laboratory scale. Mode: B, batch; SC, semi-continuous.

11 Physicochemical Treatment

Combination of Alkaline and Ultrasonic Pre-Treatment of Sludge

11.1 INTRODUCTION

The combination of ultrasonication and chemicals aims at enhancing degradation by the action of chemicals and by the generation of hydroxyl radicals from the chemicals by ultrasonication. It was mentioned that ultrasonication can also cause water sonolysis, which produces radicals. Hydroxyl radicals can react with organic scavengers in the sample and degrade them into simpler compounds. However, it should be noted that the radicals' production during sonolysis is pH dependent. Therefore, the sonochemical effect may be different at alkaline (ALK) conditions. Several authors have investigated the combination of ultrasound and alkali as shown in Table 11.1.

Compared to alkali, acids in combination with ultrasonication were less favored in the literature. Liu et al. (2008) found that the acid/ultrasound treatment gave poorer performance in solids and chemical oxygen demand (COD) solubilization compared to ALK/ultrasound treatment. However, the total solids (TS) concentration was too high and only one acid dosage was tested, which may not be representative. Şahinkaya (2015) compared pH values of 2, 4 and 6 and TS concentrations of 0.5%, 1.0% and 2.0% during sulfuric acid/ultrasound treatment. Among the tested conditions, pH 2, a power density of 1 W/mL for 10 min and a TS concentration of 1.0% were found to be the optimum condition for the combined treatment in terms of sludge disintegration. A comparison between alkali and acid with ultrasonication was not conducted, but the results indicated that the combination of ultrasound and acid could be an alternative to ultrasound and ALK treatment. The selection of a combination method should be further considered based on energy input, chemical costs and the erosion of ultrasonic (ULS) probes.

A number of scientists have proposed the Fenton process as a promising option; however, only a small number of studies have reported the improved Fenton process through a combination with other methods, i.e., ULS (Anjum et al., 2016). The Fenton process combined with the ULS pre-treatment of sludge is also advantageous, due its high destruction of microbial flocs and removal of extracellular polymeric substances (EPS). Ning et al. (2014) found that the use of the Fenton-ULS pre-treatment for 10

TABLE 11.1

Combined Ultrasonication and Alkaline Pre-Treatments of Sludge

Substrate	Ultrasonication Conditions	Alkaline Conditions	Solubilization		Anaerobic Digestion			Energy and Cost	Scale, Mode, (HRT)	References
			COD	DD (%)	CH_4	VS Removal	TCOD			
WAS (TS: 1%)	20 kHz, 2.5 min, 1 W/mL	0.05 M NaOH for 30 min	65–2700 mg/L	26.7	+15.9%	37.8%	49.2%	–	LS, B	Şahinkaya and Sevimli (2013b)
WAS (TS: 1%)	20 kHz, 1 min, 1 W/mL	0.05 M NaOH for 30 min	–	–	+33.3%	50.6% (43%)	50.4% (44.4%)	–	LS, SC (14)	Şahinkaya and Sevimli (2013b)
Pulp mill WAS (2.5%) (6.5%)	40 kHz 39,600 kJ/kg TS 16,800 kJ/kg TS	2 h 0.206 g/g TS 0.261 g/g TS	SCOD/TCOD: 6%–19% 3.6%–17.9%	–	–16% –2.5%	30% (21%) 27% (23%)	–	297 kWh/m³ (ULS) +1.75$/m³ for NaOH + 5.95$/m³ for NaOH	LS, B	Park et al. (2012)
WAS (TS: 1%)	20 kHz, 120 W, 14.4 s/mL	40 mM/L NaOH	77.9%	–	–	–	–	–	LS, B	Chiu et al. (1997)
PS+WAS (3:7, volume base)	20 kHz, 7,500 kJ/kg TS	1 h, pH=9	–	36.5%	+55.6%	+17.1%	+20.5%	Remark: Synergistic DD, RSM	LS, C	Kim et al. (2010)
WAS	20 kHz, 1.9 W/mL, 3.8 kJ/g TS	0.01 g/g TS	18.1%	–	+41.8% +31.1% +17.3%	+26.3% +39.3% +26.5%	–	–	LS, SC (10) (15) (25)	Seng et al. (2010)
WAS (TS: 3%)	25 kHz, 0.2 W/mL, 0–15,000 kJ/kgTS	0.04 M	–	30%–42%	–	–	–	–	LS, B	Jin et al. (2009b)

Note: LS: laboratory scale. Mode: batch (B), semi-continuous (SC) and continuous (C); DD; $(SCOD_T - SCOD_O)/(SCOD_{NaOH} \, SCOD_O) \times 100\%$.

min increased the production of hydroxyl radicals (2.90 mol/L) in comparison to the Fenton single treatment. An increase in the production of hydroxyl radicals will increase the oxidation of the sludge components (Gong et al., 2015). Furthermore, the optimum range of ULS density for Fenton hydroxyl radical production was found to be 0.12–0.16 W/mL (Ning et al., 2014). Sludge disintegration under the Fenton-ULS pre-treatment can also increase the release of nutrients, e.g., carbon, nitrogen and phosphorous (Gong et al., 2015), which can play a synergistic effect in the biological treatment of sludge. The combined pre-treatment of sludge using Fenton with other methods is currently gaining high acceptance, due to the use of iron being less costly and non-toxic, whereas hydrogen peroxide can yield a high level of oxidants. Furthermore, it is possible to hinder the issues associated with a single Fenton process (i.e., low acidic requirements and toxic sludge treatment in the final biological treatment) by combining the Fenton process with ultrasonication.

Alkalis are often used to control pH during the anaerobic digestion step. It has been reported as a sludge pre-treatment process since the late 1970s (Stuckey and McCarty, 1978a). Sodium hydroxide (NaOH) was reported to induce higher sludge solubilization followed by potassium hydroxide and calcium hydroxide $(Ca(OH)_2)$ sequentially at the same pH value (Kim et al., 2003). Alkali provides a large quantity of hydroxyl ions and increases sludge pH, which creates a hypotonic environment in which microorganism cells cannot withhold a proper turgor pressure and, as a result, lyze (Neyens et al., 2003a). COD solubilization has also been reported to be the consequence of protein degradation and various reactions such as saponification of uronic acids and acetyl esters, reactions occurring with free carboxylic groups and reactions of neutralization of various acids formed from material degradation. Previous studies quantified ALK pre-treatment differently. The most commonly used terms are NaOH concentration, pH and NaOH dosage. In addition, ALK treatment of organic material has been reported to induce swelling of particulate organics, making the substrate more susceptible to enzymatic attack (Vlyssides and Karlis, 2004). As a result, ALK pre-treatment is also capable of increasing waste activated sludge (WAS) biodegradability. Penaud et al. (1999) used 125 meq/L (0.125 N) NaOH and obtained a 40% increase in biodegradability.

Sole ALK pre-treatment has several disadvantages. The addition of chemicals not only means additional cost but also the potential for more sludge production (Foladori et al., 2010). Moreover, a too high concentration of sodium or potassium ion may inhibit anaerobic digestion (Appels et al., 2008). ALK pre-treatment is preferably coupled with other pre-treatment technologies, such as thermal pre-treatment (Valo et al., 2004b) and ultrasonication (Chiu et al., 1997). A combined pre-treatment often exhibits an obvious improvement for both COD solubilization and biogas production (Valo et al., 2004b; Tanaka et al., 1997). Moreover, the combination has the potential to decrease the alkali dosage while improving the performance of the pre-treatment.

It has been shown that COD solubilization by ultrasound was significantly enhanced at a higher pH (Wang et al., 2005). Kim et al. (2010) have shown the synergistic effect, i.e., the efficiency of alkali/ultrasound was more than the total efficiencies of alkalization and sonication individually. A synergistic disintegration degree (DD) was also observed in the combined treatment. The maximum synergistic DD was 18.9% when 7500 kJ/kg TS ultrasonication was performed at a pH of 12. A continuous anaerobic digestion reactor was used to assess the impact of pre-treatment on methane production. Methane produced from each gram of added COD increased from 81.9 to 127.4 mL (+55.6%) after the combined treatment (pH 9 + 7000 kJ/kg TS). VS reduction and COD removal increased by 17.1% and 20.5%, respectively. Şahinkaya and Sevimli (2013b) also observed the existence of the synergistic effect. The authors found that the DD increased with increasing pH up to pH 12 (using 0.05 N NaOH), beyond which there was little improvement. This was in accordance with Li et al. (2008) who found that the optimum dosage was 0.05 N for 30 min ALK pre-treatment. The optimum sono-alkalization conditions were determined as 1 W/mL for 1 min (ULS) and alkalization at 0.05 N NaOH dosage for 30 min. Under these conditions, the DD was 24.4% and it was found that the hydraulic retention time (HRT) during anaerobic digestion could be halved. The initial rate of methane production was also improved in other studies (Park et al., 2012; Karlsson et al., 2011). However, these authors treated pulp mill WAS and did not observe significant methane improvement over the control despite COD solubilization. This may be attributed to the high lignin content of this type of sludge and the generation of inhibitory compounds with NaOH. Furthermore, the dewaterability decreased after the pre-treatment and this was attributed to ULS (Saha et al., 2011; Park et al., 2012). Several authors have reported that the combined ULS/ alkali pre-treatment was uneconomical at the laboratory scale (Park et al., 2012; Şahinkaya and Sevimli, 2013b), but this may not be true at larger scales. Individual ultrasonication was uneconomical when operated at the laboratory scale but positive energy gain has been observed at pilot and full scales (Hogan et al., 2004; Xie et al., 2007; Pérez-Elvira et al., 2009b).

Chiu et al. (1997) found that the simultaneous combination of ALK and ULS pre-treatment was the most effective combination sequence. The simultaneous ALK treatment (0.04 M) with ultrasonication (14.4 s/mL) induced a higher initial hydrolysis rate and a similar SCOD concentration compared to the sequence when the ALK pre-treatment with the same NaOH concentration was followed by ultrasonication with a longer duration (24 s/mL). This was confirmed by Jin et al. (2009a). The SCOD concentration was 5976, 6408 and 6797 mg/L, with the sequence of ultrasound-alkali, alkali-ultrasound and simultaneous alkali + ultrasound, respectively. Synergistic COD solubilization was observed when alkali and ultrasound pretreatments were combined simultaneously (Kim et al., 2010). However, synergistic COD solubilization was also observed in the least effective sequence, i.e., ultrasound-alkali. This indicates that alkali and ultrasound pre-treatments have positive interactions on each other in sludge disintegration. A simultaneous combination of ultrasound and ALK treatment is therefore recommended to maximize such synergistic effects.

It has also been reported that ultrasound efficiency increases at higher TS levels (Park et al., 2012). However, there is no report on the optimal TS when ultrasound is combined with alkali. The disadvantage of this combination may be the premature pitting of the ultrasonication probes in the presence of strong alkali. This issue has not yet been addressed and more research should be carried out on this aspect. The performance of ultrasonication may decrease over time, especially when the ultrasonication probe is eroded. Although ultrasonication has been applied successfully on a large scale, there are no large-scale applications of ultrasonication combined with another technology. Pilot studies should investigate the combination of ultrasonication with other technologies to confirm the synergy observed in the laboratory.

11.2 PROCESS CONFIGURATION AND OPERATING CONDITIONS

ULS pre-treatment, a mechanical process, has been reported to be an effective sludge pre-treatment method (Tiehm et al., 1997). The collapse of cavitation bubbles during ultrasonication imposes substantial hydromechanical shear force on the particulate matters in sludge. This mechanical force breaks up biological flocs and ruptures microbial cells, resulting in the solubilization of intracellular and extracellular polymers.

COD solubilization during the ultrasonication process could be significantly enhanced with the addition of NaOH (Chiu et al., 1997), as shown in the process diagram in Figure 11.1. Alternatively, only WAS could be treated excluding the primary sludge; only a fraction of WAS could be treated using alkali and ultrasound; or even two different fractions of WAS could be treated in parallel treatment lines before being mixed together. The optimum combination is not known and is sludge dependent. Wang et al. (2005) confirmed higher ULS sludge disintegration was obtained at a higher pH with the multivariable linear regression method. A synergistic increase

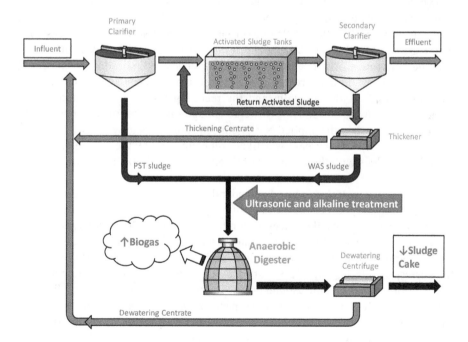

FIGURE 11.1 Proposed flow sheet for the implementation of ultrasonic and alkaline pretreatment of sludge.

in sludge DD was observed when ALK and ULS pre-treatments were applied simultaneously (Kim et al., 2010). Notwithstanding these reports, there is a relative lack of fundamental knowledge on the characteristics of the solubilized compounds following such combined pre-treatment process.

NaOH is often selected for ALK pre-treatment due to its reported higher impact on sludge (Kim et al., 2003). Data on the combination are reported in the following sections. Tests were made with sodium hydroxide pellets dissolved to make a 3 mol/L stock solution. Various NaOH concentrations were achieved by adding different volumes of stock solution to the sludge sample. The applied NaOH concentrations were 0.01, 0.02, 0.05 and 0.1 mol/L, which corresponded with NaOH dosages of 0.025, 0.05, 0.125 and 0.25 g NaOH/g TS, respectively. The sludge samples were then mixed at 200 r/min for 10 min at room temperature (25°C). ALK and ULS pre-treatment was conducted simultaneously (ALK + ULS), as it has been reported to have a higher impact on sludge compared to sequential combinations (Chiu et al., 1997; Jin et al., 2009b). ALK + ULS pre-treatment was performed by sonicating the sludge while it was being mixed at a designated NaOH concentration.

11.3   PERFORMANCE OF INDIVIDUAL PRE-TREATMENTS

The sludge pH increased from 6.5 to 7.8, 9.7, 12.2 and 12.8 after NaOH concentrations of 0.01, 0.02, 0.05 and 0.1 mol/L had been applied, respectively. As shown in Figure 11.2a, the SCOD concentrations were 1300, 1700, 3400 and 4800 mg/L, which corresponded to a DD of 0.8%, 4.0%, 19.6% and 32.8% at NaOH concentrations of 0.01, 0.02, 0.05 and 0.1 mol/L, respectively. The obvious sludge disintegration at a higher NaOH concentration (NaOH > 0.05 mol/L, pH > 12.2) was the consequence of chemical-induced cell lysis. Alkali addition increased the sludge pH value and created a hypertonic environment in microbial cells. Cell membranes could not withstand the resulting turgor pressure and so lost integrity (Neyens et al., 2003a). However, sludge disintegration was not significant (DD < 4%) when the applied NaOH concentration was lower than 0.02 mol/L (pH < 9.7). The SCOD increase was likely supplemented by the de-agglomeration of biological flocs with the increase in pH (Katsiris and Kouzeli-Katsiri, 1987; Hu et al., 2009).

The ULS pre-treatment caused considerable sludge disintegration even at low specific energy input. As shown in Figure 11.2b, 6 kJ/g TS ultrasonication was able to increase the SCOD concentration to 2400 mg/L, which corresponded to a DD of 11.7%. The SCOD concentration increased linearly with the specific energy input. The SCOD concentration reached approximately 4700 mg/L and the corresponding DD was 32.8% after 21 kJ/g TS ultrasonication. Lehne et al. (2001a) pointed out that ultrasound pre-treatment is effective in reducing floc size and causing cell lysis. However, some particulate organics can only be chemically solubilized but are relatively reluctant to mechanical attack. Sludge DD induced by mechanical treatment has been reported to be around 50% (Lehne et al., 2001a). This suggests a combined treatment technique could overcome this barrier.

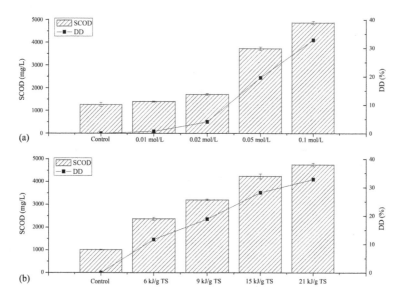

FIGURE 11.2 Change in soluble chemical oxygen demand (SCOD) and disintegration degree (DD) with applied NaOH concentrations (a) and change in SCOD and DD with ultrasound specific energy inputs (b). (Reproduced from Tian et al., 2015b, with permission from Elsevier.)

11.4 SLUDGE SOLUBILIZATION DURING COMBINED PRE-TREATMENT

Solubilized COD caused by combined treatment consists of three parts: solubilized COD caused by NaOH treatment alone, solubilized COD caused by ULS alone and solubilized COD caused by the synergistic effect. In order to investigate the influence of sequence on this combined treatment, different sequences (i.e., ULS-NaOH, NaOH-ULS and simultaneously) were tested with an NaOH dosage of 0.02 M (Figure 11.3). Simultaneous treatment as well as NaOH-ULS treatment solubilized almost the same COD (~4100 mg/L), which was more effective than ULS-NaOH treatment (~3600 mg/L). This indicates that a better performance was achieved when ultrasonication was conducted in the presence of NaOH. Thus, the following experiments were conducted with simultaneous NaOH-ULS treatment.

The change in SCOD concentration during the combined treatment is shown in Figure 11.4a. The SCOD concentration increased with the increase in both the specific energy input and the applied NaOH concentration. These results were in good accordance with a previous study when the ALK+ULS treatment was conducted in the NaOH concentration range of 0–0.04 mol/L and the ULS specific energy input range of 3.75–15 kJ/g TS (Jin et al., 2009b). Wang et al. (2005) also reported that COD solubilization induced by 30 min ULS pre-treatment increased from around 400 to 2000 mg/L when the sludge pH was increased from 6.8 to 12. The maximum SCOD concentration observed in this study was around 11,000 mg/L when

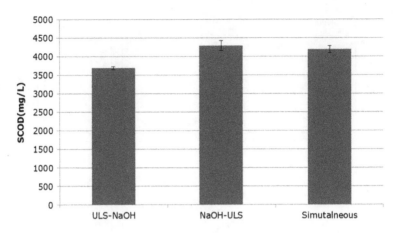

FIGURE 11.3 Effect of sequence during the combined alkali-ultrasound pre-treatment of sludge.

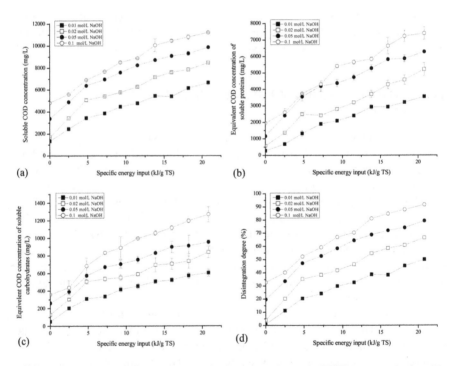

FIGURE 11.4 Change in soluble chemical oxygen demand (SCOD) concentration (a), equivalent chemical oxygen demand (COD) concentration of soluble proteins (b), equivalent COD concentration of soluble carbohydrates (c) and disintegration degree (d) with specific energy input at different applied NaOH concentrations. (Reproduced from Tian et al., 2015b, with permission from Elsevier.)

21 kJ/g TS ULS pre-treatment was combined with 0.1 mol/L ALK pre-treatment. Such a SCOD increase was the result of intracellular and extracellular organics solubilization.

The concentrations of soluble proteins and carbohydrates increased significantly in the combined treatment, as shown in Figure 11.4b and c. Proteins, the principle components of cells, were solubilized as a result of cell lysis (Wang et al., 2006b; Kim et al., 2010). The solubilization of polysaccharides in cell walls and EPS contributed to the increase in the soluble carbohydrates concentration (Wang et al., 2006b). Significant solubilization of proteins due to ALK + ULS pre-treatment was observed by Liu et al. (2008). In their study, around 67% of proteins in a WAS sample was solubilized when ULS treatment was conducted at pH 12 for 60 min. The DD increased to 50%, 67%, 80% and 91% when 21 kJ/g TS ultrasonication was conducted at NaOH concentrations of 0.01, 0.02, 0.05 and 0.1 mol/L, respectively, as shown in Figure 11.4d. Sludge disintegration of 50% was achieved when 21 kJ/g TS ULS pre-treatment was combined with 0.01 mol/L ALK pre-treatment. This was significantly higher than the DD induced by an individual 21 kJ/g TS ULS pre-treatment (DD: 32.8%). Considering that the DD caused by 0.01 mol/L ALK pre-treatment was only 0.8%, the addition of NaOH to the ultrasonication process obviously induced synergistic disintegration (i.e., synergistic COD solubilization). A similar observation was reported in a previous study (Kim et al., 2010), where 58.9% DD was obtained when a ULS treatment (7.5 kJ/g TS) was combined with ALK treatment (pH 12); whereas the DD induced by the ULS and the ALK treatment alone was only 19.5% and 21.4%, respectively. A possible reason for such synergistic disintegration was that the NaOH made the sludge structure more vulnerable to the mechanical disruption caused by ultrasound (Jin et al., 2009b; Kim et al., 2010). However, under the aforementioned combination condition, the damage to the sludge caused by the ALK pre-treatment was limited (DD: 0.8%). Therefore, apart from enhancing the mechanical disintegration due to ultrasound, it is possible that alkali addition also enhanced the sonochemical effects of the cavitation effect through the formation of radicals. Analysis of the solubilized organics was carried out to shed more light on the possible synergistic mechanisms.

The synergistic effects at different NaOH dosages and ultrasonication times are shown in Figure 11.5. It is interesting to note that the synergistic effect increased quite rapidly in the first 2 min, but no obvious change was observed after 5 min. This indicates that the addition of NaOH accelerates the subsequent ultrasonication in the first minutes, but the synergistic effect becomes negligible after 5 min.

This synergistic effect is also NaOH concentration dependent. The synergistic COD solubilization caused by 0.02, 0.05 and 0.1 M is similar (~3500 mg/L SCOD), but the synergistic effect caused by 0.01 M NaOH (~2000 mg/L SCOD) was lower than the other three dosages. This shows that 0.02 M NaOH is enough to cause the optimum synergistic effect, and a further increase in NaOH concentration does not increase synergistic COD solubilization. For an NaOH concentration lower than 0.02 M, less synergistic effect is produced when less NaOH is added. Despite this, a synergistic SCOD increase caused by 0.01 M NaOH and ultrasonication still approached 2000 mg/L, which is quite considerable.

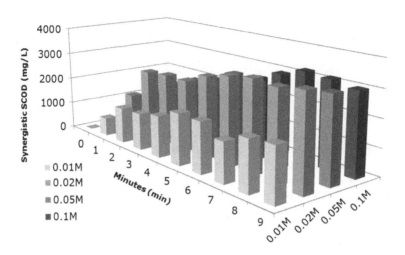

FIGURE 11.5 3-D demonstration of synergistic effect at different combinations of ultrasonication time and NaOH dosages.

It is suggested that the presence of NaOH promotes the chemical effect of ultrasound (i.e., hydroxyl radicals generation) and causes more COD solubilization. In other words, it has a "pseudo-catalytic" effect for hydroxyl radicals generation during ultrasonication. During the first 2 min of ALK addition, because of the sufficient amount of alkali, the pseudo-catalytic effect was obvious and led to a significant synergistic COD release. However, as the chemical reaction proceeded between NaOH and sludge, the synergistic effect became insignificant due to lack of "catalyst." This is because the radicals generation was a result of cavitation bubbles formation (Hua and Hoffmann, 1997). When the alkali concentration is too low, there is insufficient alkali in the vicinity of the cavitation bubbles to promote radicals generation.

Based on this theory, the conventional ultrasound/ALK treatment can be modified to a combination of stepwise ALK addition and ultrasound treatment to make good use of the synergistic effect. As in this case, the minimum NaOH dosage needed to reach the best synergistic effect was 0.02 M, and the NaOH amount of each addition was set to be 0.02 M. The second batch of ALK was added after 5 min of ultrasonication when the synergistic COD solubilization of the first batch stopped increasing. The optimized results are shown in Figure 11.6.

It is clear from Figure 11.6 that this two-step addition of ALK/ultrasound treatment was more effective than the conventional ALK/ultrasound treatment. The solubilized COD for this treatment was slightly higher than the conventional one, whereas the consumed NaOH was only 80% of the conventional treatment. The total suspended solids (TSS) and volatile suspended solids (VSS) reduction are 23.2% and 22.6%, respectively, which are similar to the conventional combined pre-treatment. These results confirm that the second addition of ALK is able to produce a better or similar performance with 20% less NaOH addition. The explanation behind this is believed to be that the pseudo-catalytic effect recurred due to the re-addition of catalyst.

The influence of power density on this treatment process was also investigated and the results are shown in Figure 11.7. A high power density produces more cavitation

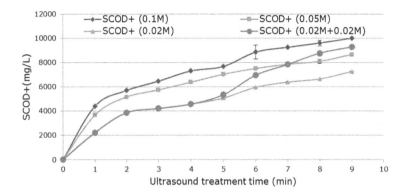

FIGURE 11.6 Comparison of conventional alkaline treatment and novel two-stage alkaline treatment in combination with ultrasound. SCOD+ represents the increase in SCOD compared to the SCOD of raw sludge.

bubbles in a specific volume of sludge (Jin et al., 2009). When no ALK was added, the performance of different power densities did not show obvious difference for the same specific energy input. Similar results were observed for an NaOH concentration of 0.02 M. When the NaOH dosage was low, the amount of NaOH in the specific volume of sludge was not sufficient to produce more radicals. When the NaOH dosage was high, more cavitation bubbles were accompanied with sufficient NaOH in the specific volume of sludge. Then, enough catalyst is available for the radicals generation, which leads to more COD solubilization.

11.5 MOLECULAR WEIGHT DISTRIBUTION OF SOLUBILIZED ORGANICS

The molecular weight (MW) of solubilized organics is important as it may impact on the downstream biological process (Eskicioglu et al., 2006c). Size exclusion

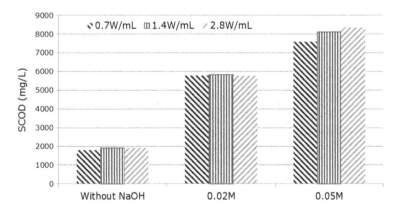

FIGURE 11.7 Influence of power density on combined alkaline/ultrasound treatment.

chromatography (SEC) has been used to measure the MW of organic substances in water samples (Her et al., 2003; Aquino et al., 2006b; Trzcinski et al., 2011). However, such an approach has not yet been reported for the assessment of the change in soluble organics due to a pre-treatment step. Therefore, an SEC measurement was conducted in this work to characterize the MW of solubilized substances after pre-treatment. Afterward, the soluble organics were fractionated into different MW ranges with ultrafiltration (UF) membranes. The COD of each MW fraction was determined to complement the SEC results. The MW distribution chromatograms of control, ALK pre-treated (0.05 mol/L), ULS pre-treated (21 kJ/g TS) and ALK + ULS pre-treated sludge are shown in Figure 11.8a. The retention time of the standard polymers are shown for comparison. High MW compounds eluted earlier and had a shorter retention time because they did not go as deep into the gel pores as low MW compounds (Trzcinski et al., 2011; Aquino et al., 2006b). As shown in Figure 11.8a, compounds with a retention time shorter than 6 min were solubilized in all the pre-treated samples. These compounds had an MW higher than 500 kDa as they eluted earlier than the largest standard polymer (MW: 500 kDa). Compounds with a retention time longer than 8 min were also solubilized (the earliest peak was from the ULS pre-treated sludge, but it was not obviously noted due to the large-scale difference compared to the ALK + ULS pre-treated sludge). The corresponding MW of these compounds was <27 kDa according to the calibration. A large UV response at a retention time around 9 min (MW: 5.6 kDa) was observed in the ALK + ULS pre-treated sample. However, such a peak was not observed in either of the individually applied ALK or ULS pre-treated sludge. Therefore, the solubilization of these substances was likely due to the synergistic effects between the ULS and ALK pre-treatments. Using SEC to characterize the solubilization products after sludge pre-treatment has not been reported previously and could provide insights into the MW distribution change due to the pre-treatment processes.

The supernatant was fractionated with UF membranes based on their MW range as determined by SEC. Apparent MW distribution in terms of total organic carbon (TOC; %, w/w) for different MW ranges was then measured to obtain quantitative results, as shown in Figure 11.8b. The TOC results showed good agreement with the SEC results, as presented in Figure 11.8a. The TOC fraction of components with an MW larger than 300 kDa increased after the pre-treatment process, indicating the solubilization of the macromolecules. This TOC fraction was only 7.8% in the control sludge and had increased to 60%, 16% and 42.3% after ALK, ULS and ALK + ULS pre-treatments, respectively. A review of the literature would suggest that UF fractionation on the supernatant of ALK + ULS-treated sludge had not been reported previously. However, Eskicioglu et al. (2006c) did report that a COD fraction with an MW over 300 kDa increased from 16.9% to 29.5% and 24.7% after conventional thermal and microwave thermal treatment, respectively. It should be noted that the COD is a general indicator that is influenced not only by organic but also inorganic interferences such as ammonium. Therefore, the results of this work confirmed that the increase had been due to organics instead of inorganics.

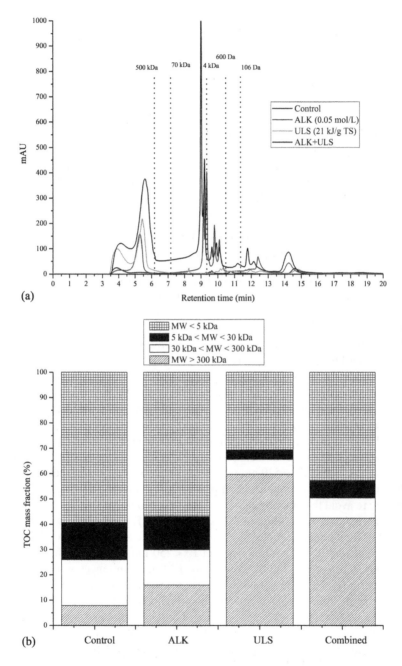

FIGURE 11.8 (a) Molecular weight distribution chromatograms of control, alkaline (ALK) and ultrasonic (ULS) and ALK+ULS pre-treated sludge in UV 254 nm signal; (b) total organic carbon (TOC) mass fraction of each molecular weight range in control and pre-treated sludge. (Reproduced from Tian et al., 2015b, with permission from Elsevier.)

In addition, it was observed that the effects of ALK treatment were different when it was applied on its own and applied together with the ULS treatment. Individual ALK treatment increased the TOC fraction of organics with an MW larger than 300 kDa from 7.8% (control) to 16% (ALK); while the ALK treatment decreased the corresponding fraction from 60% (ULS) to 42.3% (ALK + ULS) when it was combined with the ULS treatment. This was likely because the macromolecules solubilized by the ULS pre-treatment could be chemically degraded by the hydroxyl ions during the ALK + ULS treatment (Şahinkaya and Sevimli, 2013b). Examples of such degradation would be the saponification reactions of lipids and the ALK hydrolysis of proteins. Meanwhile, simpler organics were formed as degradation products. This accounted for the compounds observed at a retention time of 9 min (MW: 5.6 kDa) in the supernatant of the ALK + ULS-treated sample, as shown in Figure 11.8a.

11.6   FLUORESCENT PRODUCTS CHARACTERIZATION OF SOLUBILIZED ORGANICS

The excitation emission matrix (EEM) was an emerging technique used to characterize the solubilization products after the sludge pre-treatment process (Luo et al., 2013; Yang et al., 2013b). However, relevant research on its application to characterize the solubilization products following ALK + ULS pre-treatment has not been reported. Therefore, an EEM fluorescence spectroscopy analysis was conducted in this study to fill the information gap.

EEM fluorescence spectroscopy is a sensitive and selective technique, which is ideal for complex environmental samples (Luo et al., 2013). It detects target substances based on their excitation (Ex) and emission (Em) wavelengths without destroying samples (Luo et al., 2013). A fluorescence spectrometer (LS 55, Perkin Elmer, USA) was used to measure the fluorescence intensity (FI) of the soluble fluorescent products. The measurement procedure was previously described by Wu et al. (2011). Ex was from 230 to 520 nm with 5-nm intervals. Em was collected from 230 to 550 nm with 5-nm increments. Samples were pre-diluted 100 times with deionized water to avoid the measured FI exceeding the maximum level.

The compounds were identified based on their Ex and Em wavelengths, as reported in the literature. Peaks of simple protein-like substances such as tyrosine proteins appeared in the Ex/Em range of Ex < 250 nm, Em < 350 nm (Determann et al., 1994; Chen et al., 2003). Fulvic acid (FA)-like substances were detected in the Ex/Em range of Ex < 250 nm, Em > 380 nm (Determann et al., 1994; Chen et al., 2003; Her et al., 2003). Peaks of soluble microbial products or SMP-like substances fell in the Ex/Em range of Ex: 250–280 nm, Em < 380 nm (Determann et al., 1994; Coble, 1996; Baker, 2001; Chen et al., 2003). Humic acids or HA-like substances were detected in the Ex/Em range of Ex > 250 nm, Em > 380 nm (Determann et al., 1994; Coble, 1996; Mobed et al., 1996; Her et al., 2003).

The FIs of SMP and HA substances were measured to provide novel insights into the solubilization products after ALK + ULS pre-treatment. Fluorescent products were characterized with EEM fluorescence spectroscopy to provide more insights into the solubilized compounds. As shown in Figure 11.9, each spectrum was divided into I, II, III, IV and V, as summarized by Chen et al. (2003). According to the Ex/EM range, peaks in Regions I and II represent simple protein-like substances and peaks in Region III represent FA-like substances. However, no obvious peaks were observed in Regions I, II and III in either the control or the pre-treated sludge samples, as shown in Figure 11.9. It was possible that some of the protein-like and FA-like peaks were "over-shadowed" by other peaks or some simple proteins were aggregated with polysaccharides and detected in Region IV as SMP-like products (Ng and Ng, 2010).

The peaks of SMP-like matters (in Region IV) were observed in all the spectra, as shown in Figure 11.9. The FI of the SMP-like peak was found to increase after various pre-treatment processes. The FI values of the observed SMP-like peaks in both the control and the pre-treated sludge samples are shown in Table 11.2. The FI of the SMP-like peak increased only slightly from 463 to 531 after 0.05 mol/L ALK pre-treatment. The solubilization of the SMP-like substances was more significant after 21 kJ/g TS ULS pre-treatment and the FI of the corresponding peak increased from 463 to 768. The addition of NaOH to the ultrasonication process further enhanced the SMP-like substances solubilization. The FI of the observed SMP-like peak in ALK + ULS (0.05 mol/L + 21 kJ/g TS) pre-treated sludge was 839, which was 9.2% higher than that in the ULS pre-treated sludge (FI: 768). These results confirmed the positive interactions between the ALK and ULS pre-treatments in terms of solubilizing microbial products and this correlated well with the high concentration of soluble proteins and carbohydrates, as shown in Figure 11.4b and c.

The peaks of HA-like substances in Region V were also observed in all the spectra. The FI of the HA-like peak was found to increase after various pre-treatments; the FIs of the observed peaks are listed in Table 11.2. Individual ALK or ULS pre-treatment only slightly increased the FI of the HA-like peaks. The FI increased from 43 to 91 and 80 after 0.05 mol/L ALK pre-treatment and 21 kJ/g TS ULS pre-treatment, respectively. Interestingly, the FI of the HA-like peak was significantly high in the ALK + ULS pre-treated sludge, as highlighted by a red ellipse in Figure 11.9d. The FI of the highlighted peak was 201, as shown in Table 11.2.

TABLE 11.2

Fluorescence Intensity of the Observed Peaks in Control and Pre-Treated Sample

Sample	SMP-Like (Ex/Em)	FI	HA-like (Ex/Em)	FI
Control	280/375	463	350/400	43
ALK	280/370	531	350/450	91
ULS	280/370	768	370/425	80
Combined	290/360	839	350/440	201

Source: Reproduced from Tian et al., 2015b, with permission from Elsevier.

Note: SMP, soluble microbial products; HA, humic acid; FI, fluorescence intensity.

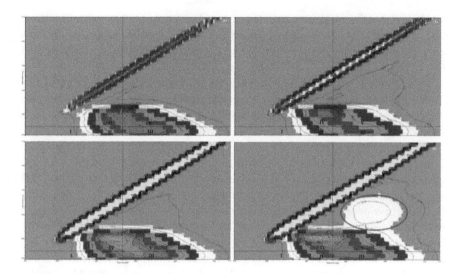

FIGURE 11.9   EEM spectra of (a) control sludge; (b) ALK pre-treated sludge (0.05 mol/L); (c) ULS pre-treated sludge (21 kJ/g TS); (d) ALK+ULS pre-treated sludge (0.05 mol/L+21 kJ/g TS). (Reproduced from Tian et al., 2015b, with permission from Elsevier.)

This was higher than that in the ALK pre-treated sludge (FI: 91), the ULS pre-treated sludge (FI: 80) and even their sum (i.e., 171). This indicated that the ALK and ULS pre-treatments had synergistic effects on the solubilization of HA-like substances, which has not been reported previously. The synergistic mechanism was proposed to be as follows. HA in sludge is normally adsorbed onto activated biomass (Esparza-Soto and Westerhoff, 2003). HA is known to be soluble in basic conditions but relatively insoluble in neutral and acidic conditions (Stevenson, 1994). When only the ALK pre-treatment was applied, most of the HAs remained attached to the biological flocs due to lack of mechanical disruption. Ultrasound disrupted the sludge matrix and mechanically set free the HAs trapped in the biological flocs. However, these HAs remained insoluble due to a neutral pH. Thus, the corresponding FI increases in both individual ALK and individual ULS pre-treatments were not obvious. When ultrasonication was performed under a basic condition, HA was mechanically set free by ultrasound and could also be solubilized in the ALK solution (pH: 12.2). The solubilized HAs then contributed to the synergistic COD solubilization, as mentioned previously. This synergistic solubilization of HA-like substances may have also influenced the subsequent anaerobic digestion.

11.7   ANAEROBIC BIODEGRADABILITY OF PRE-TREATED SLUDGE

Anaerobic digestion tests were also conducted to investigate the influence of the solubilized substances on subsequent anaerobic digestion. A BMP assay was used to test the change in sludge anaerobic biodegradability after various pre-treatment processes. As shown in Figure 11.10, the anaerobic biodegradability of all the pre-treated

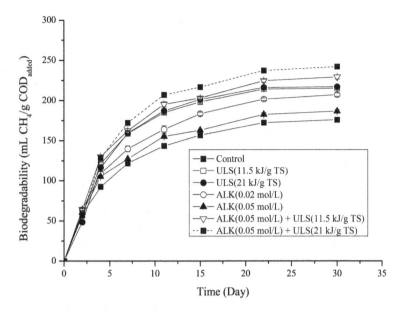

FIGURE 11.10 Biochemical methane potential results of control, ALK, ULS and ALK + ULS pre-treated sludge. (Reproduced from Tian et al., 2015b, with permission from Elsevier.)

samples was higher than the control during the first 4 days of the BMP assay, because the pre-treatment step solubilized particulate organics and so accelerated the anaerobic digestion. The sludge anaerobic biodegradability only increased by 17.8% and 5.7% after 0.02 and 0.05 mol/L ALK pre-treatment had been applied, respectively. However, the DD obtained by 0.02 and 0.05 mol/L ALK pre-treatment was 4.0% and 19.6%. This meant that the biodegradability increase (BI) after individual ALK pre-treatment was not necessarily related to the applied NaOH concentration or DD. Therefore, a further increase in sludge anaerobic biodegradability via higher NaOH dosage was not an option. Anaerobic biodegradability was also improved after ULS pre-treatment. The anaerobic biodegradability increased from 175.8 to 202.7 mL CH_4/g COD_{added} (+15.3%) after 11.5 kJ/g TS ultrasonication. However, the anaerobic biodegradability only increased to 212.2 mL CH_4/g COD_{added} (+20.7%) when the specific energy input was doubled to 21 kJ/g TS. This suggested that an increase in specific energy input could further improve the sludge anaerobic biodegradability but was relatively inefficient after a certain energy threshold.

The aforementioned limitations in individual pre-treatments were not observed after the combined pre-treatment. The ultimate sludge anaerobic biodegradability increased from 175.8 to 229.4 (+30.5%) and 242.3 (+37.8%) mLCH$_4$/g COD_{added} by combining 11.5 and 21 kJ/g TS ULS pre-treatment to 0.05 mol/L ALK pre-treatment, respectively. This showed that using NaOH to enhance ULS pre-treatment was an alternative to further improve sludge biodegradability rather than keep increasing the NaOH dosage or ULS energy. The BI in this work was slightly lower compared to the previous results from Kim et al. (2010). One possible reason is that the NaOH concentration in this work was higher compared to that in Kim et al., and the inhibition effect of sodium ions might have decreased methane production (Feijoo et al., 1995). It is also possible that the longer digestion time in this work allowed slowly degradable compounds in untreated sludge to be digested; the relative increase in the pre-treated sludge was therefore less. The results of this study were from batch reactors with a digestion time of 30 days, whereas the results of Kim et al. (2010) were based on a continuous reactor with a solids retention time of 20 days. This was confirmed by the results of Seng et al. (2010) where the methane production of ALK + ULS pre-treated sludge was 17.3%, 31.1% and 42.1% higher than the untreated sludge at a solids retention time of 25, 15 and 10 days, respectively.

Nevertheless, the contribution of synergistic effects to the anaerobic biodegradability increment was also observed. For example, the BI after ALK + ULS (0.05 mol/L + 21 kJ/g TS) pre-treatment was 37.8%, whereas the BI was only 5.7% and 20.7% after ALK and ULS pre-treatments had been applied individually under the same conditions, respectively. Obviously, the BI induced by the ALK + ULS pre-treatment (i.e., 37.8%) was significantly higher than the BI induced by individual ALK pre-treatment (i.e., 5.7%), individual ULS pre-treatment (20.7%) or the numerical summation of both terms (i.e., 26.4%). Such biodegradability improvement due to the ALK + ULS pre-treatment has not been emphasized in previous works and could be related to synergistic sludge disintegration. The solubilization of SMP-like products such as proteins and carbohydrates was significantly enhanced due to the combined pre-treatment. The solubilization of those compounds effectively accelerated the hydrolysis step, thereby enhancing methane production (Wang et al., 1999). In addition, the combined pre-treatment generated smaller organics due to the interactions between ALK and ULS treatment. This was also beneficial for the subsequent anaerobic digestion. Additionally, the synergistic solubilization of HA-like compounds might also contribute to the synergistic biodegradability improvement. HAs are generally considered recalcitrant in anaerobic digestion. However, Ho and Ho (2012) found that low concentrations of HAs (<5 g/L) could enhance methane production by serving as electron acceptors of the fatty acids degradation. Therefore, further investigation of synergistically solubilized HAs may be helpful to the mechanism investigation.

11.8   CONCLUSIONS

The SCOD concentration increased from 1,200 to 11,000 mg/L due to the ALK + ULS treatment. During the ALK + ULS pre-treatment, the hydroxyl ions were found to further degrade the macromolecules solubilized by the ULS treatment. This synergistic action generated smaller organics with an MW of around 5.6 kDa. The addition of NaOH to the ultrasonication process could enhance the solubilization of the SMP-like substances. HA was also found to be released due to the synergistic actions between alkali and ultrasound.

Individual ALK and individual ULS pre-treatments have their respective limitations in improving sludge anaerobic degradability. However, such limitation was reduced in ALK + ULS pre-treated sludge. The sludge anaerobic biodegradability increased by 30.5% and 37.8% when 0.05 mol/L ALK pre-treatment was combined with 11.5 and 21 kJ/g TS ULS pre-treatment, respectively. Therefore, ULS and ALK pre-treatment should be combined rather than individually applied for the benefit from the synergistic effects.

12 Physicochemical Treatment

Combination of Ozone and Ultrasonic Pre-Treatment of Sludge

12.1 INTRODUCTION

Sludge produced in wastewater treatment plants needs to be stabilized before it can be safely disposed. This is because of its high organic as well as pathogen content (Bitton, 2005). Anaerobic digestion is generally accepted as an appropriate way to stabilize sludge as it reduces the final amount of solids requiring disposal as well as producing methane for energy recovery (Zhang et al., 2007a). However, anaerobic digestion is a slow process and its performance is typically limited by hydrolysis of the particulate organic matters in the sludge (Pavlostathis and Giraldo-Gomez, 1991). Therefore, sludge can be pre-treated to accelerate the hydrolysis step and to enhance the overall anaerobic process before it is fed into an anaerobic digester.

Sludge pre-treatment technologies can be categorized into mechanical, thermal and chemical treatments. The aim of sludge pre-treatment is to solubilize the particulate organics, thereby making them more accessible for subsequent microbial action (Tiehm et al., 1997). Conventional pre-treatments such as thermal and alkaline treatments have been reported since the late 1970s (Stuckey and McCarty, 1978a, 3481984). Compared to these conventional pre-treatments, ultrasound and ozone pre-treatments are relatively newer technologies and have only been widely reported in the last decade. Ultrasonic sludge disintegration had been preferred at lower ultrasonic frequency (Tiehm et al., 2001). The predominant sludge disintegration mechanism in low frequency ultrasonication (ULS), a mechanical method, is of the hydro mechanical shear force caused by the collapse of cavitation bubbles (Wang et al., 2005). Ultrasound readily reduces biological floc sizes, rupturing microorganism cells as well as significantly increasing sludge biodegradability in the subsequent anaerobic digestion (Tiehm et al., 1997; Lehne et al., 2001b; Zhang et al., 2007a).

Ozonation has also been reported as an effective sludge pre-treatment technology but with a different disintegration mechanism. Ozone chemically reacts with sludge and destroys microorganism cell components (Chu et al., 2009; Yan et al., 2009). Ozone also attacks the extracellular polymeric substances (EPS) and breaks down the complex macromolecules in the soluble phase (Yan et al., 2009). In addition, ozone is able to convert refractory organic matters into biodegradable form (Volk et al., 1993; Nishijima et al., 2003). Most importantly, sludge biodegradability is reported to be remarkably improved after ozonation (Weemaes et al., 2000; Goel et al., 2003b).

Ozone is a strong oxidant, which is effective at hydrolyzing sludge as well as increasing sludge biodegradability (Goel et al., 2004). Ozone attacks the EPS of waste activated sludge (WAS) by degrading its macromolecules into smaller organics (Erden et al., 2010). Thereafter, ozone penetrates cells and causes cell lysis. Extracellular and intracellular substances can be released into the bulk for better anaerobic digestion. Besides these merits, ozone can also effectively remove resistant odors and pathogens.

Sludge ozonation is a combined chemical effect of direct and indirect ozonation. Ozone can directly react with the dissolved/particulate substances or form a secondary oxidant like hydroxyl radicals to oxidize the target substances (Staehelin and Hoigne, 1985). Scheminski et al. (2000) suggested that the direct ozonation rate is lower and reactant dependent, while the indirect hydroxyl radicals reaction is not reactant dependent. Yan et al. (2009) found that hydroxyl radical concentrations decreased with increasing ozonation time and suggested that the indirect hydroxyl oxidation was suppressed with increasing ozonation time.

Cesbron et al. (2003b) investigated the competition of particulate matter and soluble matter to ozone access and showed the importance of mass transfer. At the beginning, ozone reacts simultaneously with soluble organics and small particles. After easily oxidized reactants are used up, the oxidation of bigger particulates is possible (Cesbron et al., 2003b). They also pointed out that a screen effect will be caused on particulate matter when the reaction between the soluble fraction and ozone is high. Zhang et al. (2009) supported this conclusion by showing ozone's better performance in cell lysis than sludge matrix solubilization. Ozone could penetrate through cell walls, damage the cell membrane structure, leading to cell lysis and the release of intracellular substances. However, this mechanism is favored at higher ozone dosages because low ozone dosages may not provide enough driving force for this penetration (Zhang et al., 2009).

Ozone has been used since 1970 to reduce chemical oxygen demand (COD) in the final effluent and to degrade WAS (Moerman et al., 1994). It was reported to reduce excess sludge production and that ozonation could be toxic to non-acclimatized sludge (Sakai et al., 1997). Ozone decomposes itself into radicals and reacts with the soluble and particulate fractions, organic and mineral fractions, leaving no toxic by-products in contrast with chlorine. However, if the sludge contains bromide, then carcinogenic bromated compounds can be formed (Kurokawa et al., 1990).

Ozone is a faster process and produces less extra sludge when compared to other chemical pre-treatments. However, excess ozonation will cause COD mineralization, which converts organic carbon into inorganic carbon dioxide (Ahn et al., 2002; Lee et al., 2005). Optimum ozone dosage has been reported in the range 0.05–0.5 g O_3/g total solids (TS), but there is a phenomenon of mineralization at higher dosage. Yeom et al. (2002) showed that the rate of mineralization was 5.1%, 20.1%, 29.2% and 49.2% at ozone dosages of 0.1, 0.5, 1 and 2 g O_3/g SS. The loss of organic carbon will also decrease methane production to some extent. Weemaes et al. (2000) reported that ozone pre-treatment increased WAS biodegradability. Ozone dosages of 0.05 and 0.1 g O_3/g COD caused a 50% and 70% increase in biodegradability, respectively. However, for an ozone dosage of 0.2 g O_3/g COD, a 3-day lag phase was observed and only 30% biodegradability increase (BI) was noticed. The other drawback of ozone is its high production cost (typically 225 kWh/m^3 WAS at a dosage of 0.06 g O_3/g TS), which makes it less efficient than ULS on the same energy basis. This limits the popularization of ozone application. The improvement in solid degradation efficiencies observed after ozone pre-treatment (up to 65% solids degradation at 0.05 g O_3/g TS) (Goel et al., 2003b) was significantly higher than improvements reported for other pre-treatment options like mechanical disintegration (Kopp et al., 1997), ultrasound treatment (Tiehm et al., 1997) and mechanical jet smashing (Choi et al., 1997). A comparison with other technologies is difficult because it depends on the sludge source, energy inputs during pre-treatment and other operational parameters. The level of improvement, however, seems to be better than thermal/thermochemical treatment at low pre-treatment temperatures and comparable at higher temperatures (>100°C). Due to the high solids reduction, the use of ozone may be justified when the cost of sludge disposal is particularly high.

The combination with ULS has the potential to reduce the ozone dosage requirement and therefore operational costs. A simultaneous combination of both treatments is the most commonly reported sequence (Xu et al., 2010a; Yang et al., 2012, 2013b). However, a comparison of a combination sequence has seldom been investigated. Simultaneous ULS and ozone was shown to be more efficient in terms of COD solubilization than ozone followed by ULS (Xu et al., 2010b). As mentioned previously, ozone can induce high solids reduction, and dosing ozone prior to the ULS process would leave fewer solids as nuclei for cavitation bubbles formation and fewer solids to be mechanically disrupted by ultrasound. However, the simultaneous combination of ozonation and ULS may lead to premature pitting of the sonotrode. The application of ULS prior to ozonation may be an alternative sequence for the combination. However, there is a lack of data on the effect of sequence.

Xu et al. (2010a) first combined ULS and ozone as a sludge pre-treatment process. They found that COD solubilization and methane production were higher compared to that when ULS and ozone were individually performed. However, the synergistic COD solubilization was not obviously observed. The final soluble COD (SCOD) in ULS-treated, ozone-treated and combined-treated sludge was 797, 2347 and 3341 mg/L, respectively. They found that the COD solubilization percentage decreased as the sludge TS increased from 10 to 38 g/L. In addition, a long treatment time induced a negative effect in decreasing sludge biodegradability. Simultaneous ULS and ozone for 120 min induced a BI of around 75%, which was lower than 93% and 106% after 30 and 45 min treatment, respectively. Yang et al. (2012) suggested that ozone provides gas nuclei for the cavitation effect of ULS; meanwhile, ULS enhances the decomposition of ozone into hydroxyl ions. However, no evidence was given on these possible synergisms. Yang et al. (2013b) observed that synergistic solids solubilization could be achieved by the simultaneous combination of ULS and ozone. Additionally, they also found that the solubilization of proteins and amino acids was enhanced in combined ULS and ozone pre-treatment by employing an emission excitation matrix to categorize the solubilization productions. Yang et al. (2012) used response surface methodology to determine the optimal conditions: ozone dose of

0.158 g O_3/g dry solids (DS) and ultrasound energy density of 1.423 W/mL, but did not analyze the methane potential. However, the disintegration degrees (DD) were limited to 19.6% and 46.1% in individual ozone and ultrasound pre-treatments, while it reached up to 60.9% in combined pre-treatment.

Braguglia et al. (2012b) compared the ozonation and ULS pre-treatment from an economic view. They indicated that for the same low energy input, ozonation was less effective as ULS. Ultrasound pre-treatment (~2500 kJ/kg TS) led to a 19% volatile solids (VS) reduction and a 25% biogas production increase, whereas ozone pre-treatment (0.05 g O_3/g TS equivalent to 2000 kJ/kg TS) only led to a 7% VS reduction but no noticeable biogas production increase. When the ozone dosage was increased to 0.07 g O_3/g TS, VS reduction and biogas production increased rapidly by 27% and 17%, respectively. The literature data are still scarce on this combined pre-treatment and more research should be carried out on the toxicity, kinetic and reactor design of the pre-treatment. It seems, however, that ozone becomes useful in the synergy only after a certain dosage has been applied (Table 12.1).

12.2   PROCESS CONFIGURATION AND OPERATING CONDITIONS

Ultrasound and ozone act differently on sludge (Bougrier et al., 2006). Comparisons between sludge ULS and ozonation have been conducted by previous researchers (Bougrier et al., 2006; Braguglia et al., 2012b). Bougrier et al. (2006) found that ULS and ozonation resulted in different physical and biochemical characteristics in the same sludge. Due to the supplementary effects of these two pre-treatments, the combination of ULS and ozonation has been suggested for a larger impact (Xu et al., 2010a; Yang et al., 2012, 2013b). However, information in the published literature on such a combination is relatively scarce and the synergistic mechanisms between the two pre-treatments are still ambiguous. Eskicioglu et al. (2006b) have indicated that the molecular weight (MW) of the solubilized substances is an important indicator of the sludge pre-treatment performance. However, the size of the solubilized substances by ultrasound, ozone and their combination has not been reported previously. The optimum combination sequence was determined and changes in the sludge characteristics after individual and combined pre-treatments were measured. In addition to the conventional parameters, such as SCOD for the measurement of sludge solubilization, size exclusion chromatography measurement can be conducted to determine the change in MW distribution after each pre-treatment. Batch anaerobic digestion tests are then conducted to evaluate the influence of the pre-treatment on sludge anaerobic biodegradability and methane production kinetics.

ULS has been reported to be an effective sludge pre-treatment (i.e., treatment of pre-digestion feed sludge) technology (Tiehm et al., 1997, 3682001). Biological flocs in the sludge matrix would be mechanically disrupted, resulting in particle size reduction and the solubilization of extra/intracellular polymeric substances (Bougrier et al., 2005; Wang et al., 2006b). Consequently, methane production and solids removal efficiency during the subsequent sludge anaerobic digestion are improved (Tiehm et al., 1997, 3682001). Despite its advantages, ULS pre-treatment has limitations because it is essentially "single" effect – mechanical disintegration (Lehne et al., 2001b; Khanal et al., 2007). The enhancement of the effectiveness of

TABLE 12.1
Combined Ultrasonication and Ozone Pre-Treatment of Sludges

Substrate	Ultrasonication Conditions	Ozone Conditions	Solubilization		Anaerobic Digestion	Scale, Mode	References
			COD	DD (%)	CH₄		
WAS (TS: 19 g/L)	21 kHz, 0.26 W/mL, 5 min 30 min 45 min 120 min	0.6 g/h 5 min 30 min 45 min 120 min	5 to 60 min: 900–3300 mg/L	—	+25% +93% +106% +75%	LS, B	Xu et al. (2010a)
WAS (TS: 17.3 g/L)	20 kHz, 1.423 W/mL, 60 min	0.158 g O₃/g TS	—	60.9%	—	LS, B	Yang et al. (2012)
WAS	20 kHz, 0.3–1.5 W/mL, 60 min	0.05–0.15 g O₃/g TS	38.6% VSS red (ULS: 21.6%, Ozone: 10.9%)	—	—	LS, B	Yang et al. (2013b)

Note: LS, laboratory scale; B, batch.

ULS pre-treatment had been attempted by combining the ULS process with chemical pre-treatment methods. The combination of ULS pre-treatment with alkaline
(Chiu et al., 1997; Jin et al., 2009b; Kim et al., 2010) and acidic pre-treatments (Liu
et al., 2008; Sahinkaya, 2014) has been demonstrated to increase sludge disintegration as well as subsequent anaerobic digestion.

Apart from the aforementioned chemical methods, ozone has also been shown
feasible to enhance the ultrasonic pre-treatment (Xu et al., 2010a; Yang et al., 2012,
4222013b; Tian et al., 2014c). Xu et al. (2010a) demonstrated the feasibility of combining ultrasound and ozone to disintegrate WAS and to improve methane recovery from the subsequent anaerobic digestion. Yang et al. (2013b) observed that the
combined ultrasound and ozone pre-treatment enhanced the solubilization of amino
acids and proteins in WAS. Tian et al. (2014c) found that ozone was able to chemically degrade macromolecules solubilized by ultrasound and further increased the
sludge anaerobic biodegradability.

Figure 12.1 shows the location in the process flow sheet where the pre-treatment would take place. Alternatively, the pre-treatment could target WAS only and
exclude primary sludge (PS). It is also an option to treat a fraction of WAS to reduce
operating costs and the optimum combination should be determined for each sludge.

These previous studies focused on the characteristics of the solubilized compounds and changes in sludge properties after pre-treatment. Xu et al. (2010a) and
Tian et al. (2014c), however, investigated the influence of pre-treatment on sludge
anaerobic biodegradability in batch serum bottle tests. Information on the influence

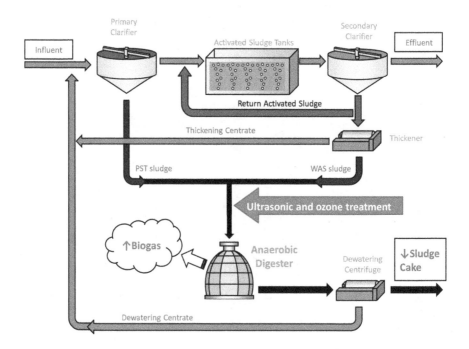

FIGURE 12.1 Proposed flow sheet with the implementation of an ultrasonic and ozone pre-
treatment of sludge.

TABLE 12.2

Characteristics of Raw Sludge

Parameter	Value Range
Total solids (g/L)	14.9–17.2
Volatile solids (g/L)	12.6–13.4
Total suspended solids (g/L)	13.7–14.1
Volatile suspended solids (g/L)	11.8–13.3
Total COD (mg/L)	16,800–25,000
Soluble COD (mg/L)	700–1,200
Equivalent COD of soluble proteins (mg/L)	<100
Equivalent COD of soluble carbohydrates (mg/L)	<50
pH	5.9–6.2

of such combined pre-treatment on solids removal efficiency and digested sludge characteristics after anaerobic digestion is not available. Additionally, the influence of the solids retention time (SRT), an important design parameter, on the anaerobic digestion of the combined pre-treated sludge has also not been reported. The following sections investigate the impact of such pre-treatment on the subsequent anaerobic digestion process with semi-continuous reactors at an SRT of 10 and 20 days.

The following data were obtained using a mixture of PS and thickened WAS (ratio around 1:1 based on DS) collected from a municipal wastewater reclamation plant. The properties of the raw sludge used in this study are listed in Table 12.2.

Ozonation was performed with an ozone generator (Wedeco, GSO 30). Pure oxygen was used as the feed gas and converted to ozone with a high-voltage converter. The power input of the ozone generator was 180 W. The applied ozone dosage was quantified according to the potassium iodide method (Konsowa, 2003). A 200 mL sludge sample (TS: 16.2–17.2 g/L) was placed in a 1 L glass bottle during the ozonation process. A stone diffuser was installed to produce fine ozone bubbles and to enhance ozone mass transfer. The applied ozone was quantified in terms of ozone dosage (g O_3/g TS). The maximum applied ozone dosage was 0.12 g O_3/g TS. The optimum combination sequence of ULS and ozonation was determined before commencing the main experiment. The ultrasonication-ozonation (ULS-ozone) sequence was performed by applying ozonation after ultrasonication and the ozonation-ultrasonication (ozone-ULS) sequence was conducted by dosing ozone prior to ultrasonication.

12.3 PERFORMANCE OF INDIVIDUAL PRE-TREATMENTS

12.3.1 COD AND BIOPOLYMERS SOLUBILIZATION

As shown in Table 12.2, soluble proteins and carbohydrates only represented a small fraction of the SCOD in the raw sludge, indicating that this initial SCOD did not originate from the degradation of microbial cells. This raw SCOD might be attributed to the soluble lipids contained in the PS, inorganic interferences (such as ammonium that contributes to COD) and/or refractory organics, which were not degraded during the activated sludge process and remained in the WAS fraction. The $SCOD_+$ increase was different during ULS and ozonation. For ultrasonicated sludge, $SCOD_+$ increased linearly with specific energy input as shown in Figure 12.2a (TS: around 17 g/L). The $SCOD_+$ was 3450 mg/L after 21 kJ/g TS ULS and values did not plateau. However, for ozonated sludge, $SCOD_+$ increased markedly to around 2400 mg/L at an ozone dosage of 0.02 g O_3/g TS and then plateaued at around 3700 mg/L at higher ozone dosages, as shown in Figure 12.2b.

Proteins and carbohydrates were responsible for around 80% of the COD solubilization measured for both pre-treatments. The equivalent COD of the solubilized proteins and carbohydrates after 21 kJ/g TS ULS was 2370 and 450 mg/L, respectively. The equivalent COD of the solubilized proteins and carbohydrates was 2500 and 560 mg/L after 0.08 g O_3/g TS ozonation, respectively. These are consistent with past results, showing that ultrasound and ozone solubilized the extracellular and intracellular biopolymers in the sludge (Wang et al., 2006b; Zhang et al., 2009).

12.3.2 CHANGE IN pH

In addition, ultrasound and ozone resulted in different changes in pH value. The sludge pH remained relatively constant at around 5.9 with increasing specific energy input, as shown in Figure 12.2c. However, the sludge pH decreased obviously with increasing ozone dosage. The sludge pH dropped from 5.9 to 5.2 after 0.04 g O_3/g TS ozonation. These results were consistent with past results and indicated that ultrasound and ozone did not disrupt the sludge in the same way (Bougrier et al., 2006). The pH drop during ozonation was due to the formation of acidic compounds. Yasui and Miyaji (1992) observed the formation of carboxylic acids during human waste ozonation. Bougrier et al. (2006) suggested that the pH decrease was due to the formation of volatile fatty acids (VFA) from the degradation of lipid compounds in sludge. Yan et al. (2009) found that the lactic acid concentration increased after ozonation due to the oxidation of the soluble macromolecules. This difference in the products of the pre-treatment process used would have an impact on subsequent anaerobic degradation.

12.4 PERFORMANCE OF COMBINED PRE-TREATMENT

12.4.1 DETERMINATION OF THE BEST COMBINATION SEQUENCE

In order to investigate the interaction between ultrasound and ozone, various combination sequences were tested. The $SCOD_+$ changes in each sequence are as shown in Figure 12.3a and b. The maximum $SCOD_+$ values obtained in the ULS-ozone and ozone-ULS sequences were around 4000 and 3800 mg/L, respectively, showing an improved COD solubilization compared to each individual treatment. Such SCOD concentration increase was slightly higher than the maximum SCOD increase (3300 mg/L) reported in a previous study treating WAS with a combined ULS and

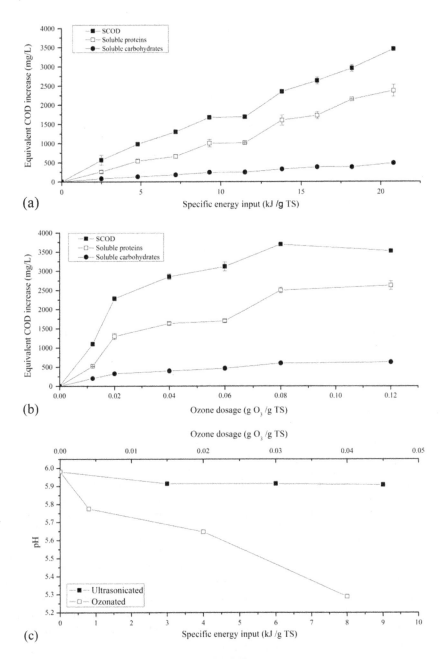

FIGURE 12.2 (a) Increase of COD, equivalent COD of proteins and carbohydrates in soluble phase with specific energy input in individual ultrasonication. (b) Increase of COD, equivalent COD of proteins and carbohydrates in soluble phase with ozone dosage in individual ozonation. (c) Change in pH with specific energy input and ozone dosage. (Reproduced from Tian et al., 2014c, with permission from Elsevier.)

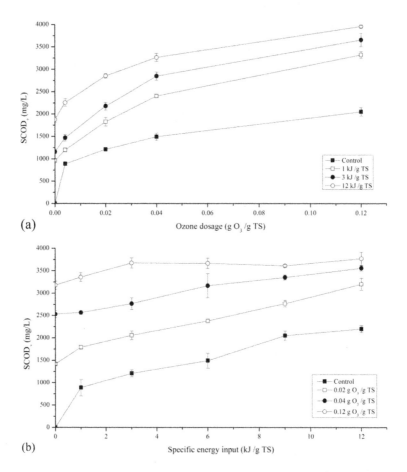

FIGURE 12.3 (a) Change in SCOD+ with ozone dosage in control and ultrasonicated feed sludge (at various specific ultrasonic energy inputs) in sequence of ULS-ozone. (b) Change in SCOD+ with specific energy input in control and ozonated feed sludge (at various ozone dosages) in sequence of ozone-ULS (SCOD+: The SCOD increase induced by pre-treatment). (Reproduced from Tian et al., 2014c, with permission from Elsevier.)

ozonation process (Xu et al., 2010a). Although WAS has been shown to be more susceptible to ULS treatment than mixtures of PS and WAS, a higher solubilization was observed in this study in comparison to Xu et al. (2010a). This is because the mixed sludge used in this study had a higher TS concentration (around 17 g/L) than the WAS (around 10 g/L) used in the previous work (Xu et al., 2010a). Therefore, more solids were available for solubilization.

Although the $SCOD_+$ values were similar between the two selected sequences in most of the combination conditions, the ozone-ULS sequence did not appear to be substantially advantageous because the $SCOD_+$ increase was marginal during the subsequent ULS step, especially at high prior ozone dosage (e.g., 0.12 g O_3/g TS), as shown in Figure 12.3b. This is because ozone significantly solubilized the solids in sludge and fewer solids were available for ultrasonic mechanical disruption. This indicated that the ozonation step should be conducted after the ULS step.

12.4.2 Sludge Solubilization

A prior ULS step did not enhance the COD solubilization induced by ozone. The COD solubilized by ozone decreased with the increase in the prior ULS energy, as shown in Figure 12.4a. For example, the $SCOD_+$ induced by 0.048 g O_3/g TS ozonation was 2600 mg/L for untreated feed sludge, but was only 1500 and 400 mg/L for feed sludge that had been pre-ultrasonicated at 9 and 21 kJ/g TS, respectively. This is because changes in SCOD can be due not only to organic solubilization but also to the degradation of organics to oxidized species such as CO_2. Previous works have shown that mineralization and degradation of the soluble organics due to ozone oxidation resulted in SCOD decrease (Ahn et al., 2002; Erden and Filibeli, 2011).

Foladori et al. (2010) suggested that volatile suspended solids (VSS) could be used as an alternative to represent the particulate organics in a sludge sample. The VSS solubilization induced by ozone for both control and pre-ultrasonicated sludge is as shown in Figure 12.4b. In contrast with the SCOD results, a greater VSS

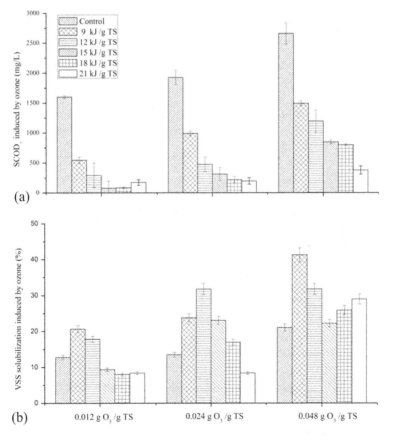

FIGURE 12.4 (a) SCOD+ induced by different ozone dosages in control sludge and ultrasonicated feed sludge (at various specific energy inputs). (b) VSS solubilization induced by different ozone dosages in control sludge and ultrasonicated feed sludge (at various specific energy inputs). (Reproduced from Tian et al., 2014c, with permission from Elsevier.)

solubilization due to ozonation was obtained when the prior ULS specific energy input was lower than 12 kJ/g TS. The highest VSS solubilization induced by ozone was 41.3% when 0.048 g O_3/g TS ozonation was applied after 9 kJ/g TS ULS. The same ozone dosage applied without prior ULS only induced 21.1% VSS solubilization. This implied that ultrasound made the organic solids easily disrupted by ozone. Agglomerations of particulates would have become smaller in size after ultrasonic dispersion resulting in a higher specific surface area, thereby affording a greater probability of contact with ozone. This is important because the half-life of ozone is only 30 min (Tian et al., 2014c), and is likely much lower in practice as it may react with non-target inorganic materials before it has the chance to do so with target organic solids. Additionally, smaller particles are more readily ozonated than bigger ones because of a lower mass transfer resistance (Cesbron et al., 2003b). Similar synergistic solids solubilization was reported in a previous study. Yang et al. (2013b) reported that combined ULS and ozonation treatment of feed sludge solubilized 6.1% more solids than the sum of solids solubilized by individual ULS and ozonation treatments.

VSS solubilization induced by ozone started to decrease when the prior ULS specific energy was higher than 12 kJ/g TS. This was possibly due to the reaction between ozone and solubilized organics released by the prior ultrasound. Cesbron et al. (2003b) showed that the soluble and particulate organics competed for ozone in a sludge ozonation system. More organics would have been released after high-energy ULS (>12 kJ/g TS). The reaction between soluble organics and ozone became significant and less ozone was therefore available for the organic solids. This negated the positive effect of the prior ultrasonic dispersion resulting in the decrease in VSS solubilization induced by ozone.

12.4.3 MOLECULAR WEIGHT DISTRIBUTION

MW distribution is a useful tool to shed more light on the soluble residual organics in digested sludge. In order to illustrate the difference in the solubilized substances between ULS and ozonation, the MW distribution chromatograms of the raw sludge, the ultrasonicated sludge and the ozonated sludge are compared in Figure 12.5a and b. No peak with the same retention time was detected in ultrasonicated and ozonated sludge, indicating solubilization of different compounds in both pre-treatments. High MW compounds (Rt < 6 min) were found to be the main solubilized compounds in both pre-treatments as shown in both UV and refractive index (RI) signals.

The corresponding MWs of these compounds were over 500 kDa according to the retention time of the largest standard polymer (500 kDa), shown as a dashed line. Such high MW compounds were reported as cell fragments and extracellular polymers in WAS with an MW as high as 10^5 kDa (Pavoni et al., 1972; Namkung and Rittmann, 1986; Schiener et al., 1998; Aquino et al., 2006b). It should be noted that RI is a universal detector that detects most of the eluted substances if these are present in sufficient concentrations, while the UV 254 nm detector provides good sensitivity toward aromatic compounds (Trzcinski et al., 2011). For both ultrasonicated and ozonated samples, the detected peaks had similar retention times in both the UV and RI detector signals, as shown in Figure 12.5a and b, suggesting that most

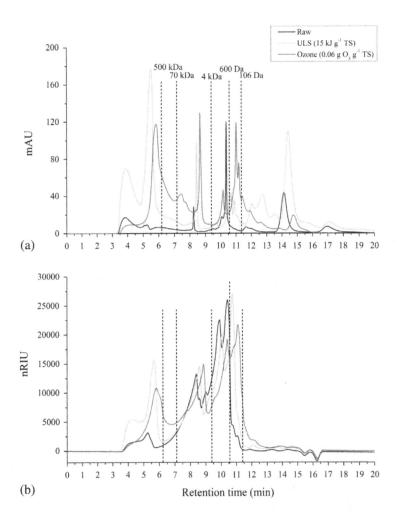

FIGURE 12.5 (a) MW distribution chromatograms of the soluble substances in raw sludge, ultrasonicated sludge (15 kJ/g TS) and ozonated sludge (0.06 g O$_3$/g TS) in (a) UV signal (254 nm) and (b) RI signal. (Reproduced from Tian et al., 2014c, with permission from Elsevier.)

of the solubilized high MW compounds were possibly aromatic compounds. Yang et al. (2013b) reported that tryptophan proteins were the main solubilization products after ULS while humic acids (HA) and fulvic acids (FAs) were the primary solubilization products after ozonation.

In addition, low MW compounds (8 min < Rt < 13.5 min) were also released in both pre-treated sludge but not as obvious as the high MW ones. The MWs of these polymers were lower than 27 kDa. Peaks with a retention time longer than 13.5 min were also detected. These peaks stand for compounds that are smaller than 106 Da (the retention time of the standard polymer is shown as the dashed line). The UV response of such compounds was lower in ozonated sludge than in raw sludge, as shown in Figure 12.5a. This decrease in the UV response of the simple organic compounds was because of the mineralization effect induced by ozone (Weemaes et al., 2000; Ahn et al., 2002). Simple organic compounds were oxidized into carbon dioxide.

The MW distribution of the soluble organics in the ultrasonicated and the ULS-ozone-treated feed sludge are shown in Figure 12.6a and b. Organic solids were mainly solubilized in the form of low MW components (8 min < Rt < 13.5 min, MW < 27 kDa), as shown by the arrows in Figure 9.4b. In addition, the response increase in high MW compounds was not as marked as was observed when sludge was only ozonated, as shown in Figure 9.3b.

Some of the high MW components released by ultrasound were found to be very sensitive to ozone attack. The peak that represented high MW components (Rt: 4 min) disappeared after ozonation was applied, as shown by an arrow in Figure 12.6a. This suggested that prior ultrasound released more organics that could be ozone scavengers. Consequently, the reaction between these soluble organics and ozone became more pronounced compared to the situation when ozone was directly applied to the raw sludge. It should be borne in mind that organic solids solubilization by ozone was also enhanced in ultrasonicated sludge ($\leq$12 kJ/g TS), as shown in Figure 12.4b. This meant that the prior ULS step enabled better utilization of ozone by the soluble and particulate organics in sludge. This is important because the utilization of the applied ozone had a significant influence on the performance of a sludge ozonation system (Chu et al., 2008; Manterola et al., 2008).

Despite the synergistic effects caused by the combined treatment, the potential for refractory compounds formation should not be neglected. Yang et al. (2013b) observed the solubilization of HA-like substances when ozone was used to pre-treat a WAS sample. Such solubilization was enhanced when the ULS and ozonation pre-treatment were combined simultaneously (Yang et al., 2013b). Therefore, it is reasonable to hypothesize that such a formation of HA-like substances may also happen when ULS and ozonation are applied sequentially. Macromolecules containing HAs (such as EPS) were first mechanically released due to the ULS and then chemically degraded to HA-like substances and other compounds due to the ozone oxidation. This might also contribute to the degradation of high MW components (Rt: 4 min), as shown in Figure 12.6b. In addition, the solubilization of the PS fraction may also contribute to such a formation of refractory products. For example, residual toilet paper fibers or lint commonly found in PS contains lignin, which could be degraded into HAs due to the ULS-ozonation pre-treatment.

12.4.4 ANAEROBIC DIGESTION OF THE PRE-TREATED SLUDGE

Biogas production and solids concentrations in the digested sludge were monitored to assess possible enhancement with such pre-treatment. The change in the anaerobic biodegradability of the raw sludge, ozonated sludge, ultrasonicated sludge and ULS-ozonated sludge during the biochemical methane potential (BMP) assay is shown in Figure 12.7.

The biodegradability of the ultrasonicated sludge increased in the first 4 days of anaerobic digestion. However, a lag phase was observed for both ozonated sludge and ULS-ozonated sludge in the first 4 days of anaerobic digestion.

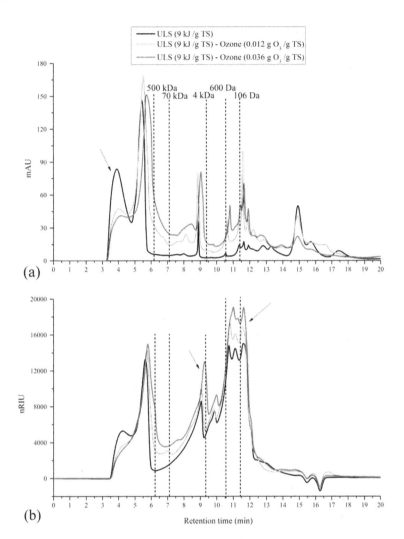

FIGURE 12.6 MW distribution chromatograms of the soluble substances in ultrasonicated sludge (9 kJ/g TS) and ultrasonicated feed sludge (9 kJ/g TS) with various subsequent ozone dosages in (a) UV signal (254 nm) and (b) RI signal. (Reproduced from Tian et al., 2014c, with permission from Elsevier.)

Their corresponding biodegradabilities in the first 4 days were lower than the raw sludge. This was because the oxidized species generated by ozone had inhibited the methanogens as these are known to thrive at very low redox potentials (Weemaes et al., 2000). After Day 7, most of the oxidized species were reduced via acidification and no further inhibition effect was observed. The biodegradability increased much faster in the ozonated and the ULS-ozonated sludge compared to the raw and ultrasonicated sludges. For all the tested samples, the increase in biodegradability became insignificant after 15 days of anaerobic digestion and reached plateau values after 22 days.

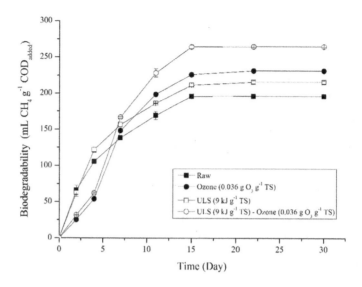

FIGURE 12.7 Results of biochemical methane potential assays for raw sludge, ozonated sludge, ultrasonicated sludge and ULS-ozonated sludge. (Reproduced from Tian et al., 2014c, with permission from Elsevier.)

The combined ULS-ozonation pre-treatment resulted in an ultimate biodegradability of 265 mL CH_4/g COD_{added}, which was 34.7% more compared to the raw sludge (196 mL CH_4/g COD_{added}). However, the sum of the ultimate BI was only 27.7% when ULS (9.9%) and ozonation (17.8%) were applied at the same conditions. Therefore, a 25.7% increase in BI could be obtained when ULS and ozonation were sequentially combined. One possibility of this synergistic increase is the synergistic VSS solubilization. Additionally, the MW reduction in the sequential combined treatment may also contribute to the synergistic BI. Eskicioglu et al. (2006b) indicated that macromolecules with an MW higher than 300 kDa were mostly complex cell fragments and HAs that were harder to anaerobically biodegrade. In this work, the soluble high MW components (MW > 500 kDa) released by ultrasound were effectively degraded by ozone and most of the organic solids were solubilized to low MW components (MW < 27 kDa), which benefited the subsequent anaerobic digestion.

However, it should be noted that the increases in methane production were only 10.9%, 6.6% and 15.4% after the ULS, ozonation and ULS-ozonation pre-treatment, respectively. These figures were lower than the increases in the anaerobic biodegradability for the ozonated and ULS-ozonated sludge. This was related to the degradation of sludge caused by the ozonation process (as indicated by the decrease in substrate total COD [TCOD] concentration). Although anaerobic biodegradability was improved by oxidizing the hardly degradable components in sludge, some biodegradable organics were also lost during ozonation.

The anaerobic digestion results of this work were compared with previous studies in Table 12.3. However, only one report on methane production increasing after combined ULS and ozonation pre-treatment could be identified. Consequently, the results from this study were also compared with those using other pre-treatment methods.

Xu et al. (2010a) showed that anaerobic biodegradability increased by 93%, 100% and 75% after 30, 45 and 120 min of simultaneous combined ULS and ozonation pre-treatment. In addition, their results also suggested that ozone overdosage could have a negative effect on methane production. This highlighted the importance of identifying the appropriate ozone dosage for the combined ULS-ozonation pre-treatment process. Seng et al. (2010) suggested methane production after ULS pre-treatment could be further enhanced by alkaline pre-treatment. Although the increases due to ozone (from 10.9% to 15.4%) and alkaline (from 12.8% to 17.3%) pre-treatments were similar, the ozone pre-treatment could significantly reduce the residual sludge amount and does not add dissolved solids (e.g., sodium ions) to the sludge, unlike the alkaline pre-treatment (Foladori et al., 2010). In the literature review, the steam-explosion method was noted to have very good performance, as shown in Table 12.3. The steam-explosion pre-treatment not only imposes mechanical shear effect on the sludge from the moisture expansion, but also thermally and chemically disintegrates the sludge with the high temperature steam

TABLE 12.3
Summary of the Anaerobic Digestion Results from this Work and Previous Studies

References	Pre-Treatment Conditions	Anaerobic Digestion Conditions	Methane Production Increase
This study (PS/WAS)	Ultrasonication (9 kJ/g TS)	Batch, 30 days	+10.9%
	Ultrasonication (9 kJ/g TS)/ozonation (0.036 g O_3/g TS)		+15.4%
Dereix et al. (2006) (TWAS/biosolids)	Steam-explosion, 220°C, 2.07 MPa	Batch, 65 days	+52%
	Steam-explosion, 260°C, 4.14 MPa		+100%
Seng et al. (2010) (WAS)	Ultrasonication (3.8 kJ/g TS)	Semi-continuous, SRT[a]: 25 days	+12.8%
	Ultrasonication (3.8 kJ/g TS)/alkaline (0.01 g NaOH/g TS)		+17.3%
Xu et al. (2010) (WAS)	Ultrasonication/ozonation (30 min)	Batch, 14 days	+93%[b]
	Ultrasonication/ozonation (45 min)		+106%[b]
	Ultrasonication/ozonation (120 min)		+75%[b]

Source: Reproduced from Tian et al., 2014c, with permission from Elsevier.

[a] Solids retention time.

[b] Increase based on anaerobic biodegradability (i.e., mL CH_4/g COD_{added}).

(Zhao et al., 2013). The steam-explosion pre-treatment resulted in a higher methane production increase (+52% and +100%) than that induced by the combined ULS-ozonation pre-treatment (+15.4%), as reported in this study. However, Dereix et al. (2006) also indicated that the extra energy recovered could not compensate for the energy cost of the pre-treatment process. This indicated that both the ULS-ozonation and steam-explosion pre-treatments required further studies if a positive energy balance is to be achieved.

12.5 ENERGY BALANCE

The energy balance is calculated in Table 12.4. The energy balance was analyzed by subtracting the energy input during the pre-treatment from the energy recovered from the combined heat and power (CHP) production of the produced methane. In order to maintain consistency, the calculation was based on 1 L of sludge. During the CHP production process, it was assumed that 30% of the methane calorific energy was converted to electricity and 50% to heat (Cho et al., 2014b).

The energy balances were 93.8, −42, −386 and −523.8 kJ for 1 L of raw sludge, ultrasonicated, ozonated sludge and ULS-ozonated sludge, respectively. This indicated that the tested pre-treatment conditions were not economically feasible, especially when ozone was used. A reduction in the ozone dosage would not only save some energy but also reduce the loss of biodegradable organics during ozonation. It should also be noted that the laboratory-scale ULS and ozonation processes could not be operated at equally high intensity and efficiency as operated in a full-scale plant (Pérez-Elvira et al., 2009a). These aforementioned factors could be responsible for the negative energy balance obtained in this study. More realistic data should be collected from pilot or full-scale applications.

TABLE 12.4
Energy Balance Analysis in 1 L of Raw and Pre-Treated Sludges

Sludge	Methane Produced (L)	Energy Recovered as Electricity[a] (kJ)	Energy Recovered as Waste Heat[a] (kJ)	Energy Input for the Pre-Treatment (kJ)	Energy Balance (kJ)
		A	B	C	D = C−A−B
Raw	3.45	35.2	58.6	−	93.8
Ultrasonicated	3.83	39	65	146	−42
Ozonated	3.68	37.5	62.5	486	−386
ULS-ozonated	3.98	40.6	67.6	632	−523.8

Source: Reproduced from Tian et al., 2014c, with permission from Elsevier.

[a]   Calorific value of methane is 33.94 kJ/L and 30% of the combustion heat was assumed to be recovered as electricity and 50% of the combustion heat was assumed to be recovered as waste heat.

12.6   KINETIC ANALYSIS OF THE ANAEROBIC DIGESTION PROCESS

A modified Gompertz equation is often applied to model methane production during the BMP assay (Zwietering et al., 1990; Yue et al., 2008). The modified Gompertz equation is

$$P = P_{max} \exp\left\{-\exp\left[\frac{R_{max}e}{P_{max}}(\lambda - t) + 1\right]\right\}$$

where

P (mL) is the cumulative methane produced from the sludge at time t;

P_{max} (mL) is the maximum cumulative methane produced from the sludge;

R_{max} (mL CH$_4$/day) is the maximum methane production rate during the BMP assay;

λ (day) is the lag phase time of the BMP assay.

The data points of cumulative methane produced (P) were fitted to the anaerobic digestion time (t) using Origin software (OriginLab, USA). The P_{max}, R_{max} and λ for each tested sample were then obtained for comparison. The impact of each pre-treatment process on the methane production kinetics was analyzed and the fitted curves are shown in Figure 12.8; the kinetic parameters obtained are summarized in Table 12.5.

All the regression coefficients of the sludges (R^2) were higher than 95%, indicating that the modified Gompertz model was suitable for the kinetic prediction in this instance. The predicted P_{max} was in good agreement with the methane production results. Negative λ values were obtained for the raw and ultrasonicated sludge. This suggested that anaerobic digestion had not been inhibited. These negative λ values were omitted because they are only mathematically possible but not possible in reality (Nevot et al., 2007). Due to the inhibition effect of ozone, the ozonated and ULS-ozonated sludge yielded positive λ values of 1.75 and 1.72, respectively. The R_{max} of raw sludge was improved from 3.53 to 4.32 and 4.21 mL CH$_4$/day after the ULS and ozonation pre-treatment, respectively. Ozonation following the ULS process further increased the R_{max} to 4.54 mL CH$_4$/day. This is because some organics in the ULS-ozonated sludge were converted to other easily biodegradable substances (e.g., VFAs) instead of methane during the inhibition period. Once the methanogens were acclimatized, these easily biodegradable organics were rapidly converted to methane, which contributed to the higher methane production rate.

12.7   OPTIMIZATION OF ULS-OZONE CONDITIONS

The combined treatment was optimized in this section. The highest ultrasonic energy input was 21 kJ/g TS and the highest ozone dosage was 0.048 g O$_3$/g TS. The SCOD concentrations after the 13 experiments were measured and are shown in Table 12.6.

The combined treatment could result in higher COD solubilization compared to individual pre-treatments (Nos. 1 and 6). The aforementioned conditions were fed into an automatic methane potential analyzer for the determination of

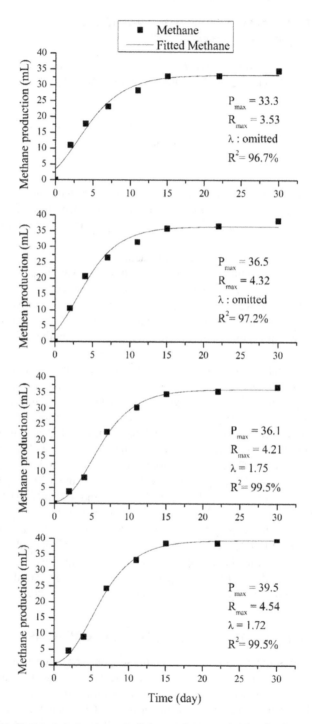

FIGURE 12.8 Data points for the cumulative methane production for (top to bottom) raw sludge, ultrasonicated sludge, ozonated sludge and ULS-ozone-treated sludge. The curve shows the best fitting to data points using Gompertz equation.

TABLE 12.5

Summary of the Methane Production Kinetics Assay

Sludge	P_{max} (mL CH_4)	R_{max} (mL CH_4/day)	λ (d)	R^2 (%)
Raw	33.3	3.53	–	96.7
ULS	36.5	4.32	–	97.2
Ozone	36.1	4.21	1.75	99.5
ULS-ozone	39.5	4.54	1.72	99.5

Source: Reproduced from Tian et al., 2014c, with permission from Elsevier.

anaerobic biodegradability. Substrate sludge (100 mL; both untreated and treated) and degassed inoculum (300 mL) were put into the Automatic Methane Potential Test System (AMPTS) system (14 bottles). One more position in the AMPTS system was for 100 mL deionized water and 300 mL degassed inoculum. This gas production was due to the autolysis of inoculum. The anaerobic biodegradability and the dewaterability of the digested sludge (in capillary suction time [CST]) are shown in Table 12.7.

A synergistic increase in anaerobic biodegradability could be observed by comparing Experiments 3, 5, 8, 11 and 13) with individual treatments (Experiments 1 and 6). The increase in the combined treatment is higher than the sum in individual treatments at the same energy input. By comparing the No. 7 with the central points, it seems that a further increase in ozone dosage did not benefit the subsequent anaerobic digestion. The ultrasonically solubilized organic carbons may be converted to carbon dioxide when the subsequent ozone dosage is too high (No. 7).

TABLE 12.6

SCOD, SCOD/TCOD Data of the Experiments

Number	ULS (A) (kJ/g TS)	Ozone (B) g (O_3/g TS)	SCOD (mg/L)	SCOD/TCOD (%)
1	10.5	0	1835.01	10.64
2	15.75	0.012	2730.64	15.51
3	10.5	0.024	2520.72	15.99
4	5.25	0.012	2106.01	11.86
5	10.5	0.024	2499.67	15.83
6	0	0.024	1833.98	11.41
7	10.5	0.048	2629.02	18.76
8	10.5	0.024	2502.24	16.97
9	21	0.024	3022.17	18.99
10	5.25	0.036	2339.54	16.15
11	10.5	0.024	2426.79	15.63
12	15.75	0.036	2714.2	19.31
13	10.5	0.024	2455.55	15.11

TABLE 12.7

Anaerobic Biodegradability and Dewaterability of the Digested Sludge

Experiment Number	ULS (A) (kJ/g TS)	Ozone (B) (g O$_3$/g TS)	Biodegradability (mLCH$_4$/g COD$_{added}$)	CST (s)
1	10.5	0	173.9 (+7.8%)	107.1
2	15.75	0.012	189.1 (+17.2%)	277.7
3	10.5	0.024	202.9 (+25.8%)	125.0
4	5.25	0.012	184.5 (+14.4%)	78.5
5	10.5	0.024	214.9 (+33.3%)	166.8
6	0	0.024	188.0 (+16.6%)	233.1
7	10.5	0.048	210.4 (+30.4%)	276.6
8	10.5	0.024	202.9 (+25.8%)	91.4
9	21	0.024	215.1 (+33.3%)	224.8
10	5.25	0.036	221.2 (+37.2%)	101.7
11	10.5	0.024	211.6 (+31.2%)	164.9
12	15.75	0.036	199.6 (+23.7%)	134.8
13	10.5	0.024	213.7 (+32.5%)	142.8
Untreated	–	–	161.3	34.0

This negated the positive effect of further ozone dosage. Additionally, prior ultrasonic energy input should not be too high. For example, the biodegradability of No. 9 was only slightly higher compared to the central points (3, 5, 8, 11 and 13) when the ultrasonic energy had been doubled. This is because when the ultrasonic energy input is too high, the VSS solubilization induced by subsequent ozonation decreases due to the competition between soluble and particulate organics for ozone.

Based on the obtained data, the ultrasonic energy input and ozone dosage should be considered as follows:

1. The prior ultrasonic energy should not be over 10.5 kJ/g TS, as this may release too much soluble compounds that may compete with the organic solids in sludge for ozone.
2. The subsequent ozone dosage should be around 0.024 g O$_3$/g TS, a higher dosage would cause a negative effect on the mineralization (conversion to CO$_2$) of ultrasonically solubilized organics.
3. However, the dosage of both treatment conditions should not be too low (as in No. 4). The improvement would not be obvious.
4. Economic reason should be considered when the increase in biodegradability is similar.

Therefore, the condition for the central point might be suitable for practical operation.

The dewaterability of treated sludge was worse than the untreated sludge. However, it should be borne in mind that the digested sludge is from a batch test (i.e., 300 mL inoculum and 100 mL substrate). The dewaterability change due to the degradation of substrate may not be significant. Nevertheless, this still could act as tentative results for dewaterability determination. Hereby, the points that had a CST over 200 s (2, 6, 7 and 9) were taken for discussion. The poor dewaterability in No. 6 might be due to the limitation of individual treatment. Excessive ultrasonic energy input or ozone dosage had been applied during the experiments of points 2, 7 and 9. This suggested that excess ultrasonic energy or ozone dosage did not efficiently improve the anaerobic degradability but deteriorated the sludge dewaterability.

12.8   CONCLUSIONS OF THE BATCH PROCESS

Ultrasound and ozone were found to disrupt the sludge differently and induce synergistic effects when sequentially combined. The ULS-ozone sequence was shown to be advantageous compared to the ozone-ULS sequence. The ULS-ozone sequence pre-treatment worked as follows:

- Prior ULS helped disperse the biological flocs and enhanced the reaction between ozone and the organic solids.
- The resulting organic solids were then solubilized by ozone in the form of low MW organics (MW < 27 kDa).
- The concentration increase of high MW compounds (MW > 500 kDa) during ozonation was also not significant.
- Ozone utilization by the ultrasonicated sludge was better compared to the non-ultrasonicated sludge. Some of the high MW compounds solubilized by ultrasound were likely ozone scavengers and hence effectively degraded by ozone.
- Synergistic sludge BI was observed due to the aforementioned synergistic actions.
- The maximum methane production rate for the ultrasonicated sludge was further improved due to the subsequent ozonation process.

12.9   PERFORMANCE OF A CONTINUOUS PROCESS

12.9.1   PROCESS CONFIGURATION AND OPERATING CONDITIONS

ULS-ozone pre-treatment was performed by sequentially applying the ULS and the ozonation treatments. The ULS energy input was selected at 9 kJ/g TS and the applied ozone dosage was 0.012 g O_3/g TS. Anaerobic digestion was conducted semi-continuously in 1.2 L glass bottles with 1 L working volume at 35°C. Seed sludge was taken from a continuous anaerobic digester with an SRT of 28 days from a local reclamation plant. Seed sludge (1 L) was fed into the reactor before starting the experiment. Sludge (100 and 50 mL) was removed daily from the reactors and replaced with the same amount of feed sludge to obtain an SRT of 10 and 20 days, respectively. The reactors operating at an SRT of 10 days were referred to as

Control10, ULS10 and ULS-Ozone10 and these received the untreated, ULS-treated and ULS-ozone-treated sewage sludge as feed, respectively. Similarly, the reactors at an SRT of 20 days were referred to as Control20, ULS20 and ULS-Ozone20. Each reactor was run for three SRTs so that process stability could be assumed. Biogas was collected with Tedlar gas bags and the volume was measured daily with a gas meter (Ritter, Germany). Feed sludge in storage was changed every 3 weeks. Each batch of feed sludge was manually adjusted to keep a consistent TS concentration of around 15 g/L. Daily biogas production was normalized by dividing the daily gas production by the amount of COD fed into the reactor.

12.9.2 BIOGAS PRODUCTION AND SOLIDS REMOVAL IN THE CONTINUOUS PROCESS

Biogas production from anaerobic reactors is shown in Figure 12.9a and b. The anaerobic biodegradability of sludge was higher at an SRT of 20 days with its longer substrate–microbe contact time. At both SRTs, the daily biogas production was higher from the reactors fed with pre-treated sludge than from the Control reactor. The average daily biogas production from each reactor is shown in Table 12.8.

These values were calculated by averaging the daily biogas production in the third SRT. At an SRT of 10 days, the daily biogas production increased from 256 to 309 (+20.7%) and 348 (+35.9%) mL biogas/g COD_{fed} because of the ULS and ULS-ozone treatments of feed sludge, respectively. At an SRT of 20 days, the daily biogas production increased from 313 to 337 (+7.7%) and 393 (+25.5%) mL biogas/g COD_{fed} due to the ULS and ULS-ozone treatments of feed sludge, respectively. These results indicated that the subsequent ozonation enhanced ULS pre-treatment in terms of increasing biogas production. Nickel and Neis (2007) found that the improvement in biogas production due to ULS treatment of feed sludge was higher when the reactor was operated at a shorter SRT. For example, biogas production increased by 16% after ULS treatment of feed sludge when the anaerobic reactor was operated at 8 days SRT; while the same ULS treatment condition only resulted in an 11% increase in biogas production when the SRT was 16 days. Similar results were obtained in this work. The ULS treatment of feed sludge increased biogas production by 20.7% and 7.7% at an SRT of 10 and 20 days, respectively. In addition, results from this work showed that the increase in biogas production after the ULS-ozone treatment of feed sludge also became more pronounced when the SRT was shortened from 20 to 10 days (from 25.5% to 35.9%), which had not been reported previously. In all the reactors, the methane content was relatively stable at around 65%. In addition, no VFAs accumulation was observed and the pH value remained near neutral throughout the anaerobic digestion test (around 7.0–7.2) in all the reactors. This suggested that neither ULS nor ULS-ozone treatments of feed sludge caused stress on the methanogenesis step in the reactors. An improvement in organic solids removal efficiency during anaerobic digestion was also observed when feed sludge was treated before anaerobic digestion. The change in VS and VSS concentrations in the digested sludge during anaerobic digestion is shown in Figure 12.10.

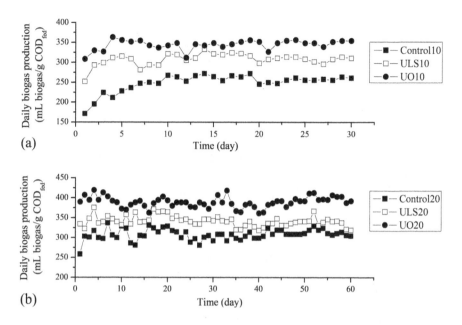

FIGURE 12.9 Daily biogas production from the reactors at (a) SRT of 10 days (Control10, ULS10, ULS-Ozone10) and (b) SRT of 20 days (Control20, ULS20, ULS-Ozone20). ULS: ultrasonication, UO: ultrasonication-ozonation. (Reproduced from Tian et al., 2015a, with permission from Elsevier.)

Solids concentration became relatively stable after 18 and 30 days of operation for reactors with an SRT of 10 and 20 days, respectively. After reaching a relatively stable level, the VS and VSS concentrations in the digested sludge were averaged for comparison. The average post-digestion VS and VSS concentrations and the corresponding VS and VSS removal efficiencies against the untreated feed sludge are as shown in Table 12.8. The control reactor had a VS removal efficiency of only 35.6% when it was operated at an SRT of 10 days. With the incorporation of ULS and ULS-ozone treatments of feed sludge, the VS removal rates increased to 38.3% and 42.1%, respectively. Solids removal efficiency was higher at the longer SRT. The VS removal rates of the Control20, ULS20 and ULS-Ozone20 reactors were 37.3%, 40.9% and 45.3%, respectively. Higher VS removal efficiency indicated more organic matters were digested and converted into biogas. Similarly, the incorporation of the pre-treatment step also improved the VSS removal efficiency, as shown in Table 12.8. The increase in VSS removal efficiency indicated that particulate organics in the treated feed sludge were better hydrolyzed for the subsequent anaerobic digestion process. It has been reported in a full-scale study that ULS treatment of feed sludge was able to slightly decrease the VSS concentration in the digested sludge from 9930 to 9810 mg/L at an SRT of 30 days (Xie et al., 2007). However, the improvement in VSS removals after anaerobic digestion due to ULS treatment of feed sludge was more obvious at a shorter SRT of 10 and 20 days in this work (e.g., from 8005 to 7640 mg/L at an SRT of 10 days). Furthermore, this

TABLE 12.8

Performance of the Semi-Continuous Anaerobic Digesters at Steady State

Reactor	SRT 10 days			SRT 20 days		
	Control10	ULS10	ULS-Ozone10	Control20	ULS20	ULS-Ozone20
Organic loading rate (g COD/L.day) (n=10 and 20)[a]	1.78±0.05	1.71±0.04	1.62±0.02	0.99±0.02	0.97±0.01	0.93±0.02
Biogas (mL/day g COD_{fed}) (n=10 and 20)[a]	256±5	309±6	348±9	313±7	337±10	393±12
Methane percentage (%) (n=5)	64.2±0.9	65.2±0.8	65.6±0.6	64.9±0.5	64.9±1.3	65.4±1.2
VS in digested sludge (mg/L) (n=5 and 8)[b]	8340±383	7990±344	7500±245	8125±252	7644±253	7081±217
VS removal efficiency (%) (n=5 and 8)[b]	35.6±4.6	38.3±4.3	42.1±3.3	37.3±3.1	40.9±3.3	45.3±3.1
VSS in digested sludge (mg/L) (n=5 and 8)[b]	8005±469	7640±155	6760±248	7619±258	7056±213	6606±206
VSS removal efficiency (%) (n=5 and 8)[b]	32.7±3.9	35.8±1.3	43.2±2.1	36.0±2.2	40.7±1.8	44.5±1.7
Total VFA in digested sludge (mg/L) (n=3)	Nd[c]	Nd	Nd	Nd	Nd	Nd

Source: Reproduced from Tian et al., 2015a, with permission from Elsevier.

[a] n=10 at SRT of 10 days and n=20 at SRT of 20 days.

[b] n=5 at SRT of 10 days and n=8 at SRT of 20 days.

[c] Not detectable (<10 mg/L).

work demonstrated that ULS-ozone treatment of feed sludge resulted in a lower VSS concentration in the digested sludge than ULS treatment, which had not been reported in previous studies.

A t-test at the significance level of 0.05 was conducted to compare the changes in biogas production and post-digestion VS concentration after the ULS and ULS-ozone treatments of feed sludge (Takashima, 2008; Rivero et al., 2006). As shown in Table 12.9, the biogas production was significantly higher and the post-digestion

TABLE 12.9
Statistical Analysis of the Biogas Production, VS Concentration in Digested Sludge and SCOD in Digested Sludge after ULS and ULS-Ozone Pre-Treatments at Different SRTs

	ULS	ULS-Ozone
	Biogas (mL/day g COD_{fed})	
Compared to control at an SRT of 10 days ($n=10$)	t: 23.398, P: 2.27×10^{-9}[a]	t: 38.203, P: 2.86×10^{-11}[a]
Compared to ULS at an SRT of 10 days ($n=10$)	–	t: 16.329, P: 5.39×10^{-8}[a]
Compared to control at an SRT of 20 days ($n=20$)	t: 12.709, P: 9.77×10^{-11}[a]	t: 33.657, P: 2.11×10^{-18}[a]
Compared to ULS at an SRT of 20 days ($n=20$)	–	t: 25.715, P: 3.16×10^{-16}[a]
	VS Concentration in Digested Sludge (mg/L)	
Compared to control at an SRT of 10 days ($n=5$)	t: −4.111, P: 0.01473[b]	t: −5.126, P: 0.0686[b]
Compared to ULS at an SRT of 10 days ($n=5$)	–	t: −3.642, P: 0.02192[b]
Compared to control at an SRT of 20 days ($n=8$)	t: −7.434, P: 1.45×10^{-3}[b]	t: −12.166, P: 5.80×10^{-6}[b]
Compared to ULS at an SRT of 20 days ($n=8$)	–	t: −7.515, P: 1.36×10^{-4}[b]
	SCOD in Digested Sludge (mg/L)	
Compared to control at an SRT of 10 days ($n=5$)	t: 8.573, P: 0.00102[a]	t: 24.305, P: 1.70×10^{-5}[a]
Compared to ULS at an SRT of 10 days ($n=5$)	–	t: 24.331, P: 1.69×10^{-5}[a]
Compared to control at an SRT of 20 days ($n=8$)	t: 4.694, P: 0.00222[a]	t: 21.649, P: 1.13×10^{-7}[a]
Compared to ULS at an SRT of 20 days ($n=8$)	–	t: 18.243, P: 3.68×10^{-7}

Source: Reproduced from Tian et al., 2015a, with permission from Elsevier.

Note: t is the t statistical value and P is the probability that the two compared values are not significantly different.

[a] The tested value was significantly higher than the reference value.

[b] The tested value was significantly lower than the reference value.

VS concentration was significantly lower than the control after the ULS and ULS-ozone treatments of feed sludge. In addition, the *t*-test results showed that ULS-ozone treatment of feed sludge resulted in statistically higher biogas production and lower VS concentration in the digested sludge than the ULS treatment. This confirmed that the application of ozonation subsequent to ULS could significantly enhance the sludge anaerobic digestion from a statistical point of view. In addition, it was noted that the daily biogas production and solids removal efficiencies of the ULS-Ozone10 reactor were better than those of the Control20 reactor. However, the daily biogas production and solids removal rates of the ULS10 reactor were lower than those of the Control20 reactor. This suggested that the ULS-ozone treatment of feed sludge could halve the SRT without affecting digestion performance; whereas the individual ULS pre-treatment was not able to provide such an advantage.

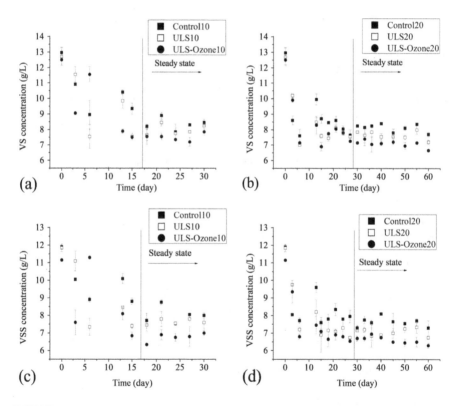

FIGURE 12.10 (a) Change in VS concentration in the digested sludge during anaerobic digestion at an SRT of 10 days. (b) Change in VS concentration in the digested sludge during anaerobic digestion at an SRT of 20 days. (c) Change in VSS concentration in the digested sludge during anaerobic digestion at an SRT of 10 days. (d) Change in VSS concentration in the digested sludge during anaerobic digestion at an SRT of 20 days (Day 0 stands for feed sludge) (solid concentration was based on at least two replicates). (Reproduced from Tian et al., 2015a, with permission from Elsevier.)

12.9.3 SCOD AND SOLUBLE BIOPOLYMERS IN THE DIGESTED SLUDGE

The SCOD concentration in the digested sludge during anaerobic digestion is shown in Figure 12.11. At both SRTs, ULS treatment of feed sludge increased the post-digestion SCOD concentration and the post-digestion SCOD increased further when ULS-ozone treatment was applied to feed sludge. The increase in SCOD in the digested sludge was also compared statistically with the t-test as shown in Table 12.9. The statistical results suggested that the SCOD in the digested sludge from the ULS and ULS-Ozone reactors was significantly higher than that from the Control reactor. Furthermore, a t-test between the SCOD in the digested sludge from the ULS and ULS-Ozone reactors showed that an increase in post-digestion SCOD due to the subsequent ozonation was also statistically significant.

As shown in Figure 12.11a and b, the SCOD in the digested sludge from ULS-Ozone reactors was around 300 and 200 mg/L higher than that from ULS reactors operating at an SRT of 10 and 20 days, respectively; while, the SCOD in the ULS-ozone-treated feed sludge was 1200 mg/L higher than the SCOD in the ULS-treated feed sludge. This indicated that much of the COD solubilized by the subsequent

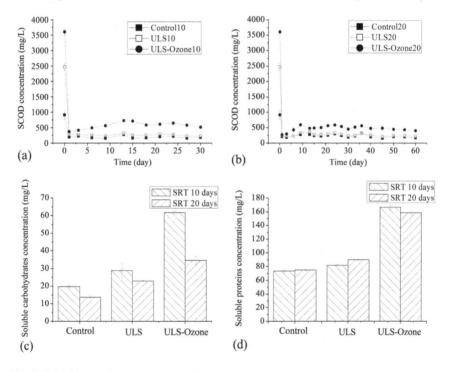

FIGURE 12.11 (a) Change in the SCOD concentration in the digested sludge during anaerobic digestion at an SRT of 10 days. (b) Change in the SCOD concentration in the digested sludge during anaerobic digestion at an SRT of 20 days (SCOD concentration was based on at least two replicates). (c) Soluble carbohydrates concentrations in the digested sludge from different anaerobic reactors ($n=3$). (d) Soluble proteins concentrations in the digested sludge from different anaerobic reactors ($n=3$). (Reproduced from Tian et al., 2015a, with permission from Elsevier.)

ozonation treatment of feed sludge was biodegraded and only a relatively small fraction accumulated in the anaerobic reactors.

It is known that biopolymers are a major component of sludge (Rittman and McCarty, 2001b). Averaged values of soluble carbohydrates and proteins concentration in the digested sludge during the last three sampling days are compared in Figure 12.11c and d. Soluble carbohydrates and proteins concentrations in the digested sludge from the ULS-Ozone reactor were much higher than the corresponding concentrations in the digested sludge from the ULS and Control reactor at both SRTs. This suggested that undigested biopolymers contributed to the higher SCOD in the digested sludge from the ULS-Ozone reactors.

The influence of SRT on the residual carbohydrates and proteins concentrations was different. As shown in Figure 12.11c, soluble carbohydrates concentrations decreased obviously when all the reactors had a longer residence time. For example, the residual soluble carbohydrates concentration decreased from 62 to 35 mg/L when the SRT of the ULS-ozone reactor increased from 10 to 20 days. This is because the solubilized carbohydrates after the treatments of feed sludge were mainly complex polysaccharides from extra and intracellular structures (Tian et al., 2014c; Wang et al., 2006b). A longer residence time was needed for sufficient degradation. However, the soluble proteins concentrations in the digested sludge did not show obvious difference between an SRT of 10 and 20 days for all the reactors, as shown in Figure 12.11d. The residual proteins were likely to be functional proteins or enzymes that could not be degraded via microbial utilization (Park et al., 2008). In addition, HAs that were generated during the anaerobic digestion could also be mistakenly detected as proteins with Lowry's method used.

12.9.4 DEWATERABILITY OF DIGESTED SLUDGE

The ULS and ULS-ozone treatments of feed sludge are often detrimental to the dewaterability of the digested sludge, as shown in Figure 12.12a. At an SRT of 10 days, the CST of the digested sludge from the Control reactor, ULS reactor and ULS-Ozone reactor was 56.9, 145.6 and 179.1 s, respectively. The dewaterability of the digested sludge further improved at a longer residence time. At an SRT of 20 days, the CST of the digested sludge from the Control reactor, ULS reactor and ULS-ozone reactor was 50.3, 101.4 and 120 s, respectively. This was because the treatment of feed sludge solubilized biopolymers that could bind with free water and worsen the sludge dewaterability (Wang et al., 2006a). Some of these biopolymers were persistent after anaerobic digestion and deteriorated the dewaterability of the digested sludge.

At the same SRT, the CST of the digested sludge from the ULS-Ozone reactor was slightly higher than that from the ULS reactor. And, the digested sludge from the Control reactor had the lowest CST in comparison to the digested sludge from the ULS and ULS-Ozone reactors. This indicated that the dewaterability of digested sludge was deteriorated by the ULS treatment of feed sludge and was further worsened by the subsequent ozonation to ULS pre-treatment. Although

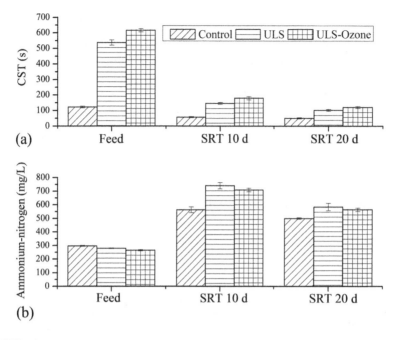

FIGURE 12.12 (a) Dewaterability ($n=3$) and (b) ammonia-nitrogen ($n=3$) in feed sludge and digested sludge from different anaerobic reactors. (Reproduced from Tian et al., 2015a, with permission from Elsevier.)

the influence of ULS-ozone treatment of feed sludge on the dewaterability of digested sludge has not been reported, the results of this work were in accordance with observations in a previous study where individual ULS and individual ozone treatments of feed sludge were found to deteriorate the dewaterability of digested sludge (Braguglia et al., 2012b).

12.9.5 AMMONIA-NITROGEN IN DIGESTED SLUDGE

Ammonia-nitrogen concentration increased after anaerobic digestion as a result of the degradation of proteinous compounds and the absence of nitrogen removal pathways (Kim et al., 2010). Averaged ammonia-nitrogen concentration in the digested sludge in the last 3 days of the anaerobic digestion tests was compared in Figure 12.12b. Digested sludge from the reactors fed with treated feed sludge had higher ammonia-nitrogen concentration than that from the control reactors. Previous studies that indicated the increase in ammonium concentration in the digested sludge could be a drawback of the pre-treatment step because pre-treatment steps released intra and extracellular proteins to be anaerobically degraded (Doğan and Sanin, 2009; Kim et al., 2010). However, it was noted that ULS-ozone treatment of feed sludge did not have such an effect on ammonia in the digested sludge of the anaerobic digester compared to ULS treatment. This might be due to the oxidative effect of ozone.

12.9.6   Molecular Weight Distribution of Organics in Digested Sludge

The MW distribution chromatograms of standard polymers are shown in Figure 12.13a. The MW chromatograms of the soluble substances in the feed sludge, in the digested sludge from reactors operating at 10 days SRT and in the digested sludge from reactors operating at 20 days SRT are shown in Figure 12.13c and d, respectively. Detected peaks were divided into five groups (A–F) in ascending order of retention time. The MWs of the components in these peaks are in descending order from A to F because larger compounds were retained for a shorter time in the column and eluted earlier.

12.9.6.1   High MW Compounds

Peaks A (Rt: 4.0 min) and B (Rt: 5.6 min) had the most obvious increase after treatments of feed sludge, as shown in Figure 12.13b. Compounds detected in these peaks were macromolecules with an MW higher than 500 kDa because the retention time of these peaks was shorter than the retention time of the largest tested standard polymer (MW: 500 kDa, Rt: 6.2 min). Therefore, these compounds were likely to be high MW extra- and intra-polymeric substances released from the sludge matrix after the treatment of feed sludge. The MW distribution of the digested sludge is shown in Figure 12.13c and d. Peak C (Rt: 6.0 min, MW > 500 kDa) instead of Peak B

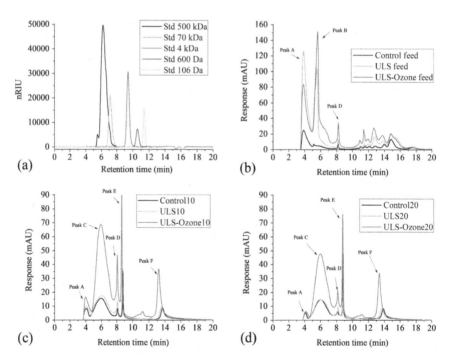

FIGURE 12.13 Molecular weight distribution chromatograms of (a) standard polymers, (b) supernatant in the feed sludge, (c) supernatant in the digested sludge from reactors operating at an SRT of 10 days and (d) supernatant in the digested sludge from reactors operating at an SRT of 20 days. (Reproduced from Tian et al., 2015a, with permission from Elsevier.)

was detected in the digested sludge together with Peak A. Compounds detected in Peak C could be generated from the hydrolysis of particulate polymers and higher MW macromolecules (Peak A) by hydrolytic bacteria because Peak C was detected only after anaerobic digestion. It should be noted that Peak C was broader than Peak B and covered the retention time of Peak B by comparing Figure 12.13c and d to b. Therefore, Peak B was possibly over-dominated by Peak C and thus not detected. As a result, soluble biopolymers released by treatments of feed sludge could also be detected in Peak C if remaining undigested.

At an SRT of 10 days, the responses of Peaks A and C in the digested sludge from the ULS-Ozone reactor were significantly higher than the corresponding responses in the digested sludge from the Control and ULS reactors, as shown in Figure 12.13c. This was due to the subsequent ozonation process because such a response increase was not observed when only ULS treatment was applied to the feed sludge. Similar observations were made on Peak D (Rt: 8.2 min) and Peak E (Rt: 8.8 min) with an MW around 19.4 and 7.7 kDa, respectively. These compounds (detected in Peaks D and E) were most likely intermediate products generated during the anaerobic degradation of macromolecules into monomers because their amounts were significantly lower in the feed sludge than those in the digested sludge.

The responses of Peaks A, C, D and E were lower at an SRT of 20 days, as compared in Figure 12.13c and d. This indicated that some of the compounds detected in these peaks were slowly biodegradable compounds. They were not biodegraded at an SRT of 10 days but could be digested at an SRT of 20 days. Some of these compounds were likely to be carbohydrates because the chemical results showed that some carbohydrates were complex polysaccharides and were not biodegradable at an SRT of 10 days but became biodegradable at an SRT of 20 days. At an SRT of 20 days, no obvious difference was observed between the MW chromatograms of the digested sludge from the Control and ULS reactors, as shown in Figure 12.13d. However, responses in Peaks A, C, D and E were significantly higher in the digested sludge from the ULS-Ozone reactor. This indicated that considerable amounts of polymeric substances remained undigested in the ULS-Ozone reactor even at an SRT of 20 days. These residual soluble polymeric compounds in the anaerobic digested sludge were possibly related to the solubilization of persistent compounds after the ULS-ozone treatment of feed sludge (Yang et al., 2013b; Tian et al., 2014c). These results correlated very well with the increase in biopolymers in the digested sludge.

12.9.6.2 Low MW Compounds

Peak F (Rt: 13.3 min, <106 Da) was detected in the supernatant of digested sludge, as shown in Figure 12.13c and d. The corresponding compounds were not monomers or other easily biodegradable components because they remained undigested at an SRT of 20 days. Therefore, they were possibly short chain alkenes or aromatics, which are anaerobic digestion by-products and could be detected by the UV 254 nm detector. The formation of these by-products is related to the chemical effects of the subsequent ozonation process because the response of Peak F was obviously higher in the digested sludge from the ULS-Ozone reactors. It is possible that the complex polymers in the untreated and ULS pre-treated sludge could only be broken down via biodegradation; while dosing ozone could provide different degradation pathways by chemically breaking down the high MW biopolymers into smaller fragments (Tian et al., 2014c). Additionally, ozone can convert some refractory compounds into biodegradable ones (Nishijima et al., 2003). These aforementioned factors could generate different substrates and the anaerobic digestion of these new substrates could contribute to the accumulation of the detected by-products.

Previous studies had focused on VSS removal and biogas production increase due to the treatment of feed sludge, but insights into the SCOD in the anaerobic digested sludge were not discussed in these studies (Tiehm et al., 2001; Doğan and Sanin, 2009). SCOD in the digested sludge could be attributed to slowly degradable components and recalcitrant anaerobic digestion by-products. MW distribution results allow better realization of the possible sources and categories of the residual components according to their MWs. This would be good supplementary information to conventional approaches in understanding the influence of treatments of feed sludge on subsequent anaerobic digestion.

12.9.7 FLUORESCENT PRODUCTS IN THE DIGESTED SLUDGE

An excitation emission matrix (EEM) fluorescence spectroscopy analysis was conducted to measure the fluorescence intensity (FI) of fluorescent compounds in the supernatant of digested sludge in the reactors. The EEM spectra of all the samples are shown in Figure 12.14a–f. The FI of the detected peaks is shown in numbered and colored contour lines for reference. According to the Ex/Em range introduced in Chapter 11, the main peaks were HA-like (as highlighted with a white arrow in Figure 12.14c) and FA-like (as highlighted with a white arrow in Figure 12.14d) substances. These substances were released from the biodegradation of the EPS, sludge pellets and refractory components (e.g., lignin) in the sewage sludge (Luo et al., 2013). Aside from these two groups, soluble microbial products (SMP)-like substances and simple protein-like matters were also detected in each spectrum. However, their FIs were relatively low and over-dominated by peaks of HA-like and FA-like substances.

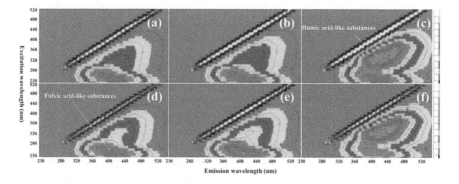

FIGURE 12.14 Excitation emission matrix fluorescent spectra of the supernatant in the digested sludge from (a) Control10, (b) ULS10, (c) ULS-Ozone10, (d) Control20, (e) ULS20 and (f) ULS-Ozone20. (Reproduced from Tian et al., 2015a, with permission from Elsevier.)

12.9.7.1   Humic Acid–Like Substances

At an SRT of 10 days, the FI of HA-like substances was similar in the digested sludge from the Control and ULS reactors, as shown in Figure 12.14a and b. In contrast, the FI of the HA-like substances was significantly higher in the digested sludge from the ULS-Ozone reactor, as shown in Figure 12.14c. This confirmed that some proteins detected with Lowry's method were attributed to humic substances. This is because the ULS-ozone treatments of feed sludge disintegrated the sludge better and solubilized more HA-containing substances in comparison to the ULS treatment. The biodegradation of these solubilized HA-containing substances resulted in a higher concentration of HAs as by-products. These HAs should contribute to the residual polymeric substances in the digested sludge from the ULS-Ozone reactors (e.g., Peak C, in Figure 12.13d), because HAs are known to be persistent and have a high MW (Li et al., 2009a; Stevenson, 1994). Such an increase in HAs during anaerobic digestion of pre-treated sludge was in good agreement with results obtained by Luo et al. (2013). They observed that the anaerobic digestion of enzymatically pre-treated WAS resulted in a higher FI of HA-like substances in the digested sludge compared to anaerobic digestion of untreated WAS (Luo et al., 2013).

By comparing the EEM spectra in Figure 12.14a and b to d and e, the FI of the HA-like matters in the digested sludge from the Control and the ULS reactors was found to increase at the longer retention time. This was likely because sludge was better digested at the longer retention time, which released more HAs as anaerobic digestion by-products. In contrast, the FI of the HA-like substances was similar in the digested sludge of the ULS-Ozone10 and the ULS-Ozone20 reactors, as shown in Figure 12.14c and f. This indicated that the HA-containing substances were mostly biodegraded and that the HAs were released into the supernatant within 10 days of anaerobic digestion for the ULS-ozone pre-treated sludge. A longer digestion time did not further increase the FI of the HA-like substances in the anaerobic digested sludge.

12.9.7.2   Fulvic Acid–Like Substances

The FI of the FA-like substances was similar in the digested sludge from the Control, ULS and ULS-ozone reactors at an SRT of 10 days, as shown in Figure 12.14a–c. By comparing Figure 12.14a and b to d and e, the FI of the FA-like substances in the digested sludge of the Control and the ULS reactors was found to increase when the SRT increased to 20 days, which was similar to the observations on the HA-like matters. However, the FI of the FA-like substances in the digested sludge of the ULS-Ozone reactor decreased when the SRT increased from 10 to 20 days, indicating that the FA-like compounds became biodegradable at a longer digestion time due to the subsequent ozonation step. This was supported by previous studies that observed the increase in the biodegradability of FAs due to the ozonation process (Volk et al., 1997; Kozyatnyk et al., 2013). Such an increase in the biodegradability of FAs is a potential advantage of the ozonation treatment of feed sludge and has not been emphasized in previous studies.

12.10   CONCLUSIONS OF THE CONTINUOUS PROCESS

This work investigated the impact of ULS-ozone treatment of pre-digestion feed sludge on sludge anaerobic digestion. The findings of this work are summarized as follows:

- The subsequent ozonation complemented the ULS treatment in improving biogas production and VS removal when the feed sludge was treated before anaerobic digestion.
- The ULS-ozone treatment of feed sludge could shorten the anaerobic digestion SRT from 20 to 10 days without an adverse impact on anaerobic digestion performance.
- The soluble polymeric substances were found to accumulate in the anaerobic digested sludge following anaerobic digestion of ULS-ozone-treated feed sludge. Such digested sludge had deteriorated dewaterability. Although some of these polymers were anaerobically degradable at 20 days SRT, most were HA-like substances and persistent.
- The biodegradability of FA-like substances was improved due to the application of ozone.

13 Comparison of the Effects of Ultrasonication, Ultrasonication-Ozonation and Ultrasonication-Alkaline Pre-Treatments of Sludge in Continuous Anaerobic Process

13.1 INTRODUCTION

The effects of ultrasonication (ULS), ultrasonication-ozonation (ULS-ozone) and ultrasonication + alkaline (ULS + ALK) treatments on sewage sludge have been introduced in previous chapters. In this chapter, the ULS, ULS-ozone and ULS + ALK-treated feed sludge were anaerobically digested in semi-continuous reactors at solids retention times (SRT) of 10 and 20 days. The biogas productions and solids removal efficiencies were measured to assess possible enhancement with such pre-treatments. Furthermore, the dewaterability of the digested sludge and soluble residual organics in the anaerobic digested sludge was also characterized to evaluate the impact of the treatment of feed sludge on the characteristics of digested sludge.

The ULS, ULS-ozone and ULS + ALK conditions tested in this chapter were selected following consideration of the results reported in previous chapters. The ULS energy input was selected at 9 kJ/g total solids (TS). The ozone dosage for ULS-ozone treatment was 0.012 g O_3/g TS. An NaOH concentration of 0.02 mol/L (NaOH dosage: 0.05 g NaOH/g TS) was selected for the ULS + ALK treatment of feed sludge because a higher NaOH concentration introduced more sodium ions, which induced negative effects on sludge anaerobic biodegradability.

13.2 BIOGAS PRODUCTION AND SOLIDS REMOVAL

Biogas production from anaerobic reactors is shown in Figure 13.1a and b. The anaerobic biodegradability of feed sewage sludge was higher at an SRT of 20 days with its longer substrate–microbe contact time. At both SRTs, the daily biogas production was higher from the reactors fed with treated feed sludge than from the Control reactor. Average daily biogas production from each reactor is as shown in Table 13.1. These values were calculated by averaging the daily biogas production in the third SRT. At an SRT of 10 days, daily biogas production increased from 256 to 309 (+20.7%), 348 (+35.9%) and 319 (+24.6%) mL biogas/g COD_{fed} as a result of the ULS, ULS-ozone and ULS + ALK treatments of feed sludge, respectively. At an SRT of 20 days, daily biogas production increased from 313 to 337 (+7.7%), 393 (+25.5%) and 365 (+16.6%) mL biogas/g COD_{fed} due to the ULS, ULS-ozone and ULS + ALK treatments of feed sludge, respectively. For all the reactors, the methane content was relatively stable at around 65%. In addition, no volatile fatty acids (VFAs) accumulation in the digested sludge was observed and the pH value remained near neutral throughout the anaerobic digestion test (around 7.0 to 7.2) in all the reactors. This suggested the ULS, ULS-ozone and ULS + ALK treatments of feed sludge did not cause stress on the methanogenesis step in the reactors.

The average volatile solids (VS) and volatile suspended solids (VSS) concentrations in the digested sludge are as shown in Table 13.1, and the corresponding VS removal efficiencies after anaerobic digestion against the untreated feed sludge are 35.6%, 38.3%, 42.1% and 41.2% for Control10, ULS10, ULS-Ozone10 and ULS + ALK10 reactors, respectively. The corresponding increase in VS removal rates after anaerobic digestion due to the ULS, ULS-ozone and ULS + ALK treatments of feed sludge were 7.6%, 18.3% and 15.7%, respectively. Solids removal efficiency was higher at longer SRTs. The VS removal rates after anaerobic digestion of the untreated ULS, ULS-ozone and ULS + ALK-treated feed sludge were 37.3%, 40.9%, 45.3% and 44.1% at an SRT of 20 days, respectively. The corresponding improvements in VS destruction after anaerobic digestion were 9.7%, 21.4% and 18.2% due to the ULS, ULS-ozone and ULS + ALK treatments of feed sludge. Higher VS removal efficiency indicated more organic matters were digested and converted into biogas. Similarly, integration of the pre-treatment step also improved the VSS removal efficiency as shown in Table 13.1. The increase in VSS removal efficiency indicated that particulate organics in the pre-treated sludge could be better hydrolyzed for the subsequent anaerobic digestion process.

The daily biogas production and solids removal efficiency was in the order of ULS-ozone > ULS + ALK > ULS > Control for the tested reactors. In addition, it was noted that the daily biogas production and solids removal efficiencies of the ULS-Ozone10 and ULS + ALK10 reactors were higher than those of the Control20 reactor. However, the daily biogas production and solids removal rates of the ULS10 reactor were lower than those of the Control20 reactor. This suggested that the ULS-ozone and ULS + ALK treatments of feed sludge could halve the SRT without affecting digestion performance; whereas, the individual ULS treatment of feed sludge was not able to provide such an advantage.

13.3 SCOD AND DEWATERABILITY IN DIGESTED SLUDGE

The average soluble chemical oxygen demand (SCOD) concentrations in the digested sludge during anaerobic digestion are as shown in Table 13.1. The treatments

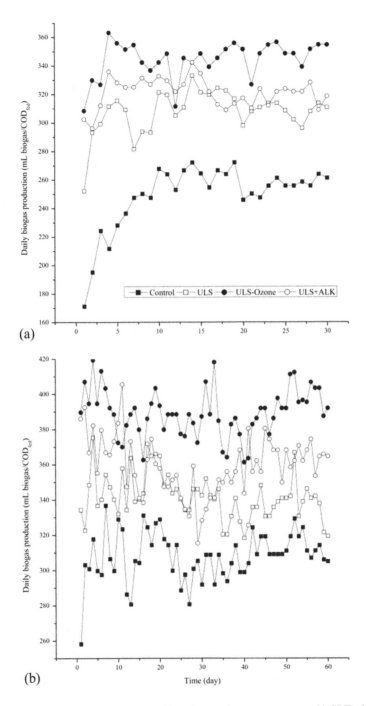

FIGURE 13.1 Daily biogas production from the continuous reactors at (a) SRT of 10 days (Control10, ULS10, ULS-Ozone10 and ULS+ALK10); (b) SRT of 20 days (Control20, ULS20, ULS-Ozone20 and ULS+ALK20).

TABLE 13.1

Performance of the Semi-Continuous Anaerobic Digesters at Steady State

Reactor	SRT 10 days				SRT 20 days			
	Control10	ULS10	ULS-Ozone10	ULS+ALK10	Control20	ULS20	ULS-Ozone20	ULS+ALK20
Biogas (mL/day g COD_{fed}) (n = 10 and 20)[a]	256±5	309±6	348±9	319±7	313±7	337±10	393±12	365±8
Methane percentage (%) (n=5)	64.2±0.9	65.2±0.8	65.6±0.6	65.1±0.8	64.9±0.5	64.9±1.3	65.4±1.2	65.0±1.3
Post-digestion TS (mg/L) (n=5 and 8)[b]	10650±602	10410±401	9720±370	10580±527	10306±180	10001±764	9375±236	10231±233
Post-digestion VS (mg/L) (n=5 and 8)[b]	8340±383	7990±344	7500±245	7620±202	8125±252	7644±253	7081±217	7243±437
Post-digestion TSS (mg/L) (n=5 and 8)[b]	9885±468	9340±171	8330±220	8970±501	9481±164	8769±153	8189±371	8613±415
Post-digestion VSS (mg/L) (n=5 and 8)[b]	8005±469	7640±155	6760±248	7038±165	7619±258	7056±213	6606±206	6706±246
Post-digestion SCOD (mg/L) (n=4)	193.6±28.1	256.8±34	588.7±53.3	296.3±30.3	181.6±24.3	227±15.1	440±33.5	245.7±19.5
Post-digestion proteins (mg/L) (n=3)	73.3±1.5	81.7±1.6	166.6±3.3	88.3±1.8	74.7±1.1	89.7±0.8	158.6±1.9	93.6±1.6
Post-digestion carbohydrates (mg/L) (n=3)	19.7±0.6	28.8±4.3	61.7±0.8	33.8±0.6	13.7±0.7	22.9±0.4	34.6±2.4	24.5±2.1
Post-digestion total VFA (mg/L) (n=3)	Nd[c]	Nd	Nd	Nd	Nd	Nd	Nd	Nd
CST in digested sludge (s) (n=3)	56.9±2.8	145.6±6.2	179.1±10	173.9±7	50.3±3.5	101.4±5.1	120±6.1	120.2±8.8

[a] n=10 at an SRT of 10 days and n=20 at an SRT of 20 days.

[b] n=5 at an SRT of 10 days and n=8 at an SRT of 20 days.

[c] Not detectable (<10 mg/L).

of feed sludge increased the post-digestion SCOD concentration at both SRTs. The post-digestion SCOD of the ULS-Ozone reactor was much higher than that of the ULS and ULS + ALK reactors. The SCOD in the digested sludge from the ULS-Ozone reactor was around 300 and 200 mg/L higher than that from the ULS reactors operating at an SRT of 10 and 20 days, respectively; while the SCOD in the ULS-ozone-treated feed sludge was 1200 mg/L higher than the SCOD in the ULS-treated feed sludge. These increases in post-digestion SCOD were much lower compared to the increase in SCOD in feed sludge due to the subsequent ozonation (around 1200 mg/L). This indicated that much of the COD solubilized by the subsequent ozonation treatment of feed sludge was biodegraded and only a relatively small fraction accumulated in the anaerobic reactors. However, the post-digestion SCOD concentration of the ULS + ALK reactor did not noticeably increase compared to the post-digestion SCOD of the ULS reactors at an SRT of 10 and 20 days. This indicated that the combination of alkaline treatment with ULS treatment did not increase the amount of soluble recalcitrant organics in the anaerobic digested sludge.

It is known that biopolymers are a major component of sludge (Rittman and McCarty, 2001a). Averaged values of soluble carbohydrates and proteins concentrations in the digested sludge during the last three sampling days are compared in Table 13.1. Soluble carbohydrates and proteins concentrations in the digested sludge from the ULS-Ozone reactor were much higher than their corresponding concentrations in the digested sludge from the ULS, ULS + ALK and Control reactors at both SRTs. This suggested that undigested biopolymers contributed to the higher SCOD in the digested sludge from the ULS-Ozone reactor.

The influence of SRT on the residual carbohydrates and proteins concentrations was different. As shown in Table 13.1, soluble carbohydrates concentrations noticeably decreased when all the reactors had a longer residence time. For example, the residual soluble carbohydrates concentration decreased from 62 to 35 mg/L when the SRT of the ULS-Ozone reactor increased from 10 to 20 days. This is because the solubilized carbohydrates after the treatments of feed sludge were mainly complex polysaccharides from extra- and intracellular structures (Tian et al., 2014c; Wang et al., 2006b). A longer residence time was needed for sufficient degradation. However, the soluble proteins concentrations in the digested sludge did not show an obvious difference between an SRT of 10 and 20 days for all the reactors. The residual proteins were likely to be functional proteins or enzymes that could not be degraded via microbial utilization (Park et al., 2008). In addition, humic acids (HAs) that were generated during anaerobic digestion could be mistakenly detected as proteins with the Lowry's method used.

The dewaterability of the digested sludge also decreased due to the treatments of feed sludge. At an SRT of 10 days, the capillary suction time (CST) of the digested sludge from the Control, ULS, ULS-Ozone and ULS + ALK reactors was 56.9, 145.6, 179.1 and 173.9 s, respectively. The dewaterability of the digested sludge further improved at a longer residence time. At an SRT of 20 days, the CST of the digested sludge from the Control, ULS, ULS-Ozone and ULS+ALK reactors was 50.3, 101.4, 120 and 120 s, respectively. This was because of the treatment of feed sludge solubilized biopolymers, which could bind with free water and worsen the sludge dewaterability (Wang et al., 2006a). Some of these biopolymers were persistent after anaerobic digestion and deteriorated the dewaterability of the digested sludge.

13.4 MOLECULAR WEIGHT DISTRIBUTION

Molecular weight (MW) distribution of organics in the digested sludge was measured with calibrated size exclusion chromatography (SEC). The MW chromatograms of the soluble substances in the digested sludge from reactors operating at 10 and 20 days SRT are as shown in Figure 13.2a and b, respectively. Detected peaks were divided into five groups (A–F) in ascending order of retention time. The MWs of the components in these peaks are in descending order from A to F because larger compounds were retained for a shorter time in the column and eluted earlier.

13.4.1 HIGH MW COMPOUNDS

At an SRT of 10 days, the responses of Peak A (Rt: 4.0 min) and Peak B (Rt: 6.2 min) in the digested sludge from the ULS-Ozone reactor were significantly higher than that from the Control, ULS and ULS+ALK reactors, as shown in Figure 13.2a. Compounds detected in these peaks were macromolecules with an MW higher than 500 kDa because the retention time of these peaks was shorter than the retention time of the largest tested standard polymer (MW: 500 kDa, Rt: 6.2 min). This indicated that anaerobic digestion of ULS-ozone-treated feed sludge resulted in the accumulation of soluble polymeric substances in the digested sludge. This was due to the subsequent ozonation process because these compounds were not observed when feed sludge was only treated with ULS.

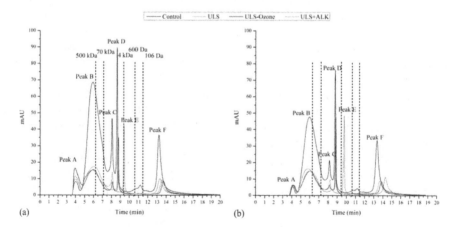

FIGURE 13.2 Molecular weight distribution chromatograms of (a) supernatant in the digested sludge from reactors operating at an SRT of 10 days; (b) supernatant in the digested sludge from reactors operating at an SRT of 20 days.

Similarly, the responses of Peak C (Rt: 8.2 min) and Peak D (Rt: 8.8 min) with an MW of around 19.4 and 7.7 kDa, respectively, were also obviously higher in the digested sludge from the ULS-Ozone reactor than those in the digested sludge from the Control, ULS and ULS + ALK reactors. However, it should be noted that Peak E (Rt: 9.4 min, MW: 3 kDa) instead of Peak D was detected in the digested sludge of the ULS + ALK reactor. This was likely related to the alkaline hydrolysis of soluble macromolecules during the ULS + ALK treatment of feed sludge. This generated new substrates compared to the ULS and ULS-ozone pre-treatments and anaerobic digestion then resulted in the formation of different residual organics in the anaerobic digested sludge.

The responses of Peaks A, C, D and E were lower at an SRT of 20 days than at an SRT of 10 days as compared in Figure 13.2a and b. This indicated that some of the compounds detected in these peaks were slowly biodegradable compounds. They were not biodegraded at an SRT of 10 days but could be degraded at an SRT of 20 days. Some of these compounds were likely to be carbohydrates because it was shown that some carbohydrates were not biodegradable at an SRT of 10 days but became biodegradable at an SRT of 20 days. At an SRT of 20 days, hardly any difference was observed between the MW chromatograms of the digested sludge from the Control, ULS and ULS + ALK reactors as shown in Figure 13.2b. However, the responses in Peaks A, B, C and D were still significantly higher in the digested sludge from the ULS-Ozone reactor. This indicated considerable amounts of polymeric substances remained undigested in the ULS-Ozone reactor even at an SRT of 20 days. These residual soluble polymeric compounds in the anaerobic digested sludge were possibly related to the solubilization of persistent compounds after the ULS-ozone treatment of feed sludge (Yang et al., 2013b; Tian et al., 2014c). These results correlated very well with the increase in biopolymers in the digested sludge.

13.4.2 LOW MW COMPOUNDS

Peak F (Rt: 13.3 min, <106 Da) was detected in the digested sludge from the ULS-Ozone reactor, as shown in Figure 13.2a and b. These compounds were not monomers or other easily biodegradable components because they remained undigested at an SRT of 20 days. Therefore, they were possibly short chain alkenes or aromatics, which are by-products from the degradation of larger molecules and could be detected by the UV_{254} nm detector. The formation of these by-products is related to the subsequent ozonation process because the response was only obviously higher in the digested sludge from the ULS-Ozone reactor. It is possible that the complex polymers in the untreated and ULS-treated feed sludge could only be broken down via biodegradation; while dosing ozone could provide different degradation pathways by chemically breaking down the high MW biopolymers into smaller fragments (Tian et al., 2014c). Besides, ozone can convert some refractory compounds into biodegradable ones (Nishijima et al., 2003). These aforementioned factors could generate different substrates and the anaerobic digestion of these new substrates could contribute to the accumulation of the detected by-products.

13.5   FLUORESCENT PRODUCTS CHARACTERIZATION

Excitation emission matrix (EEM) fluorescence spectroscopy analysis was conducted to measure the fluorescence intensity (FI) of fluorescent compounds in the supernatant of digested sludge from the reactors. The EEM spectra of all the samples are as shown in Figure 13.3a–h. The FI of the detected peaks is shown in numbered and color contour lines for reference. According to the Ex/Em range, the main peaks were HA-like (as highlighted with a white arrow in Figure 13.3c) and fulvic acid (FA)-like substances (as highlighted with a white arrow in Figure 13.3e). These substances were released from the biodegradation of the extracellular polymeric substances, sludge pellets and refractory components (e.g., lignin) in sewage sludge (Luo et al., 2013). Aside from these two groups, soluble microbial product (SMP)-like substances and simple protein-like matters were also detected in each spectrum. However, their FIs were relatively low and over-dominated by peaks of HA-like and FA-like substances.

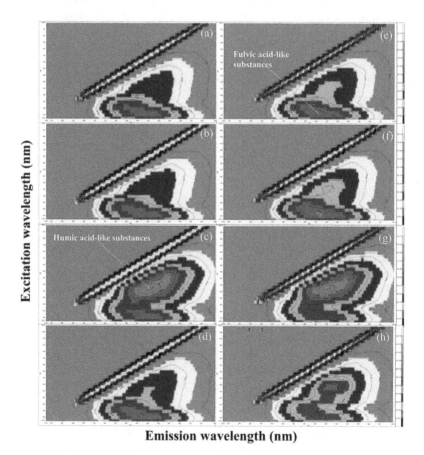

FIGURE 13.3 Excitation emission matrix fluorescent spectra of the supernatant in the digested sludge from (a) Control10, (b) ULS10, (c) ULS-Ozone10, (d) ULS + ALK10, (e) Control20, (f) ULS20, (g) ULS-Ozone20 and (h) ULS + ALK20 reactors.

13.5.1 HUMIC ACID–LIKE SUBSTANCES

At an SRT of 10 days, the FI of HA-like substances was similar in the digested sludge from the Control and ULS reactors, as shown in Figure 13.3a and b. In contrast, the FI of the HA-like substances was significantly higher in the digested sludge from the ULS-Ozone reactor, as shown in Figure 13.3c. This confirmed that some proteins detected with Lowry's method were attributed to humic substances. This is because the ULS-ozone treatment of feed sludge disintegrated the sludge better and solubilized more HA-containing substances in comparison with the ULS treatment. The biodegradation of these solubilized HA-containing substances resulted in a higher concentration of HAs as by-products. These HAs should contribute to the residual polymeric substances in the digested sludge from the ULS-Ozone reactor (e.g., Peak C in Figure 13.2b), because HAs are known to be persistent and have a high MW (Li et al., 2009a; Stevenson, 1994). Similarly, the FI of the HA-like substances was slightly higher in the digested sludge from the ULS + ALK reactor than in the digested sludge from the ULS reactor. Such an increase in HAs during anaerobic digestion of treated feed sludge was in good agreement with the results obtained by Luo et al. (2013). They observed that anaerobic digestion of enzymatically treated waste activated sludge (WAS) resulted in a higher FI of HA-like substances compared to the anaerobic digestion of untreated WAS in the digested sludge (Luo et al., 2013).

By comparing the EEM spectra in Figure 13.3a, b and d to e, f and h, the FI of the HA-like matters in the digested sludge from the Control, ULS and ULS + ALK reactors was found to increase at the longer retention time. This was likely because sludge was better digested at the longer retention time, releasing more HAs as anaerobic digestion by-products. In contrast, the FI of the HA-like substances was similar in the digested sludge from the ULS-Ozone10 and the ULS-Ozone20 reactors, as shown in Figure 13.3c and g. This indicated that the HA-containing substances were mostly biodegraded and the HAs were released into the supernatant within 10 days of anaerobic digestion for the ULS-ozone-treated feed sludge. A longer digestion time did not further increase the FI of the HA-like substances in the anaerobic digested sludge.

13.5.2 FULVIC ACID–LIKE SUBSTANCES

The FI of the FA-like substances was similar in the digested sludge from the Control, ULS and ULS-Ozone reactors at an SRT of 10 days, as shown in Figure 13.3a–d. By comparing Figure 13.3a and b with Figure 13.3e and f, the FI of the FA-like substances in the digested sludge from the Control and the ULS reactors was found to increase when the SRT increased to 20 days, which was similar to the observations on the HA-like matters. However, the FI of the FA-like substances in the digested sludge from the ULS-Ozone and ULS + ALK reactors decreased when the SRT increased from 10 to 20 days, indicating that the FA-like compounds became biodegradable at a longer digestion time. Such a biodegradability increase in the FA-like substances is related to the use of chemical treatment (e.g., ozone or alkali) to enhance the ULS treatment of feed sludge. For example, ozone was reported to increase the biodegradability of FAs in previous studies (Volk et al., 1997; Kozyatnyk et al., 2013). Such an increase in the biodegradability of FAs is a potential advantage of ULS + ALK and ULS-ozone pre-treatments and has not been emphasized in previous studies.

13.6   CONCLUSIONS

This chapter compared the impact of ULS, ULS-ozone and ULS + ALK pre-treatments on sewage anaerobic sludge digestion. The biogas production and solids removal efficiencies from the tested anaerobic reactors were in the order of ULS-Ozone > ULS + ALK > ULS > Control reactors at both SRTs of 10 and 20 days. ULS-ozone and ULS + ALK treatments of sludge could shorten the anaerobic SRT from 20 to 10 days without an adverse impact on the anaerobic digestion performance. Soluble polymeric substances were found to accumulate in the anaerobic digested sludge following anaerobic digestion of ULS-ozone-treated feed sludge. Such digested sludge had deteriorated dewaterability. Although some of these polymers were anaerobically degradable at 20 days SRT, most were HA-like substances and persistent. In addition, the biodegradability of FA-like substances was improved due to the ULS-ozone and ULS + ALK treatments of feed sludge.

14 Physicochemical Treatment

Application of Ozone, Ultrasonic and Alkaline Post-Treatment of Sludge

14.1 INTRODUCTION

The waste activated sludge (WAS) process produces between 180 and 270 kg of sludge per megaliter of wastewater treated (Metcalf and Eddy, 2014), and the biosolids management system is considered cost-intensive as it typically accounts for 25%–60% of the total operational costs of conventional activated sludge–based wastewater treatment plants (Canales et al., 1994; Verstraete and Vlaeminck, 2011). Anaerobic digestion (AD) is a commonly used technology for sewage sludge stabilization in wastewater treatment plants. Aside from solids stabilization, biogas is produced. However, AD is a slow process with a solids retention time (SRT) of around 20–30 days (Appels et al., 2008). Hydrolysis of particulate organics in sludge is the rate-limiting step in sludge digestion (Eastman and Ferguson, 1981; Pavlostathis and Giraldo-Gomez, 1991). Pre-treatment of sludge before AD is often applied to solubilize these solids to accelerate the subsequent digestion. However, Takashima et al. (1996) indicated that the pre-treatment targets not only the slowly biodegradable solids but also the easily biodegradable solids in WAS. As a result, part of the energy and chemical input during pre-treatment is wasted on solubilizing the easily biodegradable organic particulates without increasing the overall sludge biodegradability. Such wastage of pre-treatment energy and chemicals would be present when pre-treatment is applied to sludge containing primary sludge (PS) (e.g., sewage sludge), which would contain more biodegradable solids. In view of such potential inefficiency, post-treatment of the digested sludge and thereafter digesting the treated digested sludge again could be an alternative for more economical improvement of the AD performance (Nielsen et al., 2010; Li et al., 2013). Compared to pre-treatment, post-treatment would more specifically target the solids that are more difficult to biodegrade in digested sludge. However, reports on post-treatment are relatively limited in number.

14.2 PROCESS CONFIGURATION AND OPERATING CONDITIONS

The rationale for post-treatment and pre-treatment is similar wherein both aim to rupture the microbial cells and release the extra and intracellular substances. Therefore, pre-treatment methods are also potentially suitable for post-treatment. For example, the reported pre-treatment methods such as alkaline (ALK), ozone and thermal pre-treatment (Ray et al., 1990; Goel et al., 2003a; Bougrier et al., 2006) have all been reported to be feasible for post-treatment as well (Battimelli et al., 2003; Takashima, 2008; Li et al., 2013). The process configuration is show in Figure 14.1.

Ultrasonication (ULS), a proven pre-treatment method, has, however, not been reported for post-treatment. Ultrasonic pre-treatment can be enhanced when assisted with chemical methods (Kim et al., 2010; Xu et al., 2010a). This chapter discusses the feasibility of applying ULS and chemically (with alkali or ozone) assisted ULS in the treatment of digested sludge (i.e., post-treatment). A change in the physical properties of digested sludge was assessed with particle size distribution. Sludge solubilization was measured chemically in terms of chemical oxygen demand (COD), proteins and carbohydrates, and fluorescently in terms of soluble microbial products (SMP) and humic acid (HA). Furthermore, the molecular weight (MW) distribution of the soluble substances was also investigated to shed light on the solubilization products. Batch AD tests on the treated digested sludge were then conducted to evaluate the influence of post-treatment on the improvement of biodegradability if the digested sludge was digested again.

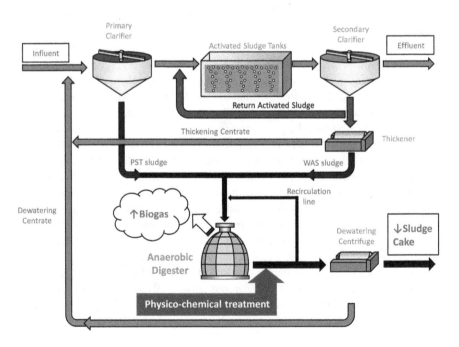

FIGURE 14.1   Flow sheet including a post-treatment of digested sludge with a recirculation line.

Based on prior tests, the specific energy input was set at 9 kJ/g TS (Tian et al., 2014c). A control sludge sample without any form of post-treatment was also analyzed along the tested post-treatments. Ozonation was performed with an ozone generator (Wedeco, GSO 30). A stone diffuser was installed to produce fine ozone bubbles and to enhance ozone mass transfer. The applied ozone dosage of 0.012 g O_3/g TS was also set based on prior tests (Tian et al., 2014c). For the ALK post-treatment, a 3 M NaOH stock solution was added to the digested sludge sample to reach a concentration of 0.02 M. The digested sludge samples were then mixed at 200 rpm for 10 min at room temperature (25°C). For chemically assisted ULS treatment, the combination sequence was determined based on earlier tests (Tian et al., 2014b,c). Combined ULS and ozone (ULS-ozone) treatment was conducted by applying ozonation after ULS treatment. The combined ultrasound and ALK (ULS + ALK) treatment was applied by ultrasonicating the sludge after NaOH addition.

14.3 PARTICLE SIZE REDUCTION FOLLOWING POST-TREATMENT

Median particle diameter results are shown in Figure 14.2. Chemical post-treatments without ULS post-treatment did not have a great impact on the digested sludge particle size. The median diameter slightly changed from 45.3 to 43.1 and 48.3 μm (statistically different at 95% confidence) after the ozone and ALK post-treatments, respectively. The slight increase in median diameter can be explained by flocs re-flocculating with the aid of electropositive organic polymers solubilized after the ALK treatment (Li et al., 2008). In contrast, the ULS post-treatment showed better capability of reducing particles sizes than the chemical methods and reduced the

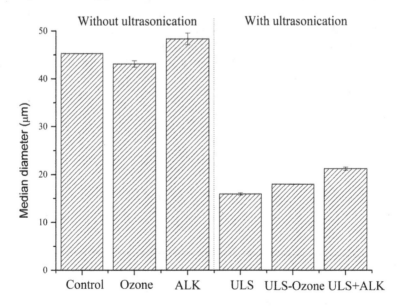

FIGURE 14.2 Effect of investigated post-treatments on the mean diameter of biological flocks in digested sludge. Error bars indicate the standard deviation of triplicate measurements. (Reproduced from Tian et al., 2016, with permission from Elsevier.)

median diameter significantly from 45.3 to 15.9 µm. The ULS treatment mechanically reduced the floc whereas the chemical methods could only chemically lyze the cells or extra-polymeric substances without causing significant change in particle sizes (Bougrier et al., 2006). When the chemical methods were used in combination with ULS, the treatment resulted in a statistically greater median diameter compared to the individual ULS post-treatment. This indicated that combining either ozone or ALK to the ULS post-treatment would not enhance the floc size reduction in digested sludge.

14.4   COD AND BIOPOLYMERS SOLUBILIZATION DURING POST-TREATMENT

SCOD concentrations of the sludge increased after the post-treatment steps, as shown in Figure 14.3a. The ULS post-treatment was obviously more effective than the ozone and ALK post-treatments in terms of COD solubilization. The ozone and ALK post-treatments only increased the SCOD concentrations from 200 ± 1.1 to 647.5 ± 9.7 and 501 ± 23.8 mg/L, respectively; whereas the ULS post-treatment was able to increase the SCOD to 1502 ± 6.5 mg/L. SCOD further increased to 2611 ± 14.2 and 2648 ± 66.1 mg/L when the ULS post-treatment was assisted with the ozone and ALK post-treatments, respectively. COD solubilization by the ozone and ALK post-treatments was significantly improved when these post-treatments were combined with the ULS post-treatment, as shown in Figure 14.3a. This suggested that the chemical post-treatments induced synergistic COD solubilization when combined with the ULS post-treatment. Although such a synergistic SCOD increase has not been reported before for the post-treatment of digested sludge, synergistic SCOD solubilization had been reported for the combined ULS and ALK pre-treatment of WAS (Kim et al., 2010). Kim et al. (2010) had suggested that the alkali addition made the cells in WAS more vulnerable to ultrasound attack and induced synergistic COD solubilization. However, synergistic SCOD solubilization

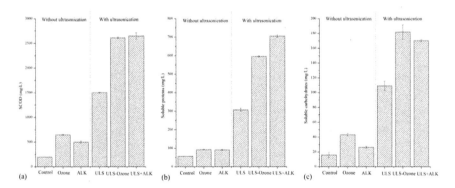

FIGURE 14.3   Change in (a) soluble chemical oxygen demand (SCOD), (b) soluble proteins and (c) soluble carbohydrates due to chemical, ultrasound and chemically assisted ultrasound post-treatments. Error bars indicate the standard deviation of triplicate measurements. (Reproduced from Tian et al., 2016, with permission from Elsevier.)

had not been observed when ULS and ozone were combined for pre-treatment, though the smaller particles created by the ULS step benefited the mass transfer of the subsequent ozonation process. Tian et al. (2014c) suggested that the reaction between ozone and some of the organics solubilized by ultrasound decreased the net SCOD during the combined ultrasound and ozone pre-treatment. Therefore, it is possible that ozone did not readily react with the organics solubilized by ultrasound, but instead specifically solubilized the particulate organics in digested sludge during the combined ultrasound and ozone post-treatment, which resulted in the synergistic COD solubilization.

The concentration of soluble biopolymers also increased after various post-treatments, as shown in Figure 14.3b and c. The increases in the concentration of soluble proteins and carbohydrates after the ozone and ALK post-treatments were relatively insignificant in comparison to the increases after the ULS post-treatment. In addition, the solubilization of proteins and carbohydrates due to the ozone or ALK post-treatment became more obvious when they were combined with the ULS post-treatment. These results indicated that synergistic interactions between the ULS and the chemical treatments helped disintegrate the biological flocs, resulting in the release of biopolymers from the sludge.

14.5 MOLECULAR WEIGHT DISTRIBUTION OF SOLUBILIZED ORGANICS

The MW distribution of the standard polymer mixture solution is shown in Figure 14.4a. The MW distributions of the samples without and with ULS post-treatment were compared separately in Figure 14.4b and c. The MW distribution results showed good agreement with the SCOD results. The ozone and ALK post-treatments were relatively incapable of solubilizing the sludge and thus the corresponding peak response increases were low, as shown in Figure 14.4b. However, the UV responses of the soluble organics were obviously higher after the ULS or chemically assisted ULS treatments, as shown in Figure 14.4b and c.

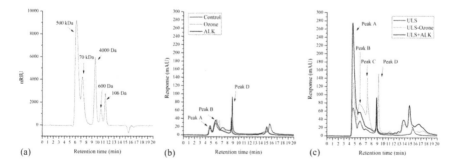

FIGURE 14.4 Molecular weight distribution chromatograms of (a) a mixture of standard polymers solution; (b) control, ozone-treated and ALK-treated digested sludge; and (c) ULS-treated, ULS-ozone-treated and ULS + ALK-treated digested sludge. (Reproduced from Tian et al., 2016, with permission from Elsevier.)

The soluble organics in the control and treated digested sludge were detected at Peaks A, B, C and D. The retention times of Peak A (Rt: 4.7 min) and Peak B (Rt: 5.8 min) were shorter than the retention time of the largest standard polymer (Rt: 6.2 min). Therefore, the MWs of the compounds detected in Peaks A and B were larger than 500 kDa. The compounds detected in Peak C (Rt: 7.2 min) had an MW around 103 kDa according to the calibration equation. Due to the high MWs, these compounds (i.e., in Peak A, B and C) were possibly from the solubilization of cellular polymeric substances and humic substances from the biological flocs in the digested sludge. The combination of the ozone or ALK post-treatment with the ULS post-treatment obviously increased the UV responses of macromolecules, as shown in Figure 14.4c. However, such increases were relatively negligible when the ozone and the ALK post-treatments were individually applied, as shown in Figure 14.4b. This correlated well with the synergistic COD solubilization and indicated that synergistically solubilized organics have an MW higher than 500 kDa. Although similar SCOD concentrations were obtained, the MW distributions of the solubilized compounds were different after the ULS-ozone and ULS+ALK post-treatments. For example, the ULS-ozone post-treatment increased the UV responses of Peaks A, B and C, while the ULS+ALK post-treatment only significantly increased the UV responses of Peak A and had relatively little impact on the UV responses of Peaks B and C. This indicated that the solubilized compounds due to the ULS-ozone and ULS+ALK treatments were different.

In addition to the aforementioned macromolecules, Peak D with a retention time of around 8.9 min (MW: 7.6 kDa) was also detected in the supernatant of all the tested digested sludge. The UV response of Peak D decreased after the ozone and ALK post-treatments. It is possible that the corresponding compounds were broken down due to the chemical attack induced by the ozone and ALK post-treatments. However, the ozone and ALK post-treatments increased the UV responses of Peak D when they were combined with the ULS post-treatment. This indicated that the synergistic effects in the ULS-ozone and ULS+ALK treatments also resulted in the solubilization of compounds detected in Peak D. However, such an increase was relatively less obvious in comparison with the increase in Peaks A, B and C.

14.6   CHARACTERIZATION OF FLUORESCENT PRODUCTS IN SOLUBILIZED ORGANICS

The excitation emission matrix (EEM) spectra of the supernatant of the control and treated digested sludge are compared in Figure 14.5a–f. The most obvious change in the fluorescence intensity (FI) increase was observed in the SMP-like and HA-like substances according to the Ex/Em ranges. The maximum FI of the SMP-like and HA-like peaks is summarized in Table 14.1.

The FI of the SMP-like substances increased after post-treatment due to the solubilization of extra and intracellular polymers. The FI increase of the SMP-like substances was relatively insignificant after the ozone and the ALK post-treatments (from 443 to 453 and 450), while the ULS post-treatment showed an obvious increase in SMP-like matters (from 443 to 620), which was in accordance with the biochemical results.

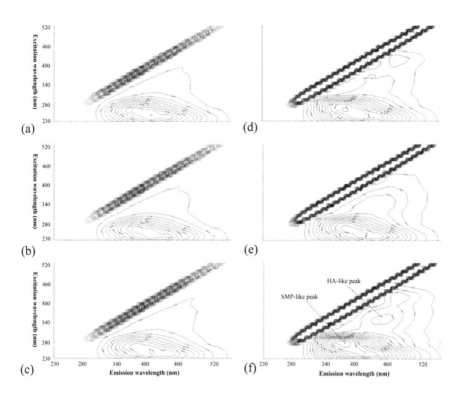

FIGURE 14.5 EEM spectra of the supernatant of (a) control, (b) ozone-treated digested sludge, (c) ALK-treated digested sludge, (d) ULS-treated digested sludge, (e) ULS-ozone-treated digested sludge and (f) ULS+ALK-treated digested sludge. (Reproduced from Tian et al., 2016, with permission from Elsevier.)

The FI of the SMP-like substances was highest (FI: 820) when the ULS post-treatment was combined with the ALK post-treatment, as shown in Table 14.1. This is in good agreement with the biopolymers results and supported the conclusion that the ALK post-treatment made the cells more vulnerable, which resulted in more cell lysis during ULS. However, the FI of the SMP-like peak in the ULS-ozone-treated sample (FI: 589) was slightly lower than the ULS-treated sample (FI: 620), despite the higher soluble biopolymers concentration. This indicated that the soluble biopolymers in the ULS-ozone post-treated sludge could not be fluorescently detected. It is also possible that the solubilized compounds did not have fluorescent characteristics or the FI was weakened at a higher redox potential due to ozonation.

The FI of the HA-like substances was also found to increase after the post-treatments. The FI increase in the HA-like substances was relatively insignificant from 71 to 73 and 91 after the ozone and ALK post-treatments, respectively, while the FI of the HA-like substances was increased to 122 after the ULS post-treatment. In addition, the FI became even higher when the ULS treatment was combined with the chemical methods. The FI of HA-like peaks increased to 193 and 225 after ULS-ozone and ULS+ALK post-treatments, respectively. Such an increase in the HA-like substances is related to the better disintegration of the biological

TABLE 14.1

Maximum Fluorescence Intensity (FI) of the Observed Peaks in the Supernatant of the Control and Post-Treated Digested Sludge

Sample	SMP-Like (Ex/Em)	FI	HA-Like (Ex/Em)	FI
Control	260/375	443	330/425	71
Ozone	260/380	453	330/425	73
ALK	260/380	450	330/340	91
ULS	270/375	620	330/420	122
ULS-ozone	270/380	589	340/425	193
ULS+ALK	280/370	820	350/445	225

Source: Reproduced from Tian et al., 2016, with permission from Elsevier.
Note: SMP, soluble microbial products; HA, humic acid; FI, fluorescence intensity.

flocs and, as a result, the HAs contained in EPS, sludge pellets and some refractory compounds were released (Yang et al., 2013b; Luo et al., 2013). However, it should not be neglected that the ULS+ALK post-treatment increased the pH of the sample, which favored the solubilization of HAs as observed previously (Tian et al., 2014b). Although HAs are known to be non-biodegradable, such solubilization of HA-like substances is often related to better disintegration of the sludge structure.

14.7 BATCH ANAEROBIC RE-DIGESTION FOLLOWING THE POST-TREATMENTS

14.7.1 BIOGAS PRODUCTION

The control, ULS, ULS-ozone and ULS+ALK post-treated digested sludge were anaerobically digested to determine an improvement in methane production due to the post-treatments. As shown in Figure 14.6a, methane production was significantly improved with the post-treatments. The methane produced increased by 28.3% from 54.7 ± 0.1 to 70.2 ± 0.1 mL CH_4 after the ULS post-treatment. Such an increase was much higher than that observed in our previous study (around 10%) when the same energy was applied to sewage sludge as a pre-treatment (Tian et al., 2014c). The sludge used for the pre-treatment contained both PS and WAS in a 1:1 ratio. Energy would have been wasted on solubilizing biodegradable solids in sludge during pre-treatment without necessarily increasing the sludge biodegradability. However, easily biodegradable components in raw sludge could be biologically degraded by AD. The post-treatment of the digested sludge could then focus on the solubilization of the less biodegradable solids. Methane production improved from 54.7 ± 0.1 to 81.1 ± 0.5 and 76.3 ± 1.1 mL CH_4 after the ULS-ozone and ULS+ALK post-treatments, respectively. This increase in methane production was because the ULS-ozone and ULS+ALK post-treatments released more organics for AD in comparison to the ULS post-treatment. The methane production from the ULS-ozone-treated sludge

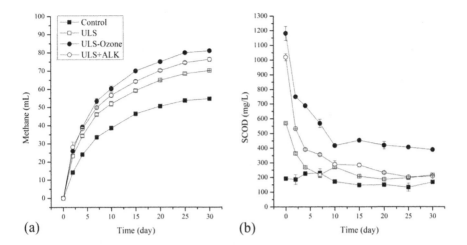

FIGURE 14.6 (a) Change in methane production from the control and post-treated digested sludge. (b) Change in SCOD during the batch anaerobic digestion of the control and post-treated digested sludge. Error bars indicate the standard deviation of triplicate tests. The error bars were omitted when smaller than the marker. (Reproduced from Tian et al., 2016, with permission from Elsevier.)

was higher than that from the ULS + ALK-treated sludge despite the similar SCOD concentration. This was possibly because ozone was not only able to solubilize organics but was also able to convert non-biodegradable solids into biodegradable ones, whereas the ALK treatment was unable to do so. The batch test results showed that the ULS and chemically assisted ULS post-treatments were able to improve the biodegradability of digested sludge. As a result, the overall methane recovery from an AD system could potentially be increased with greater solids destruction when post-treated digested sludge is re-digested –by recycling either to the original digester or to a downstream digester.

As the ULS and chemically aided ULS post-treatment had not been reported before, the results from this work were compared with previous studies using different post-treatment methods. Takashima (2008) obtained a 130%–200% improvement in CH_4 production when a 120°C thermal post-treatment was applied; while Takashima and Tanaka (2014b) observed a 172.7%–190.9% increase in sludge biodegradability when the post-treatment was conducted in a pH range of 2–6. However, the methane production increase after the ULS, ULS-ozone and ULS + ALK treatments in this work was in the range 28.3%–48.3%. In some studies, thermal treatment at medium (60°C–80°C) and high (130°C–170°C) temperatures and pressure (up to 21 bar) resulted in a biogas production increase in the range 30%–80% (Barber, 2016), but the efficiency depends on the type of sludge, the sludge rheology, the inoculum used in the AD test and the scale of the process. Compared to high temperature and pressure treatments, the methods described in this study do not require a boiler and cooling step prior to the digester and require less investment.

Aside from the different efficiencies of the compared treatment methods, such performance difference could also be attributed to the sludge age (i.e., SRT) difference of the tested digested sludge. In this work, the digested sludge was taken from a semi-continuous anaerobic reactor with an SRT of 10 days, while the digested sludge used in previous studies was taken from an anaerobic digester with an SRT of 20 days. As a result, the digested sludge in this work should contain more readily biodegradable solids, which may absorb the treatment energy without increasing the biodegradability and weakening the treatment efficiency.

In addition, it was reported that chemical methods had not enhanced the thermal post-treatment in terms of methane production. Takashima and Tanaka (2008) found that the addition of Na_2CO_3, H_2O_2 and HCl did not further improve methane production following thermal post-treatment at 170°C, while ozone only slightly improved methane production from 148 to 150 mL CH_4/g VS_{added} (+1.3%). Similarly, Nielsen et al. (2010) observed similar ultimate methane production when a 170°C thermal treatment was applied individually (377 mL CH_4/g VS_{added}) and when it was assisted with an ALK treatment at pH 10 (374 mL CH_4/g VS_{added}). However, the results of this work showed that ozone and ALK treatments were good supplements to the ULS treatment by further increasing methane production significantly. The corresponding improvements were 15.5% (ozone: from 70.2 to 81.1 mL CH_4) and 8.7% (ALK: from 70.2 to 76.3 mL CH_4). This is possibly because combined thermal and chemical treatments would create extreme conditions, which may result in the formation of recalcitrant compounds; while combined ULS and chemical treatments may not have such negative effects.

14.7.2　CHANGE IN SCOD DURING ANAEROBIC DIGESTION

The change in SCOD in serum bottles during AD is shown in Figure 14.6b. The SCOD in the control digested sludge only fluctuated slightly during the AD tests. This indicated that the COD solubilized by the hydrolytic bacteria was immediately used for biogas production. However, the SCOD dropped significantly during AD of the post-treated digested sludge. This indicated that the hydrolysis bottleneck was better overcome after post-treatment. However, it should be noted that post-treatment also resulted in higher residual SCOD at the end of the AD test ($P < 0.05$). Such an increase in the residual SCOD was also observed in ULS and ULS + ALK pre-treatments (Tiehm et al., 2001; Kim et al., 2010) and thermal post-treatment (Takashima, 2008). Such treatment of the digested sludge could have resulted in the solubilization of refractory compounds. For example, the solubilization of HA-like matters increased after ULS, ULS-ozone and ULS + ALK post-treatments. Luo et al. (2013) also pointed out that the pre-treatment step could solubilize compounds, which were further biodegraded to HAs and these HAs remained in the digested sludge due to poor biodegradability.

14.8　CONCLUSIONS OF THE BATCH RE-DIGESTION

This chapter demonstrated the possibility of using ULS treatment and chemically assisted ULS treatment to disintegrate anaerobically digested sludge to enhance methane production. Although the ozone and the ALK post-treatments were poor at disintegrating the digested sludge when applied on their own, their performance was significantly improved when combined with the ULS post-treatment. Synergistic COD and biopolymers solubilization was observed during ULS-ozone and ULS + ALK post-treatments of the digested sludge. However, the synergistically solubilized substances were different according to the MWs and the fluorescent characteristics analysis. Methane production from re-digesting the treated digested sludge was increased by 28.3%, 48.3% and 39.5% after the ULS, ULS-ozone and ULS + ALK post-treatments, respectively.

14.9 PERFORMANCE OF A CONTINUOUS PROCESS WITH POST-TREATMENT

14.9.1 INTRODUCTION

Post-treatment is realized by treating the digested sludge and recycling the treated digested sludge back to the original anaerobic reactor, as shown in Figure 14.7. The concept was first proposed by Gossett et al. (1982) who found that the performance of thermal treatment was more efficient when the substrate (i.e., municipal refuse) was first biodegraded compared to the situation where the thermal treatment was directly applied to the substrate. Pre-treatment of sludge before AD is often applied to solubilize these solids to accelerate subsequent digestion. The rationale for post-treatment and pre-treatment is similar wherein both aim to rupture the microbial cells and release the extra and intracellular substances. However, Takashima et al. (1996) indicated that pre-treatment targets not only the slowly biodegradable solids, but also the easily biodegradable solids in WAS. As a result, part of the energy and chemical input during pre-treatment would then be wasted on solubilizing the easily biodegradable organic particulates without increasing the overall sludge biodegradability. Takashima et al. (1996) suggested that the post-treatment of digested sludge

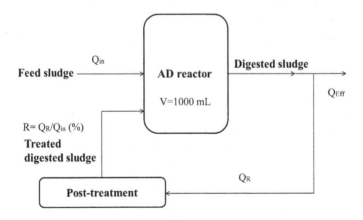

FIGURE 14.7 Schematic diagram of the anaerobic digestion process incorporating a post-treatment. Q_{in} = influent flow rate (mL/day), Q_{Eff} = effluent flow rate (mL/day), Q_R = recycled flow rate (mL/day) and V = working volume (mL). (Reproduced from Tian and Trzcinski, 2017, under open access license.)

and recycling the treated digested sludge back to the anaerobic reactor could be an alternative to pre-treatment.

As the digested sludge primarily contains slowly biodegradable and refractory solids, the energy of the post-treatment focuses on converting the non-biodegradable solids into biodegradable ones (Li et al., 2013; Nielsen et al., 2010; Takashima et al., 1996).

In some wastewater treatment plants, the highly biodegradable PS and more recalcitrant WAS are combined. In this situation, the post-treatment could be more efficient than the pre-treatment, because the solids in PS contain a higher content of biodegradable solids (Takashima, 2008). Compared to a pre-treatment, the post-treatment would more specifically target solids, which are more difficult to biode-grade in digested sludge.

However, studies on post-treatment techniques are relatively scarce in comparison with the information on pre-treatment, and most studies carried out batch AD tests. Ozone (Battimelli et al., 2003; Goel et al., 2003c), ALK (Li et al., 2013), thermal (Rivero et al., 2006; Takashima, 2008), thermal/acid (Takashima and Tanaka, 2010, 2014b) and thermal/ALK (Nielsen et al., 2010) treatments were successfully applied in sludge post-treatment. There are also papers showing that post-treatment was superior to pre-treatment in terms of improving the AD effectiveness (Rivero et al., 2006; Takashima, 2008). Battimelli et al. (2003) and Li et al. (2013) indicated that the recycle ratio of the post-treated sludge is an important operation parameter. It has impacts on the actual SRT of the anaerobic reactor, as well as the AD performance.

However, ULS, combined ULS-ozone and combined ULS + ALK post-treatments have not yet been documented in continuous reactors. Accordingly, information about the AD performance and the stress on microbial communities with post-treatment at different hydraulic retention times (HRTs) is not available. Therefore, this section aims to compare the influence of the ULS, ULS-ozone and ULS + ALK post-treatments on the AD performance of sewage sludge in semi-continuous reactors, as well as comparing the performance of pre- and post-treatment under the same conditions. The change in daily biogas production and suspended solids concentration was used to evaluate the AD performance at different HRTs and recycle ratios.

14.9.2 Pre- and Post-Treatment Conditions

The ULS treatment was performed with an ultrasonicator (Misonix, Q700, Qsonica, Newtown, USA) at 20 kHz. The power rating of the ultrasonicator is 700 W. During ULS, the temperature was monitored and maintained at about 30°C with an ice-water bath. According to previous chapters, the specific energy input was selected at 9 kJ/g TS. The ULS-ozone treatment was conducted by applying ozonation after the ULS treatment. The ozonation was performed with an ozone generator (Wedeco, GSO 30, Xylem Water Solutions Herford GmBH, Herford, Germany). A stone diffuser was installed to produce fine ozone bubbles and to enhance ozone mass transfer. The applied ozone dosage of 0.012 g O_3/g TS was selected based on Chapter 13. The ULS + ALK treatment was applied by ultrasonicating the sludge, which was mixed at 200 rpm at an NaOH (Sigma-Aldrich, St Louis, USA) concentration of 0.02 M. The NaOH concentration was reached by adding a 3 M stock solution into the

TABLE 14.2

Operational Conditions of Each Reactor

Operational Conditions	Condition I	Condition II	Condition III
Duration of the experiment (days)	15	15	31
HRT = V/Q_{in} (days)	10	10	20
Influent flow rate, Q_{in} (mL/day)	100	100	50
Recycle ratio, $R = Q_R/Q_{in}$ (%)	50	100	100
Post-treatment factor, $\alpha = Q_R/V$ (%)	5	10	5

Source: Reproduced from Tian and Trzcinski, 2017, under open access license.

sludge. The ULS + ALK post-treated digested sludge was neutralized with 6 M HCl before being recycled back to the anaerobic digester.

Before starting the post-treatment tests, all reactors were operated at 10 days HRT for 20 days to stabilize the reactor and obtain similar reactor performances. Afterward, a specific amount of sludge was treated and the recycle ratio R was calculated as follows: $R = Q_R/Q_{in}$ (%). Three different conditions were tested in the reactors, as shown in Table 14.2.

Condition I: HRT = 10 days and R = 50%
Condition II: HRT = 10 days and R = 100%
Condition III: HRT = 20 days and R = 100%

Feeding, withdrawal and recycling of sludge were conducted manually once a day. The recycle ratio (R) was calculated as the ratio of recycled sludge (Q_R) to the influent flow rate (Q_{in}). For a recycle ratio of 100%, the same volume of fresh sludge and post-treated sludge is added to the reactor, so the reactor receives half its feed as fresh sludge. The post-treatment factor (α) was calculated as the ratio of daily recycled sludge volume to the reactor working volume as defined by Li et al. (2013).

14.9.3 Biogas Production

The daily biogas production from each reactor is shown in Figure 14.8. The daily gas production from the four reactors was similar in the first 20 days stabilization period, indicating that the performance of each reactor was similar before the post-treatment was applied. The incorporation of the post-treatment improved the daily biogas production from 20 days onward.

The methane composition in biogas was around 64% in all the tests, indicating that post-treatments did not affect the methane composition. In Condition I, the biogas production due to the ULS, ULS-ozone and ULS + ALK post-treatments was, respectively, 5.2%, 7.1%, and 8.2% greater than in the control. This was achieved at R = 50% meaning that 50 mL/day of digested sludge going through post-treatment is mixed with 100 mL/day of raw feed. This post-treatment configuration would,

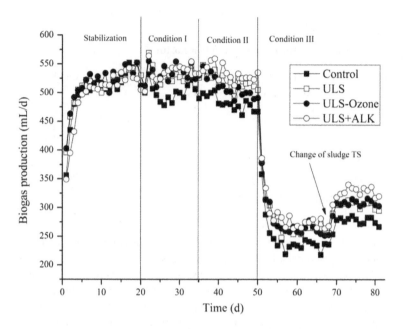

FIGURE 14.8 Daily biogas production from the control and test reactors. A fraction of the digested sludge was treated by ultrasound (ULS), ultrasound and ozone (ULS-ozone) or ultrasound and alkali (ULS + ALK). Condition I: HRT = 10 days and $R = 50\%$; Condition II: HRT = 10 days and $R = 100\%$; and Condition III: HRT = 20 days and $R = 100\%$. (Reproduced from Tian and Trzcinski, 2017, under open access license.)

therefore, consume half the energy required for the corresponding pre-treatment configuration (100 mL/day would have to be treated through any pre-treatment) while still achieving a significant biogas increase.

In Condition II, the ULS, ULS-ozone and ULS + ALK post-treatments increased the daily biogas production by 8%, 4.9%, and 11.1%, respectively. The biogas production due to the ULS (8%) and ULS + ALK (11.1%) post-treatments was higher than in Condition I (5.2% and 8.2%, respectively). This is because more digested sludge was post-treated and recycled as substrate (higher R). However, biogas production due to the ULS-ozone post-treatment decreased when the volume of recycled sludge doubled. Furthermore, a t-test confirmed that the daily biogas production from the ULS-Ozone reactor was statistically lower than that produced from the ULS reactor. Li et al. (2013) observed a decrease in biogas production when the recycled sludge (treated with 0.1 M NaOH for 30 min) exceeded 5% of the total working volume of the anaerobic digester ($\alpha > 5\%$). According to Li et al. (2013), the decrease in biogas production was related to the inactivation of anaerobic microorganisms at higher α values. However, in this study, an increase of α from 5% to 10% only decreased the biogas production from the ULS-Ozone reactor, but increased the biogas production from the ULS and ULS + ALK reactors. Therefore, the influence of α on biogas production was dependent on the selected treatment methods. In addition, no volatile fatty acids (VFAs) were detected in the effluent of any of the reactors during

Conditions I and II, suggesting that the methanogenesis step was not inhibited even at 10 days HRT.

In Condition III, the biogas increases due to the ULS, ULS-ozone and ULS + ALK post-treatments were 9.8%, 10.7%, and 17.8%. These increases were statistically greater than the corresponding increases observed in Conditions I and II. This is due to the higher HRT of 20 days applied during Condition III. On the one hand, the higher HRT provided more time for the biodegradation of the feed and post-treated sludge. On the other hand, the digested sludge contained less biodegradable solids. The post-treatment energy could solubilize biodegradable solids more slowly, which also benefited the overall AD. Future studies should focus on two-stage AD with an inter-stage physicochemical treatment.

Previous studies on batch AD assays showed that the ULS-ozone post-treatment resulted in higher ultimate methane production than the ULS + ALK post-treatment (Tian et al., 2015a,b, 2016). However, this was not the case when the post-treatment was applied in semi-continuous reactors. This is because post-treated digested sludge acted as a substrate and was given sufficient time (30 days) for degradation during the batch assay. In contrast, the HRT was much shorter in semi-continuous reactors, as shown in Table 14.2. It is known that the addition of ozone unavoidably increased the oxidation-reduction potential of the reactor and may have induced a lag phase. This shortened the degradation time under strict anaerobic conditions in one cycle and might have decreased the biogas recovery rate.

Effluent soluble COD (SCOD) increased due to the incorporation of the post-treatment and this was accompanied by higher capillary suction times (CST). These recalcitrant organics were the result of the post-treatment that solubilized some non-biodegradable biopolymers when lyzing the anaerobic microorganisms in digested sludge. In addition, HA-like substances were also formed as by-products during the AD of the solubilized macromolecules, which contributed to the effluent SCOD concentration (Luo et al., 2013). The dewaterability of the digested sludge also deteriorated after post-treatment, as shown in Table 14.3. This was related to the soluble residual biopolymers in the effluent that keep the solids from being dewatered. The ULS post-treatment was mainly responsible for the increase in effluent SCOD and CST. The combination of ALK and ULS treatments did not make the effluent SCOD and dewaterability worse. This was in accordance with a previous work by Li et al. (2013) where individual ALK post-treatment (0.1 mol/L NaOH) had negligible impacts on the SCOD and dewaterability in the digested sludge.

It should be noted that the digested sludge from the ULS-Ozone reactor had slightly higher SCOD and CST compared to the digested sludge from the ULS reactor. The application of ozonation subsequent to the ULS post-treatment increased the effluent SCOD concentration from 224 to 237 mg/L in Condition I. However, such an increase was statistically insignificant compared to the change caused by the ULS post-treatment (from 182 to 224 mg/L). The increases in biogas production were statistically significant as confirmed by the t-test provided in Table 14.4. In addition, the t-test results also showed that the biogas increase due to the ULS-ozone and ULS + ALK post-treatments was statistically higher than ULS alone, showing that the chemical methods were adding value to the ultrasound post-treatment.

TABLE 14.3

Summary of Anaerobic Reactors Performance when a Post-Treatment was Applied to Digested Sludge

Performance Parameter	Control	ULS	ULS-Ozone	ULS+ALK
Condition I: 10 days HRT, RR=50%, α=5%				
Daily biogas production (mL/d) ($n=9$)	500 ± 12	526 ± 9	525 ± 12	541 ± 6
Methane yield (mL CH_4/g VS_{added}) ($n=9$)	256 ± 5	269 ± 5	268 ± 6	277 ± 3
Effluent TS (mg/L) ($n=5$)	$11,460 \pm 481$	$10,720 \pm 309$	$10,830 \pm 292$	$11,390 \pm 392$
Effluent TSS (mg/L) ($n=5$)	$9,980 \pm 220$	$9,710 \pm 606$	$9,240 \pm 487$	$9,840 \pm 198$
Effluent VS (mg/L) ($n=5$)	$8,470 \pm 333$	$7,940 \pm 420$	$8,160 \pm 429$	$8,140 \pm 397$
Effluent VSS (mg/L) ($n=5$)	$7,840 \pm 219$	$7,530 \pm 202$	$7,280 \pm 394$	$7,630 \pm 211$
SCOD (mg/L) ($n=5$)	182 ± 6	224 ± 7	237 ± 8	220 ± 7
CST (s) ($n=3$)	64.7 ± 4.3	113.8 ± 8.3	148.7 ± 7.4	128.1 ± 6.8
Condition II: 10 days HRT, RR=100%, α=10%				
Daily biogas production (mL/day) ($n=9$)	474 ± 8	512 ± 9	498 ± 8	526 ± 7
Methane yield (mL CH_4/g VS_{added}) ($n=9$)	249 ± 4	269 ± 5	261 ± 4	276 ± 4
Effluent TS (mg/L) ($n=5$)	$10,960 \pm 378$	$10,780 \pm 275$	$10,500 \pm 252$	$11,300 \pm 362$
Effluent TSS (mg/L) ($n=5$)	$9,690 \pm 368$	$9,570 \pm 529$	$9,140 \pm 608$	$9,490 \pm 595$
Effluent VS (mg/L) ($n=5$)	$8,150 \pm 406$	$7,920 \pm 431$	$7,710 \pm 222$	$7,760 \pm 347$
Effluent VSS (mg/L) ($n=5$)	$7,740 \pm 111$	$7,700 \pm 82$	$7,310 \pm 342$	$7,390 \pm 403$
SCOD (mg/L) ($n=5$)	184 ± 11	228 ± 3	242 ± 3	234 ± 4
CST (s) ($n=3$)	63.1 ± 3.3	127.4 ± 5.4	143.9 ± 6.3	131.2 ± 5.1
Condition III: 20 days HRT, RR=100%, α=5%				
Daily biogas production (mL/day) ($n=11$)	279 ± 5	306 ± 5	309 ± 5	329 ± 7
Methane yield (mL CH_4/g VS_{added}) ($n=9$)	275 ± 5	301 ± 5	304 ± 5	324 ± 7
Effluent TS (mg/L) ($n=5$)	$11,820 \pm 480$	$11,320 \pm 649$	$11,470 \pm 160$	$12,230 \pm 850$
Effluent TSS (mg/L) ($n=5$)	$10,560 \pm 227$	$9,850 \pm 173$	$9,800 \pm 509$	$9,860 \pm 403$
Effluent VS (mg/L) ($n=5$)	$8,710 \pm 399$	$8,160 \pm 282$	$8,080 \pm 354$	$8,210 \pm 530$
Effluent VSS (mg/L) ($n=5$)	$8,460 \pm 393$	$7,750 \pm 364$	$7,540 \pm 531$	$7,700 \pm 285$
SCOD (mg/L) ($n=5$)	225 ± 6	245 ± 4	270 ± 11	246 ± 4
CST (s) ($n=3$)	74.6 ± 4.4	134.8 ± 5.7	156 ± 5	143.8 ± 5.4

Source: Reproduced from Tian and Trzcinski, 2017, under open access license.

Note: TS, total solids; TSS, total suspended solids; VS, volatile solids; VSS, volatile suspended solids; SCOD, soluble chemical oxygen demand; CST, capillary suction time; n, number of days at the end of the experiment during which the data were averaged; ULS, ultrasound post-treatment; ULS-ozone, ultrasound and ozone post-treatment; ULS+ALK, ultrasound and alkali post-treatment.

14.9.4 MICROBIAL STRESS DURING SEMI-CONTINUOUS ANAEROBIC DIGESTION

The adenosine tri-phosphate (ATP) distribution in digested sludge is shown in Figure 14.9. All the post-treatments under all HRTs and the recycle ratio tested resulted in a lower cellular ATP compared to the control despite the small proportion of post-treated sludge compared to the total reactor volume (small α values). The effect on dissolved ATP was marginal. The decrease in the cellular ATP concentration indicated the decrease in the activity of microorganisms due to the post-treatment, which was not shown in earlier studies. Interestingly, this lower cellular ATP did not prevent higher biogas production when a post-treatment was applied.

The biomass stress index (BSI), the ratio between the dissolved and total ATP concentrations, was used to quantify the stress of microbial communities in the anaerobic reactor. Surprisingly, the BSI in the ULS-Ozone reactor was slightly lower than in the ULS reactor. This meant that the use of ozone in the post-treatment did not impose further stress on the anaerobic reactor compared to ULS alone.

However, the BSI in the ULS + ALK reactor was the highest under all conditions tested. This can be due to the accumulation of dissolved solids (i.e., sodium ions). In addition, the increase in BSI was more obvious in Conditions I and II (10 days HRT). For example, the BSI increased from 40.1% to 55.7% in Condition I, but only increased from 28.4% to 32.4% in Condition III. This would suggest that the ULS + ALK post-treatment imposed more stress on the anaerobic reactor when the HRT was 10 days. Furthermore, when looking at all the BSI during the three consecutive test periods (I–III), it was found that the stress gradually decreased, which could be due to the acclimatization and adaptation of the microbial communities to the corresponding post-treatment over time. Future studies should, therefore, look at the long-term performance of such post-treatments.

14.9.5 PERFORMANCE OF THE CONTINUOUS POST-TREATMENT PROCESSES

The post-treatment also had impacts on the characteristics of the digested sludge, as shown in Table 14.3. The solids concentration (i.e., total solids [TS], total suspended solids [TSS], volatile solids [VS] and volatile suspended solids [VSS]) in the digested sludge was determined by averaging the corresponding concentrations from 5 sampling days. A t-test was conducted to statistically compare the results, as shown in Table 14.4. In many cases, the average effluent VS concentrations from the ULS, ULS-Ozone and ULS + ALK reactors were lower, but not statistically significant compared to the effluent VS concentration from the Control reactor. The post-treatment could obviously improve the biogas recovery from sludge AD while its effects on VS destruction were relatively limited (in the range 4%–7%). This was because the recycling of the post-treated sludge increased the biodegradable organic loadings of the reactor, which benefited biogas production.

The tested post-treatments showed the highest VS removals in Condition III. This was because a longer residence time was given for the hydrolysis of feed and post-treated sludge. Battimelli et al. (2003) showed that the COD and solids removal rates started to decrease when the recycle ratio (R) between the recycled sludge (treated

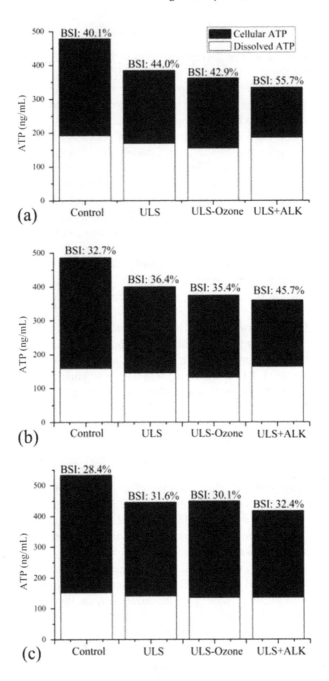

FIGURE 14.9 ATP distribution of each reactor at Condition I (a), Condition II (b) and Condition III (c). (Reproduced from Tian and Trzcinski, 2017, under open access license.)

TABLE 14.4

Statistical Analysis of the Biogas Production Increase and Effluent VS Decrease Due to Post-Treatment at Different Conditions

Statistical Parameter	Control	ULS	ULS-Ozone	ULS+ALK
Condition I: 10 days HRT, RR=50%, α=5%				
Daily biogas production increase (%)	–	5.2	7.1	8.2
t-test compared to control	–	7.83[a]	7.65[a]	11.06[a]
t-test compared to ULS	–	–	3.63[a]	7.67[a]
Decrease in effluent VS (%)	–	6.3	3.7	3.9
t-test compared to control	–	–4.17[b]	–2.48[c]	–2.8[b]
t-test compared to ULS	–	–	1.09[c]	0.83[c]
Condition II: 10 days HRT, RR=100%, α=10%				
Daily biogas production increase (%)	–	8	4.9	11.1
t-test compared to control	–	12.67[a]	9.44[a]	16.75[a]
t-test compared to ULS	–	–	–10.93[b]	5.8[a]
Decrease in effluent VS (%)	–	2.8	5.4	4.8
t-test compared to control	–	–1.76[c]	–3.96[b]	–1.77[c]
t-test compared to ULS	–	–	–2.09[c]	–0.61[c]
Condition III: 20 days HRT, RR=100%, α=5%				
Daily biogas production increase (%)	–	9.8	10.7	17.8
t-test compared to control	–	22.6[a]	29.6[a]	24.21[a]
t-test compared to ULS	–	–	1.84[c]	16.68[a]
Decrease in effluent VS (%)	–	6.3	7.2	5.7
t-test compared to control	–	–5.5[b]	–6.68[b]	–1.96[c]
t-test compared to ULS	–	–	–1.21[c]	0.21[c]

Source: Reproduced from Tian and Trzcinski, 2017, under open access license.
Note: VS, volatile solids.
[a] Significantly higher (P-value >2.306).
[b] Significantly lower (P-value <–2.306).
[c] Not significant higher (P-value between –2.306 and 2.306).

with 0.16 g O_3/g TS) and the feed sludge exceeded 25% due to the reduction in SRT. This reduction was caused by cell lysis in the recycle line due to the post-treatment. These results confirm the importance of an appropriate recycle ratio and sufficient residence time of the anaerobic reactor with the post-treatment incorporated.

The t-test results also indicated that neither the ULS-ozone nor the ULS+ALK post-treatment showed obvious increases in VS removal compared to the ULS post-treatment in the tested conditions, indicating that the chemical methods did not significantly benefit the solids removal caused by the ULS treatment. Although the effluent TS concentration decreased for the ULS and ULS-Ozone reactors as a result of the VS destruction, the effluent TS was similar to the Control reactor during Condition I. However, the effluent TS was slightly higher for the ULS + ALK reactor (from 11.82 g/L in the control to 12.23 g/L in Condition III) due to the addition of NaOH, which increased the dissolved solids concentration in the reactor over time. This increase is consistent with our NaOH dosage of 0.02 M or 800 mg/L. This is in contrast with the literature that reported a decrease in TS due to the ULS + ALK pre-treatment in batch mode: in Seng et al. (2010), the TS removal increased from 12.5% (control digester) to 17% with a chemical dose of 15 mg/g TS and then continued increasing to around 18% when the chemical dose increased to 25 mg/g TS. However, in a continuous reactor with 10 mg NaOH/g TS, the TS removal was only 2% at 25 days HRT. The authors explained that the low TS removal for chemical–ultrasound pre-treated WAS was due to the addition of NaOH, which contributed to the TS content. This work confirmed that low TS removals can be expected when ULS + ALK is used as post-treatment.

Li et al. (2013) indicated the potential risk at an α factor of 10% and 15% while the reactor was operated at 20 days HRT. This emphasized the importance of choosing an appropriate recycle ratio, especially when the ALK treatment is applied. Although inhibition due to dissolved solids (e.g., sodium ions) was not observed in Li et al. (2013) or this study, the risk of sodium inhibition is present over time. Moreover, contamination of excess sludge with sodium may require special disposal considerations.

The biogas production increase in the semi-continuous AD reactors due to pre- and post-treatments is compared in Table 14.5. The ULS-ozone pre-treatment resulted in a higher biogas production increase than the ULS-ozone post-treatment at 10 and 20 days HRT. This indicated that the ULS-ozone was more suitable for the treatment of feed sewage sludge than for the treatment of digested sludge in enhancing biogas production. This is related to the effects of the treated sludge on the AD process. The feed sludge acted as a substrate for AD, whereas the digested sludge not only acted as a substrate for the AD, but it also contained active anaerobic microorganisms that were essential for the AD (Battimelli et al., 2003; Li et al., 2013). Consequently, the post-treatment method can have negative effects such as the inactivation of anaerobic bacteria in digested sludge. Therefore, the lower increase in biogas production observed in the ULS-ozone post-treatment configuration was due to the inactivation or lysis of essential anaerobic microorganisms (e.g., hydrogenotrophic methanogens) in the digested sludge. This could have negated its positive effects on the biodegradability improvement.

Similarly, the pre-treatment configuration was more advantageous than the post-treatment configuration in terms of enhancing the biogas production at a HRT of 10 days for the ULS and ULS + ALK treatments. In contrast, the post-treatment configuration performed slightly better at a HRT of 20 days for these treatments.

The full-scale application of ultrasound to pre-treat sludge was reported to result in a 13%–58% increase in biogas and up to 22% solids destruction at an energy input of 1.44 kWh/m^3 of treated sludge (Xie et al., 2007). A small laboratory-scale probe

TABLE 14.5

Comparison of Performance of Pre-Treatment and Post-Treatment using ULS, ULS-Ozone and ULS + ALK Treatments at 10 and 20 days HRT

Performance Parameter	Treatment (HRT, R)	ULS	ULS-Ozone	ULS + ALK
Biogas increase (%)	Pre-treatment (10)	20.7	35.9	24.6
	Pre-treatment (20)	7.7	25.5	16.6
	Post-treatment (10%, 50%)	5.2	7.1	8.2
	Post-treatment (10%, 100%)	8	4.9	11.1
	Post-treatment (20%, 100%)	9.8	10.7	17.8
Solids removal (%)	Pre-treatment (10)	7.6	18.3	15.7
	Pre-treatment (20)	9.7	21.4	18.2
	Post-treatment (10%, 50%)	11.7	6.8	7.3
	Post-treatment (10%, 100%)	4.7	9.1	8
	Post-treatment (20%, 100%)	9.5	10.9	8.6
Post-digestion SCOD concentration (mg/L)	Pre-treatment (10)	194–257	194–589	194–296
	Pre-treatment (20)	182–227	182–440	182–246
	Post-treatment (10%, 50%)	182–224	182–236	182–220
	Post-treatment (10%, 100%)	184–228	184–242	184–234
	Post-treatment (20%, 100%)	225–245	225–270	225–246

Source: Reproduced from Tian and Trzcinski, 2017, under open access license.
Note: HRT, hydraulic retention time (days); *R*, recycle ratio in post-treatment (%).

was used in this study, which required a significantly higher energy input of 9000 kJ/g TS (equivalent to 25 kWh/m^3) to observe a similar performance. Nevertheless, the combination of ozone or ALK with ultrasound is unlikely to justify the additional energy demand given the increment in biogas production compared to ULS alone. Based on the laboratory data, the application of ULS post-treatment seems justified and future studies could investigate the inclusion of a ULS step in between two digesters.

In terms of solids removal, ULS-ozone and ULS-ALK achieved better results in a pre-treatment configuration regardless of the HRT and recycle ratio. However, the ULS post-treatment at 10 days HRT and 50% recycle ratio achieved better removal than in the pre-treatment (11.7% versus 7.6%). Moreover, this was achieved with only 50 mL of sludge being post-treated, whereas 100 mL was treated in the pre-treatment configuration. This indicated the potential of ULS to be used as a post-treatment using half the amount of sludge, hence, half the energy input. At 20 days HRT, both pre- and post-ULS treatment achieved about 9.5% solids removal, which was consistent with the corresponding increase in biogas production. All configurations (pre and post) resulted in an increase in the final effluent SCOD, which translated to an increase in CST.

14.9.6 Conclusions of the Continuous Processes with Post-Treatments

This section has shown that the post-treatments were able to increase the biogas production and decrease the VS in the final effluent. The maximum daily biogas increase was 17.8% when the ULS + ALK post-treatment was applied to a reactor operating at 20 days HRT and 100% recycle ratio. At 50% recycle ratio (Condition I), biogas increase in the range 5%–8% can be achieved at half the energy input required in a comparable pre-treatment configuration. Based on the results, the post-treatment of digested sludge or treating the sludge between two digesters is an interesting alternative to pre-treatments.

14.10 CONCLUSIONS AND FUTURE PROSPECTS

AD holds the potential to harvest bio energy from sewage sludge. A number of pre-treatments have been proposed to enhance this process. Currently, mechanical, thermal and chemical pre-treatments are being intensively investigated. A few patented technologies have been applied in practical sludge treatment: CambiTHP™, Biothelys® (TH), Biosonator (ultrasonication), Aspal SLUDGE™, Praxair® Lyso™ (microwave), BioCrack (electrokinetic disintegration), MicroSludge™ and Cellruptor (HPH). However, the work on biological techniques is not exhaustive and is still undergoing laboratory-scale experiments. System up-scaling is extremely necessary in order to make the biological processes practically applicable for industrial applications. In addition, to maximize resources recovery from sewage sludge, the dewatering centrate can been post-treated to produce value-added by-products (i.e., struvite, $MgNH_4PO_4 \cdot 6H_2O$) by chemical precipitation.

In addition, a systematic assessment of different pre-treatment options is quite necessary for deciding which one would be the most suitable from an industrial point of view. Making a comparison, however, is a very hard task since multiple influencing factors should be considered. The technical feasibility of one method is dependent upon not only the degree of sludge disintegration and methane conversion efficiency, but also the energy and environmental benefits. Additionally, to increase methane conversion efficiency, pre-treatment can also affect the energy required for AD, dewatering, transportation and ultimate disposal (landfill, incineration, compost or land application) as well as corresponding greenhouse gas emissions. Unfortunately, to date, most studies dealing with sludge pre-treatment have mainly focused on the former with hardly any consideration of the energy and environmental issues. Other factors that will impact the evaluation results include the operation skill of the operator, the maintenance frequency of devices, the local circumstance of labor, land prices, the market for renewable energy exchange, etc., in a real-world scenario. Apparently, for a reliable comparison, supplementary information (e.g. "standard cost–benefit optimization tool") is needed to evaluate such technologies from energy, economic and environmental perspectives.

Although the combined pre-treatment generally provides better performance compared to individual pre-treatment, special care is needed in selecting the treatment methods. When appropriate treatments are selected, performance may be enhanced due to synergistic effects. However, not all the treatments with different mechanisms have supplements or synergistic effects when combined. Using the wrong treatment combination may even result in a negative effect on sludge solubilization and the subsequent AD process. Alkali was found to be the most commonly used chemical for combined treatment. Synergistic mechanisms should be clearly understood before applying a combined treatment, because it is necessary to optimize treatment inputs and maximize the benefits. Mathematical optimization such as central composite design and multivariable linear regression can be applied for process optimization. The sequence of the combination is important in some combined treatment processes. When two proven feasible treatment methods are combined in the wrong sequence, lower or even no synergistic effect is normally observed. Despite the advantages of synergistic effects, process complexity, scaling-up issues and cost increases in combined treatments can be deterrents. Further research on a larger scale is needed to determine the optimum operating conditions, sequence order and reactor design in order to optimize the synergy. An economic assessment is needed to determine if the combination is more interesting than a single treatment. More importantly, there is a lack of a "standard cost–benefit optimization tool" to assess the technical availability of each pre-treatment method from the energy, economic and environmental perspectives, limiting a reliable comparison across the literature. The necessity to establish such standardization should become a vital research issue in the future so as to help the industry to determine the most cost-efficient co-treatment route to ensure optimal sludge conversion and energy recovery.

References

Ahn, K. H., K. Y. Park, S. K. Maeng, J. H. Hwang, J. W. Lee, K. G. Song, and S. Choi. (2002). Ozonation of wastewater sludge for reduction and recycling. *Water Science and Technology: A Journal of the International Association on Water Pollution Research* 46 (10):71–77.

Ahn, J. H., S. G. Shin, and S. Hwang. (2009). Effect of microwave irradiation on the disintegration and acidogenesis of municipal secondary sludge. *Chemical Engineering Journal* 153 (1–3):145–150.

Akin, B. (2008). Waste activated sludge disintegration in an ultrasonic batch reactor. *CLEAN – Soil, Air, Water* 36 (4):360–365.

Aldin, S. (2010). The effect of particle size on hydrolysis and modeling of anaerobic digestion. PhD thesis. Department of Chemical and Biochemical Engineering, University of Western Ontario, London, Ontario.

Alfaro, N., R. Cano, and F. Fdz-Polanco. (2014). Effect of thermal hydrolysis and ultrasounds pretreatments on foaming in anaerobic digesters. *Bioresource Technology* 170:477–482.

Ali, M. and S. Okabe. (2015). Anammox-based technologies for nitrogen removal: Advances in process start-up and remaining issues. *Chemosphere* 141:144–153.

Anjum, M., N. H. Al-Makishah, and M. A. Barakat. (2016). Wastewater sludge stabilization using pre-treatment methods. *Process Safety and Environmental Protection* 102:615–632.

APHA. (2005). *Standard Methods for the Examination of Water and Wastewater*. Washington, DC: American Public Health Association.

APHA. (2012). *Standard Methods for the Examination of Water and Wastewater*. edited by A. D. Eaton, L. S. Clesceri, A. E. Greenberg, and M. A. H. Franson. Washington, DC: American Public Health Association.

Appels, L., J. Baeyens, J. Degrève, and R. Dewil. (2008). Principles and potential of the anaerobic digestion of waste-activated sludge. *Progress in Energy and Combustion Science* 34 (6):755–781.

Appels, L., A. Van Assche, K. Willems, J. Degrève, J. Van Impe, and R. Dewil. (2011). Peracetic acid oxidation as an alternative pre-treatment for the anaerobic digestion of waste activated sludge. *Bioresource Technology* 102 (5):4124–4130.

Appels, L., S. Houtmeyers, J. Degrève, J. Van Impe, and R. Dewil. (2013). Influence of microwave pre-treatment on sludge solubilization and pilot scale semi-continuous anaerobic digestion. *Bioresource Technology* 128 (0):598–603.

Apul, O. G., and F. D. Sanin. (2010). Ultrasonic pretreatment and subsequent anaerobic digestion under different operational conditions. *Bioresource Technology* 101 (23):8984–8992.

Aquino, S. F., A. Y. Hu, A. Akram, and D. C. Stuckey. (2006). Characterization of dissolved compounds in submerged anaerobic membrane bioreactors (SAMBRs). *Journal of Chemical Technology and Biotechnology* 81 (12):1894–1904.

Ardern, E. and W.T. Locke. (1914). Experiments in the oxidation of sewage without the aid of filters. *Journal of the Society of the Chemical Industry* 33 (10), 524.

Asinari Di San Marzano, C. M., R. Binot, T. Bol, J. L. Fripiat, J. Hutschemakers, J. L. Melchior, I. Perez, H. Naveau, and E.J. Nyns. (1981). Volatile fatty acids, an important state parameter for the control of the reliability and the productivities of methane anaerobic digestions. *Biomass* 1 (1):47–59.

Ayol, A., A. Filibeli, D. Sir, and E. Kuzyaka. (2008). Aerobic and anaerobic bioprocessing of activated sludge: Floc disintegration by enzymes. *Journal of Environmental Science and Health. Part A, Toxic/Hazardous Substances and Environmental Engineering* 43 (13):1528–1535.

Baier, U. and P. Schmidheiny. (1997). Enhanced anaerobic degradation of mechanically disintegrated sludge. *Water Science and Technology* 36 (11):137–143.

Baker, A. (2001). Fluorescence excitation–emission matrix characterization of some sewage-impacted rivers. *Environmental Science and Technology* 35 (5):948–953.

Barber, W. P. (2003). Full-scale studies of part-stream ultrasound to improve sludge treatment. Paper read at the Eighth European Biosolids and Organic Residual Conference, at Wakefield.

Barber, W. P. F. (2010). The influence on digestion and advanced digestion on the environmental impacts of incinerating sewage sludge: A case study from the UK. In *Proceedings of the Water Environment Federation, Residuals and Biosolids*, pp. 865–881. Water Environment Federation.

Barber, W. P. F. (2016). Thermal hydrolysis for sewage treatment: A critical review. *Water Research* 104:53–71.

Barjenbruch, M., and O. Kopplow. (2003). Enzymatic, mechanical and thermal pre-treatment of surplus sludge. *Advances in Environmental Research* 7 (3):715–720.

Barnes, D., P. J. Bliss, B. Grauer, E. M. Kuo, K. Robbins, and G. McLean. (1983). Pretreatment of high-strength wastewaters by an anaerobic fluidized bed process. 1. Overall performance. *Environmental Technology Letters* 4 (5):195–202.

Basim, Y., N. Jaafarzadeh, and M. Farzadkia. (2016). A novel biological method for sludge volume reduction by aquatic worms. *International Journal of Environmental Science and Development* 7 (4):253–256.

Batstone, D. J., S. Tait, and D. Starrenburg. (2009). Estimation of hydrolysis parameters in full-scale anaerobic digesters. *Biotechnology and Bioengineering* 102 (5):1513–1520.

Batstone, D. J., P. D. Jensen, and H. Ge. (2011). Biochemical treatment of biosolids: Emerging drivers, trends, and technologies. *Water* 38 (3):90–93.

Batstone, D. J., T. Hülsen, C. M. Mehta, and J. Keller. (2014). Platforms for energy and nutrient recovery from domestic wastewater: A review. *Chemosphere* 140:2–11.

Battimelli, A., C. Millet, J. P. Delgenès, and R. Moletta. (2003). Anaerobic digestion of waste activated sludge combined with ozone post-treatment and recycling. *Water Science and Technology: A Journal of the International Association on Water Pollution Research* 48 (4):61–68.

Beszédes, S., S. Kertész, Z. László, G. Szabó, and C. Hodúr. (2009). Biogas production of ozone and/or microwave-pretreated canned maize production sludge. *Ozone: Science and Engineering* 31 (3):257–261.

Beszédes, S., Z. László, G. Szabó, and C. Hodúr. (2011). Effects of microwave pretreatments on the anaerobic digestion of food industrial sewage sludge. *Environmental Progress and Sustainable Energy* 30 (3):486–492.

Bi, X., C. Liu, X. Ran, W. Lv, J. Wang, and Z. Liao. (2013). Experimental studies on pretreatment of sludge by combination of thermal hydrolysis and ultrasound and sand removal by hydrocyclone to optimize wastewater and sludge treatment. Paper read at Aquaenviro's 18th European Biosolids and Organic Resources Conference and Exhibition, at Manchester, UK.

Bitton, G. (2005). *Wastewater Microbiology*. 3rd ed. Hoboken, NJ: John Wiley & Sons.

Boehnke, B., B. Diering, and S. W. Zuckut. (1997). Cost-effective wastewater treatment process for removal of organics and nutrients. *Water Engineering and Management* 144 (7):18–21.

Boehnke, B., R. Schulze-Rettmer, and S. W. Zuckut. (1998). Cost-effective reduction of high-strength wastewater by adsorption-based activated sludge technology. *Water Engineering and Management* 145 (12):31–34.

Böhnke, B. (1977). Das absorptions-belebungs-verfahren. *Korrespondenz Abwasser* 24 (77):33–42.

Bolzonella, D., P. Pavan, M. Zanette, and F. Cecchi. (2007). Two-phase anaerobic digestion of waste activated sludge: Effect of an extreme thermophilic prefermentation. *Industrial and Engineering Chemistry Research* 46 (21):6650–6655.

Bolzonella, D., C. Cavinato, F. Fatone, P. Pavan, and F. Cecchi. (2012). High rate mesophilic, thermophilic, and temperature phased anaerobic digestion of waste activated sludge: A pilot scale study. *Waste Management* 32 (6):1196–1201.

Boone, D. R. and M. P. Bryant. (1980). Propionate-degrading bacterium, *Syntrophobacter wolinii* sp. nov. gen. nov. from methanogenic ecosystems. *Applied and Environmental Microbiology* 40 (3):626–632.

Bougrier, C., H. Carrère, and J. P. Delgenès. (2005). Solubilisation of waste-activated sludge by ultrasonic treatment. *Chemical Engineering Journal* 106 (2):163–169.

Bougrier, C., C. Albasi, J. P. Delgenès, and H. Carrère. (2006). Effect of ultrasonic, thermal and ozone pre-treatments on waste activated sludge solubilisation and anaerobic biodegradability. *Chemical Engineering and Processing: Process Intensification* 45 (8):711–718.

Bougrier, C., J. P. Delgenès, and H. Carrère. (2007). Impacts of thermal pre-treatments on the semi-continuous anaerobic digestion of waste activated sludge. *Biochemical Engineering Journal* 34 (1):20–27.

Bougrier, C., J. P. Delgenès, and H. Carrère. (2008). Effects of thermal treatments on five different waste activated sludge samples solubilisation, physical properties and anaerobic digestion. *Chemical Engineering Journal* 139 (2):236–244.

Braguglia, C. M., A. Gianico, G. Mininni. (2011). Laboratory-scale ultrasound pre-treated digestion of sludge: Heat and energy balance. *Bioresource Technology* 102 (16):7567–7573.

Braguglia, C. M., M. C. Gagliano, and S. Rossetti. (2012a). High frequency ultrasound pretreatment for sludge anaerobic digestion: Effect on floc structure and microbial population. *Bioresource Technology* 110 (0):43–49.

Braguglia, C. M., A. Gianico, and G. Mininni. (2012b). Comparison between onzone and ultrasound disintegration on sludge anaerobic digestion. *Journal of Environmental Management* 95:S139–S143.

Bruus, J. H., P. H. Nielsen, and K. Keiding. (1992). On the stability of activated sludge flocs with implications to dewatering. *Water Research* 26 (12):1597–1604.

Cabirol, N., M. R. Rojas Oropeza, and A. Noyola. (2002). Removal of helminth eggs, and fecal coliforms by anaerobic thermophilic sludge digestion. *Water Science and Technology: A Journal of the International Association on Water Pollution Research* 45 (10):269–274.

Cadoret, A., A. Conrad, and J.-C. Block. (2002). Availability of low and high molecular weight substrates to extracellular enzymes in whole and dispersed activated sludges. *Enzyme and Microbial Technology* 31 (1–2):179–186.

Camacho, P., V. Geaugey, P. Ginestet, and E. Paul. (2002). Feasibility study of mechanically disintegrated sludge and recycle in the activated-sludge process. *Water Science and Technology: A Journal of the International Association on Water Pollution Research* 46 (10):97–104.

Camacho, P., P. Ginestet, and J. M. Audic. (2005). Understanding the mechanism of thermal disintegrating treatment in the reduction of sludge production. *Water Science and Technology: A Journal of the International Association on Water Pollution Research* 52 (10–11):235–245.

Canales, A., A. Pareilleux, J. L. Rols, G. Goma, and A. Huyard. (1994). Decreased sludge production strategy for domestic wastewater treatment. *Water Science and Technology* 30 (8):97–106.

Cano, R., S. I. Pérez-Elvira, and F. Fdz-Polanco. (2015). Energy feasibility study of sludge pretreatments: A review. *Applied Energy* 149:176–185.

Cao, Y., C. L. Lau, L. Lin, Y. Lee, K. S. Lee, Y. Abd Ghani, and Y. L. Wah. (2013). Mass flow and energy efficiency in a large water reclamation plant in Singapore. *Journal of Water Reuse and Desalination* 3 (4):402–409.

Carrère, H., C. Dumas, A. Battimelli, D. J. Batstone, J. P. Delgenès, J. P. Steyer, and I. Ferrer. (2010). Pretreatment methods to improve sludge anaerobic degradability: A review. *Journal of Hazardous Materials* 183 (1–3):1–15.

Carrere, H., Y. Rafrafi, A. Battimelli, M. Torrijos, J. P. Delgenes, and C. Motte. (2012). Improving methane production during the codigestion of waste-activated sludge and fatty wastewater: Impact of thermo-alkaline pretreatment on batch and semi-continuous processes. *Chemical Engineering Journal* 210 (0):404–409.

Cesbron, D., S. Déléris, H. Debellefontaine, M. Roustan, and E. Paul. (2003). Study of competition for ozone between soluble and particulate matter during activated sludge ozonation. *Chemical Engineering Research and Design* 81 (9):1165–1170.

Chai, C., D. Zhang, Y. Yu, Y. Feng, and M. S. Wong. (2015). Carbon footprint analyses of mainstream wastewater treatment technologies under different sludge treatment scenarios in China. *Water* 7 (12):918–938.

Chang, C. J., V. K. Tyagi, and S. L. Lo. (2011). Effects of microwave and alkali induced pretreatment on sludge solubilization and subsequent aerobic digestion. *Bioresource Technology* 102 (17):7633–7640.

Chauzy, J., D. Cretenot, A. Bausseron, and S. Deleris. (2008). Anaerobic digestion enhanced by thermal hydrolysis: First reference BIOTHELYS® at Saumur, France. *Water Practice and Technology* 3 (1):wpt2008004.

Chen, W., P. Westerhoff, J. A. Leenheer, and K. Booksh. (2003). Fluorescence exication-emission matrix regional integration to quantify spectra for dissolved organic matter. *Environmental Science and Technology* 37 (24):5701–5710.

Chen, Y., S. Jiang, H. Yuan, Q. Zhou, and G. Gu. (2007). Hydrolysis and acidification of waste activated sludge at different pHs. *Water Research* 41 (3):683–689.

Chen, Y. C., M. J. Higgins, S. M. Beightol, S. N. Murthy, and W. E. Toffey. (2011). Anaerobically digested biosolids odor generation and pathogen indicator regrowth after dewatering. *Water Research* 45 (8):2616–2626.

Cheng, C. J., and P. K. A. Hong. (2013). Anaerobic digestion of activated sludge after pressure assisted ozonation. *Bioresource Technology* 142:69–76.

Chiavola, A., A. Ridolfi, E. D'Amato, S. Bongirolami, E. Cima, P. Sirini, and R. Gavasci. (2015). Sludge reduction in a small wastewater treatment plant by electro-kinetic disintegration. *Water Science and Technology* 72 (3):364–370.

Chiu, Y. C., C. N. Chang, J. G. Lin, and S. J. Huang. (1997). Alkaline and ultrasonic pretreatment of sludge before anaerobic digestion. *Water Science and Technology* 36 (11):155–162.

Cho, S. K., H. J. Ju, J. G. Lee, and S. H. Kim. (2014). Alkaline-mechanical pretreatment process for enhanced anaerobic digestion of thickened waste activated sludge with a novel crushing device: Performance evaluation and economic analysis. *Bioresource Technology* 165:183–190.

Choi, H. B., K. Y. Hwang, and E. B. Shin. (1997). Effects on anaerobic digestion of sewage sludge pretreatment. *Water Science and Technology* 35 (10):207–211.

Choi, H., S. W. Jeong, and Y. J. Chung. (2006). Enhanced anaerobic gas production of waste activated sludge pretreated by pulse power technique. *Bioresource Technology* 97 (2):198–203.

Chu, C. P., B.-V. Chang, G. S. Liao, D. S. Jean, and D. J. Lee. (2001). Observations on changes in ultrasonically treated waste-activated sludge. *Water Research* 35 (4):1038–1046.

Chu, C. P., D. J. Lee, B. V. Chang, C. S. You, and J. H. Tay. (2002). "Weak" ultrasonic pretreatment on anaerobic digestion of flocculated activated biosolids. *Water Research* 36 (11):2681–2688.

Chu, L. B., S. T. Yan, X. H. Xing, A. F. Yu, X. L. Sun, and B. Jurcik. (2008). Enhanced sludge solubilization by microbubble ozonation. *Chemosphere* 72 (2):205–212.

Chu, L., S. Yan, X.H. Xing, X. Sun, and B. Jurcik. (2009). Progress and perspectives of sludge ozonation as a powerful pretreatment method for minimization of excess sludge production. *Water Research* 43 (7):1811–1822.

Clark, P. B. and I. Nujjoo. (2000). Ultrasonic sludge pre-treatment for enhanced sludge digestion. *Water and Environment Journal* 14 (1): 66–71.

Coble, P. G. (1996). Characterization of marine and terrestrial DOM in seawater using excitation-emission matrix spectroscopy. *Marine Chemistry* 51 (4):325–346.

Coelho, N. M. G., R. L. Droste, and K. J. Kennedy. (2011). Evaluation of continuous mesophilic, thermophilic and temperature phased anaerobic digestion of microwaved activated sludge. *Water Research* 45 (9):2822–2834.

Davis, M. L. (2011). *Water and Wastewater Engineering: Design Principles and Practice.* New York: McGraw-Hill.

Davies, R. (1959). Observations on the use of ultrasound waves for the disruption of microorganisms. *Biochimica et Biophysica Acta* 33 (2):481–493.

Dawson, M. K. and H. Ozgencil. (2009). Do hydrolysis processes effect digested sludge rheology? Paper read at Aquaenvio 14th European Biosolids and Organic Resources Conference and Exhibition, at Leeds, UK.

de Bok, F. A. M., C. M. Plugge, and A. J. M. Stams. (2004). Interspecies electron transfer in methanogenic propionate degrading consortia. *Water Research* 38 (6):1368–1375.

Dereix, M., W. Parker, and K. Kennedy. (2006). Steam-explosion pretreatment for enhancing anaerobic digestion of municipal wastewater sludge. *Water Environment Research: A Research Publication of the Water Environment Federation* 78 (5):474–485.

Determann, S., R. Reuter, P. Wagner, and R. Willkomm. (1994). Fluorescent matter in the eastern Atlantic Ocean. Part 1: Method of measurement and near-surface distribution. *Deep Sea Research Part I: Oceanographic Research Papers* 41 (4):659–675.

Devlin, D. C., S. R. R. Esteves, R. M. Dinsdale, and A. J. Guwy. (2011). The effect of acid pretreatment on the anaerobic digestion and dewatering of waste activated sludge. *Bioresource Technology* 102 (5):4076–4082.

Dewil, R., J. Baeyens, and R. Goutvrind. (2006). The use of ultrasonics in the treatment of waste activated sludge. *Chinese Journal of Chemical Engineering* 14 (1):105–113.

Dewil, R., L. Appels, J. Baeyens, and J. Degrève. (2007). Peroxidation enhances the biogas production in the anaerobic digestion of biosolids. *Journal of Hazardous Materials* 146 (3):577–581.

Dhar, B. R., G. Nakhla, and M. B. Ray. (2012). Techno-economic evaluation of ultrasound and thermal pretreatments for enhanced anaerobic digestion of municipal waste activated sludge. *Waste Management* 32 (3):542–549.

Diamantis, V., A. Eftaxias, B. Bundervoet, and W. Verstraete. (2014). Performance of the biosorptive activated sludge (BAS) as pre-treatment to UF for decentralized wastewater reuse. *Bioresource Technology* 156 (0):314–321.

Doğan, I., and F. D. Sanin. (2009). Alkaline solubilization and microwave irradiation as a combined sludge disintegration and minimization method. *Water Research* 43 (8):2139–2148.

Doháñyos, M., J. Zábranská, and P. Jeníček. (1997). Enhancement of sludge anaerobic digestion by using of a special thickening centrifuge. *Water Science and Technology* 36 (11):145–153.

Dolejs, P., R. Gotvald, A. M. L. Velazquez, J. Hejnic, P. Jenicek, and J. Bartacek. (2016). Contact stabilization with enhanced accumulation process for energy recovery from sewage. *Environmental Engineering Science* 33 (11):873–881.

Dolfing, J. and W. G. B. M. Bloeman. (1985). Activity measurements as a tool to characterize the microbial composition of methanogenic environments. *Journal of Microbiological Methods* 4 (1):1–12.

Doulah, M. S. (1977). Mechanism of disintegration of biological cells in ultrasonic cavitation. *Biotechnology and Bioengineering* 19 (5):649–660.

Dumas, C., S. Perez, E. Paul, and X. Lefebvre. (2010). Combined thermophilic aerobic process and conventional anaerobic digestion: Effect on sludge biodegradation and methane production. *Bioresource Technology* 101 (8):2629–2636.

Dytczak, M. A., K. L. Londry, H. Siegrist, and J. A. Oleszkiewicz. (2007). Ozonation reduces sludge production and improves denitrification. *Water Research* 41 (3):543–550.

Eastman, J. A., and J. F. Ferguson. (1981). Solubilization of particulate organic carbon during the acid phase of anaerobic digestion. *Journal of the Water Pollution Control Federation* 53 (3):352–366.

Ebenezer, A. V., P. Arulazhagan, S. Adish Kumar, I. Yeom, and J. Rajesh Banu. (2015). Effect of deflocculation on the efficiency of low-energy microwave pretreatment and anaerobic biodegradation of waste activated sludge. *Applied Energy* 145:104–110.

Ehlinger, F., J. M. Audic, and G. M. Faup. (1987). Relationship between the concentration of acetate in the feed and the composition of a biofilm in an anaerobic filter. *Environmental Technology Letters* 8 (1–12):197–207.

Erden, G. (2013). Combination of alkaline and microwave pretreatment for disintegration of meat processing wastewater sludge. *Environmental Technology* 34 (5–8):711–718.

Erden, G. and A. Filibeli. (2010a). Ultrasonic pre-treatment of biological sludge: Consequences for disintegration, anaerobic biodegradability, and filterability. *Journal of Chemical Technology and Biotechnology* 85 (1):145–150.

Erden, G. and A. Filibeli. (2010b). Improving anaerobic biodegradability of biological sludges by Fenton pre-treatment: Effects on single stage and two-stage anaerobic digestion. *Desalination* 251 (1–3):58–63.

Erden, G. and A. Filibeli. (2011). Ozone oxidation of biological sludge: Effects on disintegration, anaerobic biodegradability, and filterability. *Environmental Progress and Sustainable Energy* 30 (3):377–383.

Erden, G., O. Demir, and A. Filibeli. (2010). Disintegration of biological sludge: Effect of ozone oxidation and ultrasonic treatment on aerobic digestibility. *Bioresource Technology* 101 (21):8093–8098.

Ersahin, M. E., H. Ozgun, R. K. Dereli, I. Ozturk, K. Roest, and J. B. van Lier. (2012). A review on dynamic membrane filtration: Materials, applications and future perspectives. *Bioresource Technology* 122:196–206.

Eskicioglu, C., K. J. Kennedy, and R. L. Droste. (2006). Characterization of soluble organic matter of waste activated sludge before and after thermal pretreatment. *Water Research* 40 (20):3725–3736.

Eskicioglu, C., R. L. Droste, and K. J. Kennedy. (2007a). Performance of anaerobic waste activated sludge digesters after microwave pretreatment. *Water Environment Research: A Research Publication of the Water Environment Federation* 79 (11):2265–2273.

Eskicioglu, C., K. J. Kennedy, and R. L. Droste. (2007b). Enhancement of batch waste activated sludge digestion by microwave pretreatment. *Water Environment Research: A Research Publication of the Water Environment Federation* 79 (11):2304–2317.

Eskicioglu, C., N. Terzian, K. J. Kennedy, R. L. Droste, and M. Hamoda. (2007c). Athermal microwave effects for enhancing digestibility of waste activated sludge. *Water Research* 41 (11):2457–2466.

Eskicioglu, C., A. Prorot, J. Marin, R. L. Droste, and K. J. Kennedy. (2008). Synergetic pretreatment of sewage sludge by microwave irradiation in presence of H2O2 for enhanced anaerobic digestion. *Water Research* 42 (18):4674–4682.

Eskicioglu, C., K. J. Kennedy, and R. L. Droste. (2009). Enhanced disinfection and methane production from sewage sludge by microwave irradiation. *Desalination* 248 (1–3):279–285.

Esparza-Soto, M. and P. Westerhoff. (2003). Biosorption of humic and fulvic acids to live activated sludge biomass. *Water Research* 37 (10):2301–2310.

Everett, J. G. (1972). Dewatering of wastewater sludge by heat treatment. *Journal - Water Pollution Control Federation* 44 (1):92–100.

Fang, W., P. Zhang, G. Zhang, S. Jin, D. Li, M. Zhang, and X. Xu. (2014). Effect of alkaline addition on anaerobic sludge digestion with combined pretreatment of alkaline and high pressure homogenization. *Bioresource Technology* 168:167–172.

Feijoo, G., M. Soto, R. Méndez, and J. M. Lema. (1995). Sodium inhibition in the anaerobic digestion process: Antagonism and adaptation phenomena. *Enzyme and Microbial Technology* 17 (2):180–188.

Feng, X., J. Deng, H. Lei, T. Bai, Q. Fan, and Z. Li. (2009). Dewaterability of waste activated sludge with ultrasound conditioning. *Bioresource Technology* 100 (3):1074–1081.

Fernandez-Polanco, D. and H. Tatsumi. (2016). Optimum energy integration of thermal hydrolysis through pinch analysis. *Renewable Energy*. http://dx.doi.org/10.1016/j.renene.2016.01.038.

Ferrer, I., S. Ponsá, F. Vázquez, and X. Font. (2008). Increasing biogas production by thermal (70°C) sludge pre-treatment prior to thermophilic anaerobic digestion. *Biochemical Engineering Journal* 42 (2):186–192.

Ferrer, I., E. Serrano, S. Ponsa, F. Vazquez, and X. Font. (2009). Enhancement of thermophilic anaerobic sludge digestion by 70°C pre-treatment: Energy considerations. *Journal of Residuals Science and Technology* 6 (1):11–18.

Foladori, P., G. Andreottola, and G. Ziglio. (2010). *Sludge Reduction Technologies in Wastewater Treatment Plants*. London: IWA Publishing.

Frijns, J. and C. Uijterlinde. (2010). Energy efficiency in the European water industry. A compendium of best practices and case studies. Report prepared for GWRC by J. Frijns (KWR) and C. Uiterlinde (STOWA).

García-Galán, M.J., S. González Blanco, R. López Roldán, S. Díaz-Cruz, and D. Barceló. (2012). Ecotoxicity evaluation and removal of sulfonamides and their acetylated metabolites during conventional wastewater treatment. *Science of The Total Environment* 437: 403–412.

Gavala, H. N., U. Yenal, I. V. Skiadas, P. Westermann, and B. K. Ahring. (2003). Mesophilic and thermophilic anaerobic digestion of primary and secondary sludge. Effect of pretreatment at elevated temperature. *Water Research* 37 (19):4561–4572.

Ge, H., P. D. Jensen, and D. J. Batstone. (2010). Pre-treatment mechanisms during thermophilic-mesophilic temperature phased anaerobic digestion of primary sludge. *Water Research* 44 (1):123–130.

Ge, H., P. D. Jensen, and D. J. Batstone. (2011). Temperature phased anaerobic digestion increases apparent hydrolysis rate for waste activated sludge. *Water Research* 45 (4):1597–1606.

Ge, H., D. J. Batstone, and J. Keller. (2013). Operating aerobic wastewater treatment at very short sludge ages enables treatment and energy recovery through anaerobic sludge digestion. *Water Research* 47 (17):6546–6557.

Gessesse, A., T. Dueholm, S. B. Petersen, and P. H. Nielsen. (2003). Lipase and protease extraction from activated sludge. *Water Research* 37 (15):3652–3657.

Ghosh, S., and D. L. Klass. (1978). Two-phase anaerobic digestion. *Process Biochemistry* 13 (4):15.

Gianico, A., C. M. Braguglia, R. Cesarini, and G. Mininni. (2013). Reduced temperature hydrolysis at 134°C before thermophilic anaerobic digestion of waste activated sludge at increasing organic load. *Bioresource Technology* 143 (0):96–103.

Goel, R., T. Tokutomi, and H. Yasui. (2003a). Anaerobic digestion of excess activated sludge with ozone pretreatment. *Water Science and Technology: A Journal of the International Association on Water Pollution Research* 47 (12):207–214.

Goel, R., H. Yasui, and C. Shibayama. (2003b). High-performance closed loop anaerobic digestion using pre/post sludge ozonation. *Water Science and Technology: A Journal of the International Association on Water Pollution Research* 47 (12):261–267.

Goel, R., K. Komatsu, H. Yasui, and H. Harada. (2004). Process performance and change in sludge characteristics during anaerobic digestion of sewage sludge with ozonation. *Water Science and Technology: A Journal of the International Association on Water Pollution Research* 49 (10):105–113.

Gong, C., J. Jiang, and D. Li. (2015). Ultrasound coupled with Fenton oxidation pre-treatment of sludge to release organic carbon, nitrogen and phosphorus. *Science of the Total Environment* 532:495–500.

Gorczyca, B. (2000). Porosity and structure of alum coagulated flocs. PhD thesis. Department of Chemical Engineering and Applied Chemistry, University of Toronto, Toronto.

Gossett, J. M., D. C. Stucky, W. F. Owen, and P. L. McCarty. (1982). Heat treatment and anaerobic digestion of refuse. *Journal of the Environmental Engineering Division* 108 (3):437–454.

Graff, K. (1988). *Independent Study on High Power Ultrasonic*. Lecture Notes. WE 795. Columbus, OH: The Ohio State University.

Griffin, M. E., K. D. McMahon, R. I. Mackie, and L. Raskin. (1998). Methanogenic population dynamics during start-up of anaerobic digesters treating municipal solid waste and biosolids. *Biotechnology and Bioengineering* 57 (3): 342–355.

Grönroos, A., H. Kyllönen, K. Korpijärvi, P. Pirkonen, T. Paavola, Jari Jokela, and J. Rintala. (2005). Ultrasound assisted method to increase soluble chemical oxygen demand (SCOD) of sewage sludge for digestion. *Ultrasonics Sonochemistry* 12 (1–2):115–120.

Guellil, A., F. Thomas, J. C. Block, J. L. Bersillon, and P. Ginestet. (2001). Transfer of organic matter between wastewater and activated sludge flocs. *Water Research* 35 (1):143–150.

Gujer, W. and A. J. B. Zehnder. (1983). Conversion processes in anaerobic digestion. *Water Science and Technology* 15 (8–9):127–167.

Gurieff, N., J. Bruus, S. Hoejsgaard, J. Boyd, and M. Kline. (2011). Maximizing energy efficiency and biogas production: EXELYS™—continuous thermal hydrolysis. *Proceedings of the Water Environment Federation* 2011 (17):642e656.

Haider, S., K. Svardal, P. A. Vanrolleghem, and H. Kroiss. (2003a). The effect of low sludge age on wastewater fractionation (S_S, S_1). *Water Science and Technology* 47 (11):203–209.

Han, Y., Shihwu Sung, and R. R. Dague. (1997). Temperature-phased anaerobic digestion of wastewater sludges. *Water Science and Technology* 36 (6–7):367–374.

Harper, S. R., and F. G. Pohland. (1986). Recent developments in hydrogen management during anaerobic biological waste-water treatment. *Biotechnology and Bioengineering* 28 (4):585–602.

Harrison, S. T. L. (1991). Bacterial-cell disruption: A key unit operation in the recovery of intracellular products. *Biotechnology Advances* 9 (2):217–240.

Hartmann, H., and B. K. Ahring. (2005). A novel process configuration for anaerobic digestion of source-sorted household waste using hyper-thermophilic posttreatment. *Biotechnology and Bioengineering* 90 (7):830–837.

Hasegawa, S., N. Shiota, K. Katsura, and A. Akashi. (2000). Solubilization of organic sludge by thermophilic aerobic bacteria as a pretreatment for anaerobic digestion. *Water Science and Technology: A Journal of the International Association on Water Pollution Research* 41 (3):163–169.

Haug, R. T., D. C. Stuckey, G. M. Gosett, and P. L. McCarty. (1978). Effect of thermal pre-treatment on digestibility and dewaterability of organic sludges. *Journal (Water Pollution Control Federation)* 50:73–84.

Haug, R. T., T. J. Lebrun, and L. D. Totorici. (1983). Thermal pretreatment of sludges: A field demonstration. *Journal (Water Pollution Control Federation)* 55 (1):23–34.

Her, N., G. Amy, D. McKnight, Jinsik Sohn, and Yeomin Yoon. (2003). Characterization of DOM as a function of MW by fluorescence EEM and HPLC-SEC using UVA, DOC, and fluorescence detection. *Water Research* 37 (17):4295–4303.

Hernández Leal, L., H. Temmink, G. Zeeman, and C. J. N. Buisman. (2010). Bioflocculation of grey water for improved energy recovery within decentralized sanitation concepts. *Bioresource Technology* 101 (23):9065–9070.

Higgins, M., S. Beightol, U. Mandahar, S. Xiao, L. Hung-Wei, T. Le, J. Mah, B. Pathak, J. Novak, A. Al-Omari, and S. Murthy. (2015). Effect of thermal hydrolysis temperature on anaerobic digestion, dewatering and filtrate characteristics. Paper read at the Proceedings of WEFTEC 2015, at New Orleans.

Hiraoka, M., N. Takeda, S. l. Sakai, and A. Yasuda. (1985). Highly efficient anaerobic digestion with thermal pre-treatment. *Water Science and Technology* 17 (4–5):529–539.

Ho, L., and G. Ho. (2012). Mitigating ammonia inhibition of thermophilic anaerobic treatment of digested piggery wastewater: Use of pH reduction, zeolite, biomass and humic acid. *Water Research* 46 (14):4339–4350.

Hogan, F., S. Mormede, P. Clark, and M. Crane. (2004). Ultrasonic sludge treatment for enhanced anaerobic digestion. *Water Science and Technology: A Journal of the International Association on Water Pollution Research* 50 (9):25–32.

Hong, S. M., J. K. Park, and Y. O. Lee. (2004). Mechanisms of microwave irradiation involved in the destruction of fecal coliforms from biosolids. *Water Research* 38 (6):1615–1625.

Hong, S. M., J. K. Park, N. Teeradej, Y. O. Lee, Y. K. Cho, and C. H. Park. (2006). Pretreatment of sludge with microwaves for pathogen destruction and improved anaerobic digestion performance. *Water Environment Research: A Research Publication of the Water Environment Federation* 78 (1):76–83.

Houtmeyers, S., J. Degrève, K. Willems, R. Dewil, and L. Appels. (2014). Comparing the influence of low power ultrasonic and microwave pre-treatments on the solubilisation and semi-continuous anaerobic digestion of waste activated sludge. *Bioresource Technology* 171:44–49.

Hu, Y., C. Zhang, C. Zhang, X. Tan, H. Zhu, and Q. Zhou. (2009). Effect of alkaline pretreatment on waste activated sludge solubilization and anaerobic digestion. In *3rd International Conference on Bioinformatics and Biomedical Engineering*. Beijing: IEEE.

Hua, I., and M. R. Hoffmann. (1997). Optimization of ultrasonic irradiation as an advanced oxidation technology. *Environmental Science and Technology* 31 (8):2237–2243.

Hughes, D. E. and W. L. Nyborg. (1962). Cell disruption by ultrasound. *Science* 138 (3537):108–114.

Hung-Wei, L., S. Xiao, T. Le, A. Al-Omari, M. Higgins, G. Boardman, J. Novak, and S. Murthy. (2014). Evaluation of solubilization characteristics of thermal hydrolysis process. In *Proceedings of the Water Environment Federation*, pp. 6312–6336(25). Water Environment Federation.

Jang, J. H., and J.-H. Ahn. (2013). Effect of microwave pretreatment in presence of NaOH on mesophilic anaerobic digestion of thickened waste activated sludge. *Bioresource Technology* 131:437–442.

Jang, H. M., H. U. Cho, S. K. Park, J. H. Ha, and J. M. Park. (2014). Influence of thermophilic aerobic digestion as a sludge pre-treatment and solids retention time of mesophilic anaerobic digestion on the methane production, sludge digestion and microbial communities in a sequential digestion process. *Water Research* 48:1–14.

Jenicek, P., J. Kutil, O. Benes, V. Todt, J. Zabranska, and M. Dohanyos. (2013). Energy self-sufficient sewage wastewater treatment plants: Is optimized anaerobic sludge digestion the key? *Water Science and Technology: A Journal of the International Association on Water Pollution Research* 68 (8):1739–1744.

Jenkins, D., and M. G. Richard, and G.T. Daigger. (2003). *Manual on the Causes and Control of Activated Sludge Bulking, Foaming, and other Solids Separation Problems*, 3rd ed. London: IWA Publishing.

Jiang, J., S. Yang, M. Chen, and Q. Zhang. (2009). Disintegration of sewage sludge with bifrequency ultrasonic treatment. *Water Science and Technology: A Journal of the International Association on Water Pollution Research* 60 (6):1445–1453.

Jin, Y., H. Li, R. B. Mahar, Z. Wang, and Y. Nie. (2009). Combined alkaline and ultrasonic pretreatment of sludge before aerobic digestion. *Journal of Environmental Sciences* 21 (3):279–284.

Jin, N., B. Jin, N. Zhu, H. Yuan, and J. Ruan. (2015). Disinhibition of excessive volatile fatty acids to improve the efficiency of autothermal thermophilic aerobic sludge digestion by chemical approach. *Bioresource Technology* 175:120–127.

Jolis, D. and M. Marneri. (2006). Thermal hydrolysis of secondary scum for control of biological foam. *Water Environment Research: A Research Publication of the Water Environment Federation* 78 (8):835–841.

Jung, J., X.H. Xing, and K. Matsumoto. (2002). Recoverability of protease released from disrupted excess sludge and its potential application to enhanced hydrolysis of proteins in wastewater. *Biochemical Engineering Journal* 10 (1):67–72.

Jung, Y., H. Ko, B. Jung, and N. Sung. (2011). Application of ultrasonic system for enhanced sewage sludge disintegration: A comparative study of single- and dual-frequency. *KSCE Journal of Civil Engineering* 15 (5):793–797.

Karlsson, A., X. B. Truong, J. Gustavsson, B. H. Svensson, F. Nilsson, and J. Ejlertsson. (2011). Anaerobic treatment of activated sludge from Swedish pulp and paper mills: Biogas production potential and limitations. *Environmental Technology* 32 (13–14):1559–1571.

Kaspar, H. F. and K. Wuhrmann. (1978). Kinetic parameters and relative turnovers of some important catabolic reactions in digesting sludge. *Applied and Environmental Microbiology* 36 (1):1–7.

Katsiris, N., and A. Kouzeli-Katsiri. (1987). Bound water content of biological sludges in relation to filtration and dewatering. *Water Research* 21 (11):1319–1327.

Kayhanian, M. (1999). Ammonia inhibition in high-solids biogasification: An overview and practical solutions. *Environmental Technology* 20 (4):355–365.

Keiding, K. and P. H. Nielsen. (1997). Desorption of organic macromolecules from activated sludge: Effect of ionic composition. *Water Research* 31 (7):1665–1672.

Kepp, U., and O. E. Solheim. (2000). Thermo dynamical assessment of the digestion process. Paper read at the 5th European Biosolids and Organic Residiuals Conference, at Wakefield, UK.

Kepp, U., I. Machenbach, N. Weisz, and O. E. Solheim. (2000). Enhanced stabilisation of sewage sludge through thermal hydrolysis: Three years of experience with full scale plant. *Water Science and Technology* 42 (9):89–96.

Khanal, S. K., D. Grewell, Shihwu Sung, and J. van Leeuwen. (2007). Ultrasound applications in wastewater sludge pretreatment: A review. *Critical Reviews in Environmental Science and Technology* 37 (4):277–313.

Kim, D.-J. and Y. Youn. (2011). Characteristics of sludge hydrolysis by ultrasound and thermal pretreatment at low temperature. *Korean Journal of Chemical Engineering* 28 (9):1876–1881.

Kim, M., Y. H. Ahn, and R. E. Speece. (2002). Comparative process stability and efficiency of anaerobic digestion; mesophilic vs. thermophilic. *Water Research* 36 (17):4369–4385.

Kim, J., C. Park, T. H. Kim, M. Lee, S. Kim, S. W. Kim, and J. Lee. (2003). Effects of various pretreatments for enhanced anaerobic digestion with waste activated sludge. *Journal of Bioscience and Bioengineering* 95 (3):271–275.

Kim, D. H., E. Jeong, S.-E. Oh, and H.-S. Shin. (2010). Combined (alkaline+ultrasonic) pretreatment effect on sewage sludge disintegration. *Water Research* 44 (10):3093–3100.

Konsowa, A. H. (2003). Decolorization of wastewater containing direct dye by ozonation in a batch bubble column reactor. *Desalination* 158 (1–3):233–240.

Kopp, J., J. Müller, N. Dichtl, and J. Schwedes. (1997). Anaerobic digestion and dewatering characteristics of mechanically disintegrated excess sludge. *Water Science and Technology* 36 (11):129–136.

Kozyatnyk, I., J. Świetlik, U. Raczyk-Stanisławiak, A. Dąbrowska, N. Klymenko, and J. Nawrocki. (2013). Influence of oxidation on fulvic acids composition and biodegradability. *Chemosphere* 92 (10):1335–1342.

Kumar, M. S. K., T. K. Kumar, P. Arulazhagan, S. A. Kumar, I. T. Yeom, and J. R. Banu. (2015). Effect of alkaline and ozone pretreatment on sludge reduction potential of a membrane bioreactor treating high-strength domestic wastewater. *Desalination and Water Treatment* 55:1127–1134.

Kurokawa, Y., A. Maekawa, M. Takahashi, and Y. Hayashi. (1990). Toxicity and carcinogenicity of potassium bromate-a new renal carcinogen. *Environmental Health Perspectives* 87:309–335.

Lafitte-Trouqué, S., and C. F. Forster. (2002). The use of ultrasound and γ-irradiation as pretreatments for the anaerobic digestion of waste activated sludge at mesophilic and thermophilic temperatures. *Bioresource Technology* 84 (2):113–118.

Lancaster, R. (2015). Thermal hydrolysis at Davyhulme WWtW one year on. Paper read at the WEF Residuals and Biosolids 2015, at Washington DC.

Laureni, M., P. Falås, O. Robin, A. Wick, D. G. Weissbrodt, J. L. Nielsen, T. A. Ternes, E. Morgenroth, and A. Joss. (2016). Mainstream partial nitritation and anammox: Long-term process stability and effluent quality at low temperatures. *Water Research* 101:628–639.

Law, Y., L. Ye, Q. Wang, S. Hu, M. Pijuan, and Z. Yuan. (2015). Producing free nitrous acid – a green and renewable biocidal agent – from anaerobic digester liquor. *Chemical Engineering Journal* 259:62–69.

Lee, I. S. and B. E. Rittmann. (2011). Effect of low solids retention time and focused pulsed pretreatment on anaerobic digestion of waste activated sludge. *Bioresource Technology* 102 (3):2542–2548.

Lee, J. W., H. Y. Cha, K. Y. Park, K. G. Song, and K. H. Ahn. (2005). Operational strategies for an activated sludge process in conjunction with ozone oxidation for zero excess sludge production during winter season. *Water Research* 39 (7):1199–1204.

Lehne, G., A. Müller, and J. Schwedes. (2001). Mechanical disintegration of sewage sludge. *Water Science and Technology: A Journal of the International Association on Water Pollution Research* 43 (1):19–26.

Levine, A. D., G. Tchobanoglous, and T. Asano. (1985). Characterization of the size distribution of contaminants in wastewater: Treatment and reuse implications. *Journal (Water Pollution Control Federation)* 57 (7):805–816.

Li, Y. Y. and T. Noike. (1992). Upgrading of anaerobic digestion of waste activated sludge by thermal pretreatment. *Water Science and Technology* 26 (3–4):857–866.

Li, A. J., T. Zhang, and X. Y. Li. (2010). Fate of aerobic bacterial granules with fungal contamination under different organic loading conditions. *Chemosphere* 78 (5):500–509.

Li, H., Y. Jin, R. B. Mahar, Z. Wang, and Y. Nie. (2008). Effects and model of alkaline waste activated sludge treatment. *Bioresource Technology* 99 (11):5140–5144.

Li, H., Y. Jin, and Y. Nie. (2009a). Application of alkaline treatment for sludge decrement and humic acid recovery. *Bioresource Technology* 100 (24):6278–6283.

Li, H., Y. Jin, M. Rasool Bux, Z. Wang, and Y. Nie. (2009b). Effects of ultrasonic disintegration on sludge microbial activity and dewaterability. *Journal of Hazardous Materials*. 161:1421–1426.

Li, H., C. C. Li, W. J. Liu, and S. X. Zou. (2012). Optimized alkaline pretreatment of sludge before anaerobic digestion. *Bioresource Technology* 123:189–194.

Li, H., S. Zou, C. Li, and Y. Jin. (2013). Alkaline post-treatment for improved sludge anaerobic digestion. *Bioresource Technology* 140:187–191.

Lim, C., S. Zhang, Y. Zhou, and W. J. Ng. (2015). Enhanced carbon capture biosorption through process manipulation. *Biochemical Engineering Journal* 93 (0):128–136.

Lin, X. A., C. G. Lee, E. S. Casale, and J. C. H. Shih. (1992). Purification and characterization of a keratinase from a feather-degrading Bacillus licheniformis strain. *Applied and Environmental Microbiology* 58 (10):3271–3275.

Lin, J.-G., C.-N. Chang, and S.-C. Chang. (1997). Enhancement of anaerobic digestion of waste activated sludge by alkaline solubilization. *Bioresource Technology* 62 (3):85–90.

Lin, J. G., Y. S. Ma, A. C. Chao, and C. L. Huang. (1999). BMP test on chemically pretreated sludge. *Bioresource Technology* 68:187–192.

Lin, Y. Q., D. H. Wang, S. Q. Wu, and C. M. Wang. (2009). Alkali pretreatment enhances biogas production in the anaerobic digestion of pulp and paper sludge. *Journal of Hazardous Materials* 170 (1):366–373.

Lin, Y., H. Zheng, and M. Juan. (2012). Biohydrogen production using waste activated sludge as a substrate from fructose-processing wastewater treatment. *Process Safety and Environmental Protection* 90 (3):221–230.

Liu, C., B. Xiao, A. Dauta, G. Peng, S. Liu, and Z. Hu. (2009). Effect of low power ultrasonic radiation on anaerobic biodegradability of sewage sludge. *Bioresource Technology* 100 (24):6217–6222.

Liu, C., Y. Yang, Q. Wang, M. Kim, Q. Zhu, D. Li, and Z. Zhang. (2012a). Photocatalytic degradation of waste activated sludge using a circulating bed photocatalytic reactor for improving biohydrogen production. *Bioresource Technology* 125:30–36.

Liu, C., W. Shi, M. Kim, Y. Yang, Z. Lei, and Z. Zhang. (2013). Photocatalytic pretreatment for the redox conversion of waste activated sludge to enhance biohydrogen production. *International Journal of Hydrogen Energy* 38 (18):7246–7252.

Liu, C., P. Zhang, C. Zeng, G. Zeng, G. Xu, and Y. Huang. (2015). Feasibility of bioleaching combined with Fenton oxidation to improve sewage sludge dewaterability. *Journal of Environmental Sciences* 28:37–42.

Liu, J. C., C. H. Lee, J. Y. Lai, K. C. Wang, Y. C. Hsu, and B. V. Chang. (2001). Extracellular polymers of ozonized waste activated sludge. *Water Science and Technology: A Journal of the International Association on Water Pollution Research* 44 (10):137–142.

Liu, S., N. Zhu, L. Y. Li, and H. Yuan. (2011). Isolation, identification and utilization of thermophilic strains in aerobic digestion of sewage sludge. *Water Research* 45 (18):5959–5968.

Liu, S., N. Zhu, and L. Y. Li. (2012b). The one-stage autothermal thermophilic aerobic digestion for sewage sludge treatment: Stabilization process and mechanism. *Bioresource Technology* 104:266–273.

Liu, X., H. Liu, J. Chen, G. Du, and J. Chen. (2008). Enhancement of solubilization and acidification of waste activated sludge by pretreatment. *Waste Management* 28 (12):2614–2622.

Liu, X., W. Wang, X. Gao, Y. Zhou, and R. Shen. (2012c). Effect of thermal pretreatment on the physical and chemical properties of municipal biomass waste. *Waste Management* 32 (2):249–255.

Long, J. H. and C. M. Bullard. (2014). Waste activated sludge pretreatment to boost volatile solids reduction and digester gas production: Market and technology assessment. *Florida Water Resources Journal* J44–50.

Lormier, J. P. (1990). Sonochemistry: The general principles. In *Sonochemistry: The Use of Ultrasound in Chemistry*, edited by T. Mason. Cambridge, UK: Royal Society of Chemistry.

Lotti, T., R. Kleerebezem, Z. Hu, B. Kartal, M. K. de Kreuk, C. van Erp Taalman Kip, J. Kruit, T. L. Hendrickx, and M. C. M. van Loosdrecht. (2015). Pilot-scale evaluation of anammox-based mainstream nitrogen removal from municipal wastewater. *Environmental Technology* 36 (9–12):1167–1177.

Lu, J. Q., H. N. Gavala, I. V. Skiadas, Z. Mladenovska, and B. K. Ahring. (2008). Improving anaerobic sewage sludge digestion by implementation of a hyper-thermophilic prehydrolysis step. *Journal of Environmental Management* 88 (4):881–889.

Lu, L., D. Xing, and N. Ren. (2012). Pyrosequencing reveals highly diverse microbial communities in microbial electrolysis cells involved in enhanced H2 production from waste activated sludge. *Water Research* 46 (7):2425–2434.

Luo, K., Q. Yang, X. M. Li, H. B. Chen, X. Liu, G. J. Yang, and G. M. Zeng. (2013). Novel insights into enzymatic-enhanced anaerobic digestion of waste activated sludge by three-dimensional excitation and emission matrix fluorescence spectroscopy. *Chemosphere* 91 (5):579–585.

Makinia, J., K. H. Rosenwinkel, and L. C. Phan. (2006). Modification of ASM3 for the determination of biomass adsorption/storage capacity in bulking sludge control. *Water Science and Technology: A Journal of the International Association on Water Pollution Research* 53 (3):91–99.

Malliaros, C., and A. Guitonas. (1997). Pre-treatment and elimination systems of toxic industrial waste and sludges. The case study of the Department of Attika. *Water Science and Technology* 36 (2–3):91–100.

Manterola, G., I. Uriarte, and L. Sancho. (2008). The effect of operational parameters of the process of sludge ozonation on the solubilisation of organic and nitrogenous compounds. *Water Research* 42 (12):3191–3197.

Mao, T., S. Y. Hong, K. Y. Show, J. H. Tay, and D. J. Lee. (2004). A comparison of ultrasound treatment on primary and secondary sludges. *Water Science and Technology: A Journal of the International Association on Water Pollution Research* 50 (9):91–97.

Martín, M. Á., I. González, A. Serrano, and J. Á. Siles. (2015). Evaluation of the improvement of sonication pre-treatment in the anaerobic digestion of sewage sludge. *Journal of Environmental Management* 147:330–337.

McCarty, P. L. (1964). Anaerobic waste treatment fundamentals part I: Chemistry and microbiology. *Public Works* 95 (12):107–112.

McCarty, P. L., J. Bae, and J. Kim. (2011). Domestic wastewater treatment as a net energy producer: Can this be achieved? *Environmental Science and Technology* 45 (17):7100–7106.

McInerney, M. J., M. P. Bryant, and N. Pfennig. (1979). Anaerobic bacterium that degrades fatty acids in syntrophic association with methanogens. *Archives of Microbiology* 122 (2):129–135.

Meerburg, F. A., N. Boon, T. Van Winckel, J. A. R. Vercamer, I. Nopens, and S. E. Vlaeminck. (2015). Toward energy-neutral wastewater treatment: a high-rate contact stabilization process to maximally recover sewage organics. *Bioresource Technology* 179:373–381.

Mehdizadeh, S. N., C. Eskicioglu, J. Bobowski, and T. Johnson. (2013). Conductive heating and microwave hydrolysis under identical heating profiles for advanced anaerobic digestion of municipal sludge. *Water Research* 47 (14):5040–5051.

Mei, X., Z. Wang, X. Zheng, F. Huang, J. Ma, Jixu Tang, and Z. Wu. (2014). Soluble microbial products in membrane bioreactors in the presence of ZnO nanoparticles. *Journal of Membrane Science* 451:169–176.

Merry, J. and P. Fountain. (2014). Innovative design aspects of advanced digestion at beckton and crossness. In: *Proceedings of Aquaenviro's 19th European Biosolids and Organic Resources Conference and Exhibition*, Manchester, UK.

Merry, J. and B. Oliver. (2015). A comparison of real ad plant performance: howdon, bra sands, cardiff and afan. In: *Proceedings of Aquaenviro's 20th European Biosolids and Organic Resources Conference and Exhibition*, Manchester, UK.

Metcalf, E., G. Tchobanoglous, H. David Stensel, R. Tsuchihashi, and F. Burton. (2014). *Wastewater Engineering: Treatment and Resource Recovery*. 5th ed. New York: McGraw-Hill.

Mills, N., Martinicca, H., Fountain, P., Shana, A., Ouki, S., Thorpe, R., 2013. Second generation thermal hydrolysis process. In: *18th European Biosolids And Biowastes Conference*, Manchester, UK.

Mills, N., Pearce, P., Farrow, J., Thorpe, R.B., Kirkby, N.F., 2014. Environmental and economic life cycle assessment of current and future sewage sludge to energy technologies. Waste Manag. 34 (1), 185e195.

Mills, N., H. Martinicca, P. Fountain, A. Shana, S. Ouki, and R. Thorpe. (2013). Second generation thermal hydrolysis process. In: *18th European Biosolids and Biowastes Conference*, Manchester, UK.

Mobed, J. J., S. L. Hemmingsen, J. L. Autry, and L. B. McGown. (1996). Fluorescence characterization of IHSS humic substances: Total luminescence spectra with absorbance correction. *Environmental Science and Technology* 30 (10):3061–3065.

Moerman, W. H., D. R. Bamelis, H. L. Vergote, P. M. Van Holle, F. P. Houwen, and W. H. Verstraete. (1994). Ozonation of activated sludge treated carbonization wastewater. *Water Research* 28 (8):1791–1798.

Mosey, F. E. (1983). Mathematical modelling of the anaerobic digestion process: Regulatory mechanisms for the formation of short-chain volatile acids from glucose. *Water Science and Technology* 15 (8–9):209–232.

Mottet, A., J. P. Steyer, S. Déléris, F. Vedrenne, J. Chauzy, and H. Carrère. (2009). Kinetics of thermophilic batch anaerobic digestion of thermal hydrolysed waste activated sludge. *Biochemical Engineering Journal* 46 (2):169–175.

Müller, J. A. (2001). Prospects and problems of sludge pre-treatment processes. *Water Science and Technology: A Journal of the International Association on Water Pollution Research* 44 (10):121–128.

Müller, J., G. Lehne, J. Schwedes, S. Battenberg, R. Näveke, J. Kopp, N. Dichtl, A. Scheminski, R. Krull, and D. C. Hempel. (1998). Disintegration of sewage sludges and influence on anaerobic digestion. *Water Science and Technology* 38 (8–9):425–433.

Mustranta, A. and L. Viikari. (1993). Dewatering of activated sludge by an oxidative treatment. *Water Science and Technology* 28 (1):213–221.

Nabarlatz, D., F. Stüber, J. Font, A. Fortuny, A. Fabregat, and C. Bengoa. (2012). Extraction and purification of hydrolytic enzymes from activated sludge. *Resources, Conservation and Recycling* 59 (0):9–13.

Naddeo, V., V. Belgiorno, M. Landi, T. Zarra, and R. M. A. Napoli. (2009). Effect of sonolysis on waste activated sludge solubilisation and anaerobic biodegradability. *Desalination* 249 (2):762–767.

Nah, I. K. Y. Whan Kang, K.-Y. Hwang, and W.-K. Song. (2000). Mechanical pretreatment of waste activated sludge for anaerobic digestion process. *Water Research* 34 (8):2362–2368.

Namkung, E., and B. E. Rittmann. (1986). Soluble microbial products (SMP) formation kinetics by biofilms. *Water Research* 20 (6):795–806.

Navarro, N. M., T. Chave, P. Pochon, I. Bisel, and S. I. Nikitenko. (2011). Effect of ultrasonic frequency on the mechanism of formic acid sonolysis. *Journal of Physical Chemistry. B* 115 (9):2024–2029.

Neis, U., K. Nickel, and A. Tiehm. (2000). Enhancement of anaerobic sludge digestion by ultrasonic disintegration. *Water Science and Technology* 42 (9):73–80.

Neis, U., K. Nickel, and A. Lundén. (2008). Improving anaerobic and aerobic degradation by ultrasonic disintegration of biomass. *Journal of Environmental Science and Health. Part A, Toxic/Hazardous Substances and Environmental Engineering* 43 (13):1541–1545.

Nevot, M., V. Deroncele, M. J. Montes, and E. Mercade. (2007). Effect of incubation temperature on growth parameters of Pseudoalteromonas antarctica NF3 and its production of extracellular polymeric substances. *Journal of Applied Microbiology* 105:255–263.

Neyens, E. and J. Baeyens. (2003). A review of thermal sludge pre-treatment processes to improve dewaterability. *Journal of Hazardous Materials* 98 (1–3):51–67.

Neyens, E., J. Baeyens, and C. Creemers. (2003a). Alkaline thermal sludge hydrolysis. *Journal of Hazardous Materials* 97 (1–3):295–314.

Neyens, E., J. Baeyens, M. Weemaes, and B. De heyder. (2003b). Pilot-scale peroxidation (H_2O_2) of sewage sludge. *Journal of Hazardous Materials* 98 (1–3):91–106.

Neyens, E., J. Baeyens, M. Weemaes, and B. D. De heyder. (2003c). Hot acid hydrolysis as a potential treatment of thickened sewage sludge. *Journal of Hazardous Materials* 98 (1–3):275–293.

Neyens, E., J. Baeyens, R. Dewil, and B. De heyder. (2004). Advanced sludge treatment affects extracellular polymeric substances to improve activated sludge dewatering. *Journal of Hazardous Materials* 106 (2–3):83–92.

Ng, T. C. A. and H. Y. Ng. (2010). Characterisation of initial fouling in aerobic submerged membrane bioreactors in relation to physico-chemical characteristics under different flux conditions. *Water Research* 44 (7):2336–2348.

Nges, I. A. and J. Liu. (2009). Effects of anaerobic pre-treatment on the degradation of dewatered-sewage sludge. *Renewable Energy* 34 (7):1795–1800.

Nges, I. A. and J. Liu. (2010). Effects of solid retention time on anaerobic digestion of dewatered-sewage sludge in mesophilic and thermophilic conditions. *Renewable Energy* 35 (10):2200–2206.

Nickel, K. and U. Neis. (2007). Ultrasonic disintegration of biosolids for improved biodegradation. *Ultrasonics Sonochemistry* 14 (4):450–455.

Nie, Y., Z. Qiang, W. Ben, and J. Liu. (2014). Removal of endocrine-disrupting chemicals and conventional pollutants in a continuous-operating activated sludge process integrated with ozonation for excess sludge reduction. *Chemosphere* 105:133–138.

Nielsen, H. B., A. Thygesen, A. B. Thomsen, and J. Ejbye Schmidt. (2010). Anaerobic digestion of waste activated sludge: Comparison of thermal pretreatments with thermal interstage treatments. *Journal of Chemical Technology and Biotechnology* 86:238–245.

Ning, X., H. Chen, J. Wu, Y. Wang, J. Liu, and M. Lin. (2014). Effects of ultrasound assisted Fenton treatment on textile dyeing sludge structure and dewaterability. *Chemical Engineering Journal* 242:102–108.

Nishijima, W., Fahmi, T. Mukaidani, and M. Okada. (2003). DOC removal by multi-stage ozonation-biological treatment. *Water Research* 37 (1):150–154.

Noike, T., G. Endo, J. E. Chang, J. I. Yaguchi, and J. Matsumoto. (1985). Characteristics of carbohydrate degradation and the rate-limiting step in anaerobic digestion. *Biotechnology and Bioengineering* 27 (10):1482–1489.

Novak, J. T., and D. A. Carlson. (1970). Kinetics of anaerobic long chain fatty acid degradation. *Journal (Water Pollution Control Federation)* 42 (11):1932–1970.

Odegaard, H. (2004). Sludge minimization technologies: An overview. *Water Science and Technology: A Journal of the International Association on Water Pollution Research* 49 (10):31–40.

Oles, J., N. Dichtl, and H.-H. Niehoff. (1997). Full scale experience of two stage thermophilic mesophilic sludge digestion. *Water Science and Technology* 36 (6–7):449–456.

Onyeche, T. I. (2007). Economic benefits of low pressure sludge homogenization for wastewater treatment plants. In *IWA Specialist Conferences Moving Forward Wastewater Biosolids Sustainability*. Moncton, New Brunswick, Canada.

Onyeche, T. I., O. Schläfer, H. Bormann, C. Schröder, and M. Sievers. (2002). Ultrasonic cell disruption of stabilised sludge with subsequent anaerobic digestion. *Ultrasonics* 40 (1–8):31–35.

Oosterhuis, M., D. Ringoot, A. Hendriks, and P. Roeleveld. (2014). Thermal hydrolysis of waste activated sludge at Hengelo wastewater treatment plant, The Netherlands. *Water Science and Technology: A Journal of the International Association on Water Pollution Research* 70 (1):1–7.

Owen, W. F., D. C. Stuckey, J. B. Healy Jr., L. Y. Young, and P. L. McCarty. (1979). Bioassay for monitoring monitoring biochemical methane potential and anaerobic toxicity. *Water Research* 13 (6):485–492.

Palmeiro-Sánchez, T., A. Val del Río, A. Mosquera-Corral, J. L. Campos, and R. Méndez. (2013). Comparison of the anaerobic digestion of activated and aerobic granular sludges under brackish conditions. *Chemical Engineering Journal* 231:449–454.

Panter, K. and H. Kleiven. (2005). Ten years' experience of full scale thermal hydrolysis projects, Aquaenviro. In: *10th European Biosolids and Biowaste Conference*, Wakefield, UK.

Park, B., J. H. Ahn, J. Kim, and S. Hwang. (2004). Use of microwave pretreatment for enhanced anaerobiosis of secondary sludge. *Water Science and Technology: A Journal of the International Association on Water Pollution Research* 50 (9):17–23.

Park, C., R. F. Helm, and J. T. Novak. (2008). Investigating the fate of activated sludge extracellular proteins in sludge digestion using sodium dodecyl sulfate polyacrylamide gel electrophoresis. *Water Environment Research: A Research Publication of the Water Environment Federation* 80 (12):2219–2227.

Park, N. D., S. S. Helle, and R. W. Thring. (2012). Combined alkaline and ultrasound pretreatment of thickened pulp mill waste activated sludge for improved anaerobic digestion. *Biomass and Bioenergy* 46:750–756.

Parmar, N., A. Singh, and O. P. Ward. (2001). Enzyme treatment to reduce solids and improve settling of sewage sludge. *Journal of Industrial Microbiology and Biotechnology* 26 (6):383–386.

Pavlostathis, S. G. and J. M. Gossett. (1985). Alkaline treatment of wheat straw for increasing anaerobic biodegradability. *Biotechnology and Bioengineering* 27 (3):334–344.

Pavlostathis, S. G. and E. Giraldo-Gomez. (1991). Kinetics of anaerobic treatment. *Water Science and Technology* 24 (8):35–59.

Pavoni, J. L., M. W. Tenney, and W. F. Echelberger Jr. (1972). Bacterial exocellular polymers and biological flocculation. *Journal – Water Pollution Control Federation* 44 (3):414–429.

Penaud, V., J. P. Delgenès, and R. Moletta. (1999). Thermo-chemical pretreatment of a microbial biomass: Influence of sodium hydroxide addition on solubilization and anaerobic biodegradability. *Enzyme and Microbial Technology* 25 (3–5):258–263.

Pérez-Elvira, S., M. Fdz-Polanco, F. I. Plaza, G. Garralón, and F. Fdz-Polanco. (2009). Ultrasound pre-treatment for anaerobic digestion improvement. *Water Science and Technology: A Journal of the International Association on Water Pollution Research* 60 (6):1525–1532.

Perrault, A., R. B. Thorpe, A. Cataldi, N. Mills, and R. Barua. (2015). Steam explosion of sludge in the THP process. Paper read at the Sludgetech Conference, at Guilford, Surrey, UK.

Pham, T. T. H., S. K. Brar, R. D. Tyagi, and R. Y. Surampalli. (2010). Influence of ultrasonication and Fenton oxidation pre-treatment on rheological characteristics of wastewater sludge. *Ultrasonics Sonochemistry* 17 (1):38–45.

Phothilangka, P., M. A. Schoen, and B. Wett. (2008). Benefits and drawbacks of thermal prehydrolysis for operational performance of wastewater treatment plants. *Water Science and Technology: A Journal of the International Association on Water Pollution Research* 58 (8):1547–1553.

Pickworth, B., J. Adams, K. Panter, and O. E. Solheim. (2006). Maximising biogas in anaerobic digestion by using engine waste heat for thermal hydrolysis pretreatment of sludge. *Water Science and Technology* 54 (5): 101–108.

Pijuan, M., Q. Wang, L. Ye, and Z. Yuan. (2012). Improving secondary sludge biodegradability using free nitrous acid treatment. *Bioresource Technology* 116:92–98.

Pilli, S., P. Bhunia, S. Yan, R. J. LeBlanc, R. D. Tyagi, and R. Y. Surampalli. (2011). Ultrasonic pretreatment of sludge: A review. *Ultrasonics Sonochemistry* 18 (1):1–18.

Pilli, S., S. Yan, R. D. Tyagi, and R. Y. Surampalli. (2015). Overview of Fenton pre-treatment of sludge aiming to enhance anaerobic digestion. *Reviews in Environmental Science and Bio/Technology* 14 (3):453–472.

Pilli, S., S. Yan, R. D. Tyagi, and R. Y. Surampalli. (2016). Anaerobic digestion of ultrasonicated sludge at different solids concentrations: Computation of mass-energy balance and greenhouse gas emissions. *Journal of Environmental Management* 166:374–386.

Pook, M., N. Mills, M. Heitmann, K. Panter, and P. Walley. (2013). Exploring the upper limits of thermal hydrolysis at Chertsey STW. In: *Proceedings of Aqua-enviro 18th European Biosolids and Organic Residuals Conference and Exhibition*, Manchester, UK.

Pritchard, D. L., N. Penney, M. J. McLaughlin, H. Rigby, and K. Schwarz. (2010). Land application of sewage sludge (biosolids) in Australia: Risks to the environment and food crops. *Water Science and Technology: A Journal of the International Association on Water Pollution Research* 62 (1):48–57.

Pullammanappallil, P. C., D. P. Chynoweth, G. Lyberatos, and S.A. Svoronos. (2001). Stable performance of anaerobic digestion in the presence of a high concentration of propionic acid. *Bioresource Technology* 78(2):165–169.

Qiao, W., W. Wang, R. Xun, W. Lu, and K. Yin. (2008). Sewage sludge hydrothermal treatment by MW irradiation combined with alkali addition. *Journal of Materials Science* 43 (7):2431–2436.

Qiao, W., C. Peng, W. Wang, and Z. Zhang. (2011). Biogas production from supernatant of hydrothermally treated municipal sludge by upflow anaerobic sludge blanket reactor. *Bioresource Technology* 102 (21):9904–9911.

Rabinowitz, B. and R. Stephenson. (2005). Improving anaerobic digester efficiency by homogenization of waste activated sludge. Paper read at the Proceedings of the 78th Annals Conference of the Water Environment Federation, at Washington, DC.

Rai, C. L. and P. G. Rao. (2009). Influence of sludge disintegration by high pressure homogenizer on microbial growth in sewage sludge: An approach for excess sludge reduction. *Clean Technologies and Environmental Policy* 11 (4):437–446.

Rajan, R. V., J.-G. Lin, and B. T. Ray. (1989). Low-level chemical pretreatment for enhanced sludge solubilization. *Research Journal of the Water Pollution Control Federation* 61:1678–1683.

Ratsak, C. H. (1994). Grazer induced sludge reduction in wastewater treatment. PhD thesis. Vrije Universiteit, Amsterdam.

Rawlinson, D., S. Halliday, S. Garbutt, and I. Jobling. (2009). Advanced digestion plant at Bran Sands design and construct experiences. In: *Proceedings of Aquaenviro's 14th European Biosolids and Organic Resources Conference and Exhibition*, Leeds, UK.

Ray, B., T. Jih-Gaw Lin, and R. V. Rajan. (1990). Low-level alkaline solubilization for enhanced anaerobic digestion. *Research Journal of the Water Pollution Control Federation* 62 (1):81–87.

Riau, V., M. A. De la Rubia, and M. Pérez. (2015). Upgrading the temperature-phased anaerobic digestion of waste activated sludge by ultrasonic pretreatment. *Chemical Engineering Journal* 259:672–681.

Riesz, P., D. Berdahl, and C. L. Christman. (1985). Free radical generation by ultrasound in aqueous and nonaqueous solutions. *Environmental Health Perspectives* 64:233–252.

Rittman, B. E., and P. L. McCarty. (2001). *Environmental Biotechnology: Principles and Applications*. London: McGraw-Hill Int. Editions.

Rittmann, B. E., H. S. Lee, H. S. Zhang, J. Alder, J. E. Banaszak, and R. Lopez. (2008). Full-scale application of focused-pulsed pre-treatment for improving biosolids digestion and conversion to methane. *Water Science and Technology* 58 (10):1895–1901.

Rivero, J. A., N. Madhavan, M. T. Suidan, P. Ginestet, and J. Audic. (2006). Enhancement of anaerobic digestion of excess municipal sludge with thermal and/or oxidative treatment. *Journal of Environmental Engineering* 132 (6):638–644.

Rocher, M., G. Goma, A. P. Begue, L. Louvel, and J. L. Rols. (1999). Towards a reduction in excess sludge production in activated sludge process, biomass physicochemical treatment and biodegradation. *Applied Microbiology and Biotechnology* 51 (6):883–890.

Roest, K., B. Daamen, M. S. de Graaff, L. Hartog, M. H. Zandvoort, C. A. Uijterlinde, S. Dilven, J. B. van Lier, and M. C. M. van Loosdrecht. (2012). Energy production from wastewater: Dynamic filtration of A-stage sludge. Paper read at the IWA World Congress on Water, Climate and Energy, 13–18 May, at Dublin, Ireland.

Roxburgh, R., R. Sieger, B. Johnson, B. Rabinowitz, S. Goodwin, G. Crawford, , and G. Daigger. (2006). Sludge minimization technologies: Doing more to get less. Paper Read at WEFTEC®.06, at Dallas, Texas.

Saha, M., C. Eskicioglu, and J. Marin. (2011). Microwave, ultrasonic and chemo-mechanical pretreatments for enhancing methane potential of pulp mill wastewater treatment sludge. *Bioresource Technology* 102 (17):7815–7826.

Sahinkaya, S. (2014). Disintegration of municipal waste activated sludge by simultaneous combination of acid and ultrasonic pretreatment. *Process Safety and Environment Protection* http://dx.doi.org/doi:10.1016/j.psep.2014.04.002.

Şahinkaya, S. (2015). Disintegration of municipal waste activated sludge by simultaneous combination of acid and ultrasonic pretreatment. *Process Safety and Environmental Protection* 93:201–205.

Şahinkaya, S. and M. F. Sevimli. (2013a). Sono-thermal pre-treatment of waste activated sludge before anaerobic digestion. *Ultrasonics Sonochemistry* 20 (1):587–594.

Şahinkaya, S. and M. F. Sevimli. (2013b). Synergistic effects of sono-alkaline pretreatment on anaerobic biodegradability of waste activated sludge. *Journal of Industrial and Engineering Chemistry* 19 (1):197–206.

Şahinkaya, S., E. Kalıpci, and S. Aras. (2015). Disintegration of waste activated sludge by different applications of Fenton process. *Process Safety and Environmental Protection* 93:274–281.

Sakai, Y., T. Fukase, H. Yasui, and M. Shibata. (1997). An activated sludge process without excess sludge production. *Water Science and Technology* 36 (11):163–170.

Sakai, Y., T. Aoyagi, N. Shiota, A. Akashi, and S. Hasegawa. (2000). Complete decomposition of biological waste sludge by thermophilic aerobic bacteria. *Water Science and Technology* 42 (9):81–88.

Saktaywin, W., H. Tsuno, H. Nagare, T. Soyama, and J. Weerapakkaroon. (2005). Advanced sewage treatment process with excess sludge reduction and phosphorus recovery. *Water Research* 39 (5):902–910.

Salsabil, M. R., A. Prorot, M. Casellas, and C. Dagot. (2009). Pre-treatment of activated sludge: Effect of sonication on aerobic and anaerobic digestibility. *Chemical Engineering Journal* 148 (2–3):327–335.

Scheminski, A., R. Krull, and D. C. Hempel. (2000). Oxidative treatment of digested sewage sludge with ozone. *Water Science and Technology* 42 (9):151–158.

Schiener, P., S. Nachaiyasit, and D. C. Stuckey. (1998). Production of soluble microbial products (SMP) in an anaerobic baffled reactor: Composition, biodegradability, and the effect of process parameters. *Environmental Technology* 19 (4):391–399.

Seng, B., S. K. Khanal, and C. Visvanathan. (2010). Anaerobic digestion of waste activated sludge pretreated by a combined ultrasound and chemical process. *Environmental Technology* 31 (3):257–265.

Serrano, A., J. A. Siles, M. C. Gutiérrez, and M. Á. Martín. (2015). Improvement of the bio-methanization of sewage sludge by thermal pre-treatment and co-digestion with straw-berry extrudate. *Journal of Cleaner Production* 90:25–33.

Shao, L. M., X. Y. Wang, H. C. Xu, and P. J. He. (2012). Enhanced anaerobic digestion and sludge dewaterability by alkaline pretreatment and its mechanism. *Journal of Environmental Sciences* 24 (10):1731–1738.

Shao, L., T. Wang, T. Li, F. Lü, and P. He. (2013). Comparison of sludge digestion under aero-bic and anaerobic conditions with a focus on the degradation of proteins at mesophilic temperature. *Bioresource Technology* 140:131–137.

Sheng, G. P., H. Q. Yu, and X. Y. Li. (2010). Extracellular polymeric substances (EPS) of micro-bial aggregates in biological wastewater treatment systems: A review. *Biotechnology Advances* 28 (6):882–894.

Sheridan, J., and B. Cutis. (2004). Case book: Revolutionary technology cuts biosolids pro-duction and costs. *Pollution Engineering* 36 (5).

Shiota, N., A. Akashi, and S. Hasegawa. (2002). A strategy in wastewater treatment pro-cess for significant reduction of excess sludge production. *Water Science and Technology: A Journal of the International Association on Water Pollution Research* 45 (12):127–134.

Shizas, I. and D. M. Bagley. (2004). Experimental determination of energy content of unknown organics in municipal wastewater streams. *Journal of Energy Engineering* 130 (2):45–53.

Show, K.-Y., T. Mao, and D.-J. Lee. (2007). Optimisation of sludge disruption by sonication. *Water Research* 41 (20):4741–4747.

Skiadas, I. V., H. N. Gavala, J. Lu, and B. K. Ahring. (2005). Thermal pre-treatment of pri-mary and secondary sludge at 701°C prior to anaerobic digestion. *Water Science and Technology: A Journal of the International Association on Water Pollution Research* 52 (1–2):161–166.

Smith, M. R. and R. A. Mah. (1978). Growth and methanogenesis by Methanosarcina Strain 227 on acetate and methanol. *Applied and Environmental Microbiology* 36 (6):870–879.

Solley, D., S. Hu, C. Hertle, D. Batstone, T. Karastergiou-Hogan, Q. Rider, and J. Keller. (2015). Identifying novel wastewater treatment options through optimal technology integration. *Water Practice and Technology* 10 (3):496–504.

Song, L. J., N. W. Zhu, H. P. Yuan, Y. Hong, and J. Ding. (2010). Enhancement of waste activated sludge aerobic digestion by electrochemical pre-treatment. *Water Research* 44 (15):4371–4378.

Souza, T. S. O., L. C. Ferreira, I. Sapkaite, S. I. Pérez-Elvira, and F. Fdz-Polanco. (2013). Thermal pretreatment and hydraulic retention time in continuous digesters fed with sew-age sludge: Assessment using the ADM1. *Bioresource Technology* 148:317–324.

Speece, R. E. (2008). *Anaerobic Biotechnology and Odor/Corrosion Control for Municipalities and Industries.* Nashville, TN: Archae Press.

Staehelin, J. and J. Hoigne. (1985). Decomposition of ozone in water in the presence of organic solutes acting as promoters and inhibitors of radical chain reactions. *Environmental Science and Technology* 19 (12):1206–1213.

Stephenson, R., J. Shaw, S. Laliberte, and P. Elson. (2003). Sludge buster! The Microsludge™ process to destroy biosolids. Paper read at the Second Canadian Organic Residuals Recycling Conference, Longueuil, Quebec, Canada.

Stevenson, F. J. (1994). *Humus Chemistry: Genesis, Compositions Reactions.* 2nd ed. New York: John Wiley & Sons.

Stuckey, D. C. and P. L. McCarty. (1978). Thermochemical pretreatment of nitrogenous materials to increase methane yield. Biotechnology and Bioengineering Symposium 8:219–233.

Stuckey, D. C. and P. L. McCarty. (1984). The effect of thermal pretreatment on the anaerobic biodegradability and toxicity of waste activated sludge. *Water Research* 18 (11):1343–1353.

Takashima, M. (2008). Examination on process configurations incorporating thermal treatment for anaerobic digestion of sewage sludge. *Journal of Environmental Engineering* 134 (7):543–549.

Takashima, M. and Y. Tanaka. (2008). Comparison of thermo-oxidative treatments for the anaerobic digestion of sewage sludge. *Journal of Chemical Technology and Biotechnology* 83 (5):637–642.

Takashima, M. and Y. Tanaka. (2010). Application of acidic thermal treatment for one- and two-stage anaerobic digestion of sewage sludge. *Water Science and Technology: A Journal of the International Association on Water Pollution Research* 62 (11):2647–2654.

Takashima, M. and Y. Tanaka. (2014). Acidic thermal post-treatment for enhancing anaerobic digestion of sewage sludge. *Journal of Environmental Chemical Engineering* 2 (2):773–779.

Takashima, M., Y. Kudoh, and N. Tabata. (1996). Complete anaerobic digestion of activated sludge by combining membrane separation and alkaline heat post-treatment. *Water Science and Technology* 34 (5–6):477–481.

Tanaka, S., T. Kobayashi, K.-I. Kamiyama, and M. L. N. Signey Bildan. (1997). Effects of thermochemical pretreatment on the anaerobic digestion of waste activated sludge. *Water Science and Technology* 35 (8):209–215.

Tao, J., S. Wu, L. Sun, X. Tan, S. Yu, and Z. Zhang. (2012). Composition of waste sludge from municipal wastewater treatment plant. *Procedia Environmental Sciences* 12: 964–971.

Tchobanoglous, G., F. L. Burton, and H. D. Stensel. (2003). *Wastewater Engineering: Treatment and Reuse*. 4th ed. New York: McGraw-Hill.

Thomas, L., G. Jungschaffer, and B. Sprössler. (1993). Improved sludge dewatering by enzymatic treatment. *Water Science and Technology* 28 (1):189–192.

Tian, X. and A. Trzcinski. (2017). Effects of physico-chemical post-treatments on the semi-continuous anaerobic digestion of sewage sludge. *Environments* 4 (3):49.

Tian, X., A. Trzcinski, W. Chong, L. Lin, and W. J. Ng. (2014a). Insights on the solubilization products after combined alkaline and ultrasonic pre-treatment of sewage sludge. *Journal of Environmental Sciences* 27:97–105.

Tian, X., C. Wang, A. P. Trzcinski, L. Lin, and W. J. Ng. (2014b). Interpreting the synergistic effect in combined ultrasonication-ozonation sewage sludge pre-treatment. *Chemosphere* 140:63–71.

Tian, X., A. P. Trzcinski, L. L. Lin, and W. J. Ng. (2015a). Impact of ozone assisted ultrasonication pre-treatment on anaerobic digestibility of sewage sludge. *Journal of Environmental Sciences* 33:29–38.

Tian, X., C. Wang, A. P. Trzcinski, L. Lin, and W. J. Ng. (2015b). Insights on the solubilization products after combined alkaline and ultrasonic pre-treatment of sewage sludge. *Journal of Environmental Sciences* 29:97–105.

Tian, X., A. P. Prandota Trzcinski, L. L. Leonard, and W. J. Ng. (2016). Enhancing sewage sludge anaerobic "re-digestion" with combinations of ultrasonic, ozone and alkaline treatments. *Journal of Environmental Chemical Engineering* 4 (4, Part A):4801–4807.

Tiehm, A., K. Nickel, and U. Neis. (1997). The use of ultrasound to accelerate the anaerobic digestion of sewage sludge. *Water Science and Technology* 36 (11):121–128.

Tiehm, A., K. Nickel, M. Zellhorn, and U. Neis. (2001). Ultrasonic waste activated sludge disintegration for improving anaerobic stabilization. *Water Research* 35 (8):2003–2009.

Traore, A. S., M. L. Fardeau, C. E. Hatchikian, J. Le Gall, and J. P. Belaich. (1983). Energetics of growth of a defined mixed culture of *Desulfovibrio vulgaris* and *Methanosarcina barkeris*: Interspecies hydrogen transfer in batch and continuous cultures. *Applied and Environmental Microbiology* 46 (5):1152–1156.

Tronson, R., M. Ashokkumar, and F. Grieser. (2002). Comparison of the effects of water-soluble solutes on multibubble sonoluminescence generated in aqueous solutions by 20- and 515-kHz pulsed ultrasound. *Journal of Physical Chemistry B* 106 (42):11064–11068.

Trzcinski, A. P. and D. C. Stuckey. (2012). Determination of the hydrolysis constant in the biochemical methane potential test of municipal solid waste. *Environmental Engineering Science* 29 (9):848–854.

Trzcinski, A. P., N. Ofoegbu, and D. C. Stuckey. (2011). Post-treatment of the permeate of a submerged anaerobic membrane bioreactor (SAMBR) treating landfill leachate. *Journal of Environmental Science and Health. Part A, Toxic/Hazardous Substances and Environmental Engineering* 46 (13):1539–1548.

Trzcinski, A. P., X. Tian, C. Wang, L. L. Lin, and W. J. Ng Prandota, X. Tian, C. Wang, and W. J. Ng. (2015). Combined ultrasonication and thermal pre-treatment of sewage sludge for increasing methane production. *Journal of Environmental Science and Health, Part A* 50 (2):213–223.

Trzcinski, A. P., L. Ganda, C. Kunacheva, D. Q. Zhang, L.L. Lin, G. Tao, Y. Lee, and W.J. Ng. (2016). Characterization and biodegradability of sludge from a high rate A-stage contact tank and B-stage membrane bioreactor of a pilot-scale AB system treating municipal wastewaters. *Water Science and Technology* 74 (7):1716.

Trzcinski P. A., C. Wang, D. Zhang, W. S. Ang, L. L. Lin, T. Niwa, Y. Fukuzaki, and W. J. Ng. (2017). Performance of A-stage process treating combined municipal-industrial wastewater. *Water Science and Technology* 75 (1):228.

Tyagi, V. K. and S. L. Lo. (2012). Enhancement in mesophilic aerobic digestion of waste activated sludge by chemically assisted thermal pretreatment method. *Bioresource Technology* 119:105–113.

Uma Rani, R., S. Adish Kumar, S. Kaliappan, I. Yeom, and J. Rajesh Banu. (2013). Impacts of microwave pretreatments on the semi-continuous anaerobic digestion of dairy waste activated sludge. *Waste Management* 33 (5):1119–1127.

Val del Río, A., N. Morales, E. Isanta, A. Mosquera-Corral, J. L. Campos, J. P. Steyer, and H. Carrère. (2011). Thermal pre-treatment of aerobic granular sludge: Impact on anaerobic biodegradability. *Water Research* 45 (18):6011–6020.

Valo, A., H. Carrère, and J. P. Delgenès. (2004). Thermal, chemical and thermo-chemical pretreatment of waste activated sludge for anaerobic digestion. *Journal of Chemical Technology and Biotechnology* 79 (11):1197–1203.

van den Berg, L., G. B. Patel, D. S. Clark, and C. P. Lentz. (1976). Factors affecting rate of methane formation from acetic-acid by enriched methanogenic cultures. *Canadian Journal of Microbiology* 22 (9):1312–1319.

van Dijk, L., and A. de Man. (2010). TurboTec®: Continue thermische slib hydrolyse voor lagere zuiveringskosten en meer energie op de rwzi. *Neerslag* 5.

van Haandel, A. and J. van der Lubbe. (2007). *Handbook Biological Wastewater Treatment - Design and Optimisation of Activated Sludge Systems*. London: IWA Publishing.

van Loosdrecht, M. C. M., and D. Brdjanovic. (2014). Water treatment. Anticipating the next century of wastewater treatment. *Science* 344 (6191):1452–1453.

Versprille, A. I., B. Zuurveen, and Th. Stein. (1985). The A-B process: A novel two stage wastewater treatment system. *Water Science and Technology* 17 (2–3):235–246.

Verstraete, W. and S. E. Vlaeminck. (2011). Zero waste water: Short-cycling of wastewater resources for sustainable cities of the future. *International Journal of Sustainable Development and World Ecology* 18 (3):253–264.

Verstraete, W., P. V. Van de Caveye, and V. Diamantis. (2009). Maximum use of resources present in domestic "used water". *Bioresource Technology* 100 (23):5537–5545.

Vesilind, P. A. (2003). *Wastewater Treatment Plant Design*. London: Water Environment Federation, IWA Publishing.

Vlyssides, A. G. and P. K. Karlis. (2004). Thermal-alkaline solubilization of waste activated sludge as a pre-treatment stage for anaerobic digestion. *Bioresource Technology* 91 (2):201–206.

Vogelaar, J. C. T., A. De Keizer, S. Spijker, and G. Lettinga. (2005). Bioflocculation of mesophilic and thermophilic activated sludge. *Water Research* 39 (1):37–46.

Volk, C., P. Renner, H. Paillard, J. C. Joret. (1993). Effects of ozone on the production of biodegradable dissolved organic carbon (BDOC) during water treatment. *Ozone: Science and Engineering* 15 (5):390–404.

Volk, C., P. Roche, J.-C. Joret, and H. Paillard. (1997). Comparison of the effect of ozone, ozone-hydrogen peroxide system and catalytic ozone on the biodegradable organic matter of a fulvic acid solution. *Water Research* 31 (3):650–656.

Wahidunnabi, A. K. and C. Eskicioglu. (2014). High pressure homogenization and two-phased anaerobic digestion for enhanced biogas conversion from municipal waste sludge. *Water Research* 66:430–446.

Wang, F., Y. Wang, and M. Ji. (2005). Mechanisms and kinetics models for ultrasonic waste activated sludge disintegration. *Journal of Hazardous Materials* 123 (1–3):145–150.

Wang, F., M. Ji, and S. Lu. (2006a). Influence of ultrasonic disintegration on the dewaterability of waste activated sludge. *Environmental Progress* 25 (3):257–260.

Wang, F., S. Lu, and M. Ji. (2006b). Components of released liquid from ultrasonic waste activated sludge disintegration. *Ultrasonics Sonochemistry* 13 (4):334–338.

Wang, Q., C. Noguchi, Y. Hara, C. Sharon, K. Kakimoto, and Y. Kato. (1997). Studies on anaerobic digestion mechanism: Influence of pretreatment temperature on biodegradation of waste activated sludge. *Environmental Technology* 18 (10):999–1008.

Wang, Q. H., M. Kuninobu, K. Kakimoto, H. I. Ogawa, and Y. Kato. (1999). Upgrading of anaerobic digestion of waste activated sludge by ultrasonic pretreatment. *Bioresource Technology* 68 (3):309–313.

Wang, Q., L. Ye, G. Jiang, P. D. Jensen, D. J. Batstone, and Z. Yuan. (2013a). Free nitrous acid (FNA)-based pre-treatment enhances methane production from waste activated sludge. *Environmental Science and Technology* 47 (20):11897–11904.

Wang, Q., L. Ye, G. Jiang, and Z. Yuan. (2013b). A free nitrous acid (FNA)-based technology for reducing sludge production. *Water Research* 47 (11):3663–3672.

Wang, Q., G. Jiang, L. Ye, and Z. Yuan. (2014a). Enhancing methane production from waste activated sludge using combined free nitrous acid and heat pre-treatment. *Water Research* 63:71–80.

Wang, Q., L. Ye, G. Jiang, S. Hu, and Z. Yuan. (2014b). Side-stream sludge treatment using free nitrous acid selectively eliminates nitrite oxidizing bacteria and achieves the nitrite pathway. *Water Research* 55:245–255.

Wang, Q., W. Wei, Y. Gong, Q. Yu, Q. Li, J. Sun, and Z. Yuan. (2017). Technologies for reducing sludge production in wastewater treatment plants: State of the art. *Science of the Total Environment* 587–588 (Supplement c):510–521.

Weemaes, M., H. Grootaerd, F. Simoens, and W. Verstraete. (2000). Anaerobic digestion of ozonized biosolids. *Water Research* 34 (8):2330–2336.

WEF (Water Environment Federation). (1987). *Anaerobic Sludge Digestion Water Pollution Control Federation//Manual of Practice*, 2 Sub ed. ISBN 10-0943244137.

Wei, Y. S., R. T. Van Houten, A. R. Borger, D. H. Eikelboom, and Y. B. Fan. (2003). Minimization of excess sludge production for biological wastewater treatment. *Water Research* 37 (18):4453–4467.

Wett, B., K. Buchauer, and C. Fimml. (2007). Energy self-sufficiency as a feasible concept for wastewater treatment systems. Paper read at the IWA Leading Edge Technology Conference, September, at Singapore.

Wett, B., M. Hell, J. Jimenez, I. Takacs, C. Bott, and S. Murthy. (2014). Control measures for improved high-rate carbon removal (A-stage). In *Singapore International Water Week*. Singapore: IWA.

Winter, P. and P. Pearce. (2010). Parallel digestion of secondary and primary sludge. In: *15th European Biosolids and Biowastes Conference*, Leeds, UK.

Wilson, C. A. and J. T. Novak. (2009). Hydrolysis of macromolecular components of primary and secondary wastewater sludge by thermal hydrolytic pretreatment. *Water Research* 43 (18):4489–4498.

Wong, W. T., W. I. Chan, P. H. Liao, K. V. Lo, and D. S. Mavinic. (2006). Exploring the role of hydrogen peroxide in the microwave advanced oxidation process: Solubilization of ammonia and phosphates. *Journal of Environmental Engineering and Science* 5 (6):459–465.

Wonglertarak, W. and B. Wichitsathian. (2014). Alkaline pretreatment of waste activated sludge in anaerobic digestion. *Journal of Clean Energy Technologies* 2:118–121.

WRC. (1984). Theory, design and operation of nutrient removal activated sludge processes. Water Research Commission, Pretoria, South Africa.

Wu, B., T. Kitade, C. Tzyy Haur, T. Uemura, and G. Fane Anthony. (2011). Impact of membrane bioreactor operating conditions on fouling behavior of reverse osmosis membranes in MBR-RO processes. *Desalination* 311:37–45.

Xie, R., Y. Xing, Y. A. Ghani, K. Ooi, and S. Ng. (2007). Full-scale demonstration of an ultrasonic disintegration technology in enhancing anaerobic digestion of mixed primary and thickened secondary sewage sludge. *Journal of Environmental Engineering and Science* 6 (5):533–541.

Xu, C. C. and J. Lancaster. (2009). *Treatment of Secondary Sludge for Energy Recovery.* New York: Nova Science Publishers.

Xu, G., S. Chen, J. Shi, S. Wang, and G. Zhu. (2010). Combination treatment of ultrasound and ozone for improving solubilization and anaerobic biodegradability of waste activated sludge. *Journal of Hazardous Materials* 180 (1–3):340–346.

Xu, J., H. Yuan, J. Lin, and W. Yuan. (2013). Evaluation of thermal, thermal-alkaline, alkaline and electrochemical pretreatments on sludge to enhance anaerobic biogas production. *Journal of the Taiwan Institute of Chemical Engineers* 45:2531–2536.

Yan, L. and H.P.F. Herbert. (2003). Influences of extracellular polymeric substances (EPS) on flocculation, settling, and dewatering of activated sludge. *Critical Reviews in Environmental Science and Technology* 33 (3):237–237.

Yan, S., K. Miyanaga, X. H. Xing, and Y. Tanji. (2008). Succession of bacterial community and enzymatic activities of activated sludge by heat-treatment for reduction of excess sludge. *Biochemical Engineering Journal* 39 (3):598–603.

Yan, S. T., L. B. Chu, X. H. Xing, A. F. Yu, X. L. Sun, and B. Jurcik. (2009). Analysis of the mechanism of sludge ozonation by a combination of biological and chemical approaches. *Water Research* 43 (1):195–203.

Yang, Q., K. Luo, X. M. Li, D. B. Wang, W. Zheng, G. M. Zeng, and J. J. Liu. (2010). Enhanced efficiency of biological excess sludge hydrolysis under anaerobic digestion by additional enzymes. *Bioresource Technology* 101 (9):2924–2930.

Yang, Q., J. Yi, K. Luo, X. Jing, X. Li, Y. Liu, and G. Zeng. (2013a). Improving disintegration and acidification of waste activated sludge by combined alkaline and microwave pretreatment. *Process Safety and Environmental Protection* 91 (6):521–526.

Yang, S. S., W. Q. Guo, G. L. Cao, H. S. Zheng, and N. Q. Ren. (2012). Simultaneous waste activated sludge disintegration and biological hydrogen production using an ozone/ultrasound pretreatment. *Bioresource Technology* 124 (0):347–354.

Yang, S. S., W. Q. Guo, Z. H. Meng, X. J. Zhou, X. C. Feng, H. S. Zheng, B. Liu, N. Q. Ren, and Y. S. Cui. (2013b). Characterizing the fluorescent products of waste activated sludge in dissolved organic matter following ultrasound assisted ozone pretreatments. *Bioresource Technology* 131 (0):560–563.

Yasui, H. and Y. Miyaji. (1992). A novel approach to removing refractory organic compounds in drinking water. *Water Science and Technology* 26 (7–8):1503–1512.

Yeneneh, A. M., T. K. Sen, S. Chong, H. M. Ang, and A. Kayaalp. (2013). Effect of combined microwave-ultrasonic pretreatment on anaerobic biodegradability of primary, excess activated and mixed sludge. *Water Energy Environment* 2:7–11.

Yeneneh, A. M., A. Kayaalp, T. K. Sen, and H. M. Ang. (2015). Effect of microwave and combined microwave-ultrasonic pretreatment on anaerobic digestion of mixed real sludge. *Journal of Environmental Chemical Engineering* 3 (4):2514–2521.

Yeom, I. T., K. R. Lee, K. H. Ahn, and S. H. Lee. (2002). Effects of ozone treatment on the biodegradability of sludge from municipal wastewater treatment plants. *Water Science and Technology: A Journal of the International Association on Water Pollution Research* 46 (4–5):421–425.

Yin, G. Q., P. H. Liao, and K. V. Lo. (2007). An ozone/hydrogen peroxide/microwave-enhanced advanced oxidation process for sewage sludge treatment. *Journal of Environmental Science and Health, Part A* 42 (8):1177–1181.

Yu, G. H., P. J. He, L. M. Shao, and Y. S. Zhu. (2008). Extracellular proteins, polysaccharides and enzymes impact on sludge aerobic digestion after ultrasonic pretreatment. *Water Research* 42 (8–9):1925–1934.

Yu, J., X. Lan, S. Sreerama, R. Wikramanayake, R. Knauf, W. Liu, E. Jordan, and J. Hu. (2009). An energy saving solution for water reclamation: A hybrid process combining bio-sorption and anaerobic digestion. Process concept and bio-sorption using anaerobic/aerobic sludge mixtures. Paper read at the IWA Leading Edge Technology Conference, at Singapore.

Yu, J., X. Lan, S. Sreerama, R. Wikramanayake, R. Knuaf, W. Liu, E. Jordan, and J. Hu. (2014). An energy saving solution for water reclamation: A hybrid process combining bio-sorption and anaerobic digestion- process concept and bio-sorption using anaerobic/aerobic sludge mixtures. In *Singapore International Water Week*. Singapore: IWA.

Yue, Z.-B., R. H. Liu, H.-Q. Yu, H.-Z. Chen, Y. Bin, H. Harada, and Y.-Y. Li. (2008). Enhanced anaerobic ruminal degradation of bulrush through steam explosion pretreatment. *Industry and Engineering Chemistry Research* 47 (16):5899–5905.

Zábranská, J., J. Štěpová, R. Wachtl, P. Jeníček, and M. Dohányos. (2000). The activity of anaerobic biomass in thermophilic and mesophilic digesters at different loading rates. *Water Science and Technology* 42 (9):49–56.

Zábranská, J., M. Dohányos, P. Jeníček, and J. Kutil. (2006). Disintegration of excess activated sludge: Evaluation and experience of full-scale applications. *Water Science and Technology: A Journal of the International Association on Water Pollution Research* 53 (12):229–236.

Zanoni, A. E. and D. L. Mueller. (1982). Calorific value of wastewater plant sludges. *Journal of Environmental Engineering Division* 108 (1):187–195.

Zeeman, G., W. T. M. Sanders, K. Y. Wang, and G. Lettinga. (1997). Anaerobic treatment of complex wastewater and waste activated sludge: Application of an upflow anaerobic solid removal (UASR) reactor for the removal and pre-hydrolysis of suspended COD. *Water Science and Technology* 35 (10):121–128.

Zehnder, A. J. B. (1988). *Biology of Anaerobic Microorganisms*. New York: Wiley-Liss.

Zhang, D., Y. Chen, Y. Zhao, and X. Zhu. (2010). New sludge pretreatment method to improve methane production in waste activated sludge digestion. *Environmental Science and Technology* 44 (12):4802–4808.

Zhang, G., J. Yang, H. Liu, and J. Zhang. (2009). Sludge ozonation: Disintegration, supernatant changes and mechanisms. *Bioresource Technology* 100 (3):1505–1509.

Zhang, P., G. Zhang, and W. Wang. (2007a). Ultrasonic treatment of biological sludge: Floc disintegration, cell lysis and inactivation. *Bioresource Technology* 98 (1):207–210.

Zhang, S., Panyue Zhang, G. Zhang, J. Fan, and Y. Zhang. (2012a). Enhancement of anaerobic sludge digestion by high-pressure homogenization. *Bioresource Technology* 118 (0):496–501.

Zhang, S., H. Guo, L. Du, J. Liang, X. Lu, N. Li, and K. Zhang. (2015a). Influence of NaOH and thermal pretreatment on dewatered activated sludge solubilisation and subsequent anaerobic digestion: Focused on high-solid state. *Bioresource Technology* 185:171–177.

Zhang, T., Q. Wang, L. Ye, D. Batstone, and Z. Yuan. (2015b). Combined free nitrous acid and hydrogen peroxide pre-treatment of waste activated sludge enhances methane production via organic molecule breakdown. *Scientific Reports* 5:16631.

Zhang, Y., P. Zhang, G. Zhang, W. Ma, H. Wu, and B. Ma. (2012b). Sewage sludge disintegration by combined treatment of alkaline+high pressure homogenization. *Bioresource Technology* 123 (0):514–519.

Zhang, Z. B., J. F. Zhao, S. Q. Xia, C. Q. Liu, and X. S. Kang. (2007b). Particle size distribution and removal by a chemical-biological flocculation process. *Journal of Environmental Sciences* 19 (5):559–563.

Zhao, W., Y. P. Ting, J. P. Chen, C. H. Xing, and S. Q. Shi. (2000). Advanced primary treatment of waste water using a bio-flocculation-adsorption sedimentation process. *Acta Biotechnologica* 20 (1):53–64.

Zhao, P., S. Ge, and K. Yoshikawa. (2013). An orthogonal experimental study on solid fuel production from sewage sludge by employing steam explosion. *Applied Energy* 112 (0):1213–1221.

Zhao, Y., N. Ren, and A. Wang. (2008). Contributions of fermentative acidogenic bacteria and sulfate-reducing bacteria to lactate degradation and sulfate reduction. *Chemosphere* 72(2):233–242.

Zhen, G., X. Lu, H. Kato, Y. Zhao, and Y.-Y. Li. (2017). Overview of pretreatment strategies for enhancing sewage sludge disintegration and subsequent anaerobic digestion: Current advances, full-scale application and future perspectives. *Renewable and Sustainable Energy Reviews* 69:559–577.

Zhou, X., Q. L. Wang, and G. M. Jiang. (2015). Enhancing methane production from waste activated sludge using a novel indigenous iron activated peroxidation pretreatment process. *Bioresource Technology* 182:267–271.

Zwietering, M. H., I. Jongenburger, F. M. Rombouts, and K. van't Riet. (1990). Modeling of the bacterial growth curve. *Applied and Environmental Microbiology* 56 (6):1875–1881.

Index